Advanced Engineering Mathematics: A Complete Approach

Advanced Engineering Mathematics: A Complete Approach

Contributors

Nai-Chia Shih, Chun-Chun Wu et al.

AURIS
Reference

www.aurisreference.com

Advanced Engineering Mathematics: A Complete Approach

Contributors: Nai-Chia Shih, Chun-Chun Wu et al.

Published by Auris Reference Limited

www.aurisreference.com

United Kingdom

Advanced Engineering Mathematics: A Complete Approach

ISBN: 978-1-78154-821-9

British Library Cataloguing in Publication Data
A CIP record for this book is available from the British Library

Printed in the United Kingdom

Exclusively distributed by CBS Publishers & Distributors Pvt. Ltd.

Sales & Distribution Rights only for India, Pakistan, Bangladesh, Sri Lanka, Nepal and Bhutan. This book is not to be sold outside these territories.

Contents

List of Abbreviations

CCM	Chance constraints model
IFLPP	Intuitionistic fuzzy linear programming
DNLS	Discrete nonlinear Schrödinger
BECs	Bose-Einstein condensates
BEM	Boundary element method
FEM	Finite element method
DSF	Discretized scale-frequency
HUP	Heisenberg uncertainty principle
CDFs	Cumulative distribution functions
PG	Poisson-gamma
MSE	Mean squared error
PM	Permanent magnet
PWM	Pulse width modulated
ASD	Adjustable speed drive
FC	Feedback current
SI	System identification
PCD	Powell Conjugate Direction

List of Contributors

Nai-Chia Shih
Department of Applied Mathematics, Feng-Chia University, Taichung, Taiwan

Chun-Chun Wu
Department of Applied Mathematics, Feng-Chia University, Taichung, Taiwan

Jui-Sen Yang
Institute of Marine Biology, National Taiwan Ocean University, Keelung, Taiwan

Dipankar Chakraborty
Department of Mathematics, Heritage Institute of Technology, Anandapur, Kolkata, West Bengal 700107, India

Dipak Kumar Jana
Department of Applied Science, Haldia Institute of Technology, Haldia, Purba Midnapur, West Bengal 721657, India

Tapan Kumar Roy
Department of Mathematics, Indian Institute of Engineering Science and Technology, Shibpur, Howrah, West Bengal 711103, India

Bin Xu
Department of Civil Engineering, Nanchang Institute of Technology, Nanchang, Jiangxi 330029, China

Man-Qing Xu
Department of Civil Engineering, Nanchang Institute of Technology, Nanchang, Jiangxi 330029, China

Nassar H. S. Haidar
Center for Research in Applied Mathematics & Statistics (CRAMS), AUL, Lebanon

Wayne S. Kendal
Division of Radiation Oncology, University of Ottawa, 501 Smyth Road, Ottawa, ON K1H 8L6, Canada

Bent Jørgensen
Department of Mathematics and Computer Science, University of Southern Denmark, Campusvej 55, Odense M DK-5230, Denmark

Daniel Law
Department of Mathematics and Statistics, Amherst College, Amherst, MA 01002, USA

Jennie D'Ambroise
Department of Mathematics and Statistics, Amherst College, Amherst, MA 01002, USA

Panayotis G. Kevrekidis
Department of Mathematics and Statistics, University of Massachusetts, Amherst, MA 01003, USA

Detlef Kip [3]
Faculty of Electrical Engineering, Helmut Schmidt University, Hamburg 22043, Germany

Richard A. Guinee
Department of Electrical and Electronic Engineering, Cork Institute of Technology, Cork, Ireland

Arindam Banerjee,
Department of ECE, JIS College of Engineering, Kalyani, India

Aniruddha Ghosh,
Department of ECE, JIS College of Engineering, Kalyani, India

Mainuck Das
Department of ECE, JIS College of Engineering, Kalyani, India

Alecsandru Simion,
"Gh. Asachi" Technical University of Iaşi, Electrical Engineering Faculty, Romania

Leonard Livadaru
"Gh. Asachi" Technical University of Iaşi, Electrical Engineering Faculty, Romania

Adrian Munteanu
"Gh. Asachi" Technical University of Iaşi, Electrical Engineering Faculty, Romania

Preface

Engineering mathematics is a branch of applied mathematics concerning mathematical methods and techniques that are typically used in engineering and industry. The text Advanced Engineering Mathematics: A Complete Approach covers the entire sequence of mathematical topics including geometry, matrices, integrals, vector calculus, complex variables, probability/statistics, numerical methods, and more. Mathematical approach to the platonic solid structure of MS2 particles has been presented in first chapter. A new approach to solve intuitionistic fuzzy optimization problem using possibility, necessity, and credibility measures has been focused in second chapter. In third chapter, a numerical method for evaluating the vertical vibration isolation effect of pile rows embedded in a viscoelastic half space subjected to a moving load has been developed on the basis of the Cole-Cole model and Muki's method. Mathematical model of the three-phase induction machine for the study of steady-state and transient duty under balanced and unbalanced states has been introduced in fourth chapter. In fifth chapter, we present a computational analysis of the deviations between the actual positions of the prime numbers and their predicted positions from Riemann's counting formula, focused on the variance function of these deviations from sequential enumerative bins. Asymmetric wave propagation through saturable nonlinear oligomers has been discussed in sixth chapter. Seventh chapter addresses the topographical examination of various mean squared error (MSE) cost surface structures and selecting the most suitable MSE fitness function for accurate brushless motor drive (BLMD) dynamical parameter system identification (SI) of BLMD shaft load inertia and viscous damping for electric vehicle controlled propulsion. Novel high speed energy efficient square root architecture has been reported in eighth chapter. Mathematical intervening principle based on "yin yang wu xing" theory in traditional Chinese mathematics has been presented in ninth chapter. In last chapter, we use wavelet method, for estimating the density function for censoring data.

Chapter 1

MATHEMATICAL APPROACH TO THE PLATONIC SOLID STRUCTURE OF MS2 PARTICLES

Nai-Chia Shih[1], Chun-Chun Wu[1], Jui-Sen Yang[2]

[1]Department of Applied Mathematics, Feng-Chia University, Taichung, Taiwan

[2]Institute of Marine Biology, National Taiwan Ocean University, Keelung, Taiwan

ABSTRACT

Bacteriophage MS2 is a viral particle whose symmetrical capsid consists of 180 copies of asymmetrical coat proteins with triangulation number T = 3. The mathematical theorems in this study show that the phage particles in three-dimension (3D) might be an icosahedron, a dodecahedron, or a pentakis dodecahedron. A particle with 180 coat protein subunits and T = 3 requires some geometrical adaptations to form a stable regular polyhedron, such as an icosahedron or a dodecahedron. However, with mathematical reasons electron micrographs of the phage MS2 show that 180 coat proteins are packed in an icosahedron. The mathematical analysis of electron micrographs in this study may be a useful tool for surveying the platonic solid structure of a phage or virus particle before performing 3D reconstruction.

INTRODUCTION

Many viruses and phages are icosahedral particles. However, spherical viruses may be polyhedral, forming dodecahedra or pentakis dodecahedra [1]. Bacteriophage MS2 is an icosahedral bacteriophage with a diameter of 27 - 34 nm [2] [3].

When an icosahedron is transferred to a flat sheet [4], its hexagon units consist of 6 regular triangles. A convex angle can be formed by 5 angles of regular triangles, but not 6 angles, which become a plane. The 20 triangular faces created by the vertices of 12 pentagonal cones and 30 edges can form an icosahedron. The coat protein of MS2 forms a shell that protects the phage nucleic acid and acts as a translational repressor [5]. The tertiary structure of the coat proteins is asymmetrical. A particle with nucleotides packed with asymmetrical proteins requires a low free-energy to achieve a stable condition. Packing as a helix or a regular polyhedron is a way of getting a symmetrical solid with asymmetrical subunits. In icosahedral particles, proteins are packed on the faces and directed to the vertices and the particles become symmetrical [3] [6]. In the MS2 capsid, one triangle of the icosahedron contains 3 asymmetrical subunits [6] [7]. This study uses mathematical analysis to identify the reasonable Platonic solids for packing a symmetrical capsid with asymmetrical subunits. Results show that the MS2 particles with 180 coat proteins and triangulation number $(T) = 3$ might form an icosahedron, a dodecahedron, or a pentakis dodecahedron.

The overall shape of MS2 is spherical, but is difficult to see the three-dimensional (3D) figure in electron microscopy (EM) two-dimensional (2D) images before 3D reconstruction. Therefore, this study also introduces a mathematical method to predict particle solid 3D figures with 2D EM images before performing 3D reconstruction.

THEORY AND CALCULATION

The 3D structure of MS2 should be an isohedron or a regular polyhedron if the particles need to pack the asymmetrical proteins to make a stable symmetrical structure. The capsid of phage MS2 contains 180 identical copies of a coat protein with a T = 3 isohedral (such as icosahedral) shell [3] [7]. MS2 particles may be constructed with 60 triangles, 45 quadrilaterals, 36 pentagons, or 30 hexagons. This study examines particle construction using the following theorems and mathematical analysis:

Theorem 1. If a polyhedron has 180 subunits, where 180/n-polygons have n sides, n should be a positive integer and equal or less than 5.

Proof. A polyhedron is constructed with n-side polygons, the polyhedron has 180/n faces and the number of edges should be

$$180/n \times n \div 2 = 90 \quad.$$

By the Euler theorem, $V + F - E = 2$, where V, F, and E denote the number of vertices, faces, and edges, respectively,

$$V = 2 + E - F = 2 + 90 - 180/n = 92 - 180/n \quad.$$

A vertex of solid angle needs at least trimer, then

$$3V \leq 180 \Rightarrow 3 \times (92 - 180/n) \leq 180 \Rightarrow n \leq 5.625 \quad.$$

Thus, the maximum of n is 5, where the polygons are pentagons, quadrilaterals, or triangles. Therefore, a polyhedron with 180 structure protein subunits should not be constructed with 30 hexagons.

Theorem 2. If a polyhedron with 180 subunits is constructed with 36 pentagons, the polyhedron should not be an isohedron or a regular polyhedron.

Proof. If 36 pentagons can construct an isohedron or a regular polyhedron, the number of faces is 36. If two faces form dihedral angles, then the number of edges = $36 \times 5 \div 2 = 90$.

By the Euler theorem, $V + F - E = 2 \Rightarrow$

$$V = 2 + E - F = 2 + 90 - 36 = 56$$

If vertices are constructed with x n-polymers and y m-polymers, where x, y, n, and m are positive numbers, and $m > n$, then $nx + my = 180$.

Since $x + y = 56$, that is,

$$nx + ny = 56n \Rightarrow 56n \leq 180 \Rightarrow n \leq 3.213$$

the maximum of n is 3.

Thus, $3x + my = 180$.

The shell of an isohedron or a regular polyhedron consists of m-polymer units, and y m-polymers consist of 36 faces.

That is

$$my = 36$$

The solution of 3 simultaneous equations

$$3x + my = 180,$$

$$x + y = 56,$$

$$my = 36,$$

is $x = 48$, $y = 8$, $m = 4.5$.

This is a contradiction with positive integer m.

Therefore, this theorem suggests that 36 pentagons cannot form an isohedron or a regular polyhedron.

Theorem 3. If a polyhedron with 180 subunits is constructed with an odd number of quadrilaterals, the polyhedron should not be an isohedron or a regular polyhedron.

Proof. Suppose that an isohedron or a regular polyhedron can be constructed with an odd number, 2n + 1, of quadrilaterals.

The number of faces is then 2n + 1, and the number of edges $=(2n+1)\times 4 \div 2 = 4n+2$.

By the Euler theorem, $V + F - E = 2 \Rightarrow$
$V = 2 + E - F = 2 + 4n + 2 - 2n - 1 = 2n + 3$.

The number of vertices is odd, which is a contradiction because an isohedron or a regular polyhedron is symmetrical, and the number of verticies should be even. Thus, an odd number (such as 45) of quadrilaterals cannot form an isohedron or a regular polyhedron.

According to Theorems 1, 2, and 3, 45 quadrilaterals, 36 pentagons, or 30 hexagons (the polygons with more than 3 edges) cannot form an isohedron or a regular polyhedron with 180 subunits. Only 60 triangles can form a regular polyhedron for packing 180 subunits.

Lemma 4. If a convex solid angle is constructed with n equilateral triangles, n should be 3, 4, or 5.

Proof. In 3D models, a convex solid angle has at least by 3 faces.

For a solid angle, the total angle with 3 equilateral triangular angles is 180°. The total angle with 4 equilateral triangular angles is 240°. The total angle with 5 equilateral triangular angles is 300°. The total angle with 6 equilateral triangular angles is 360°, which is a cyclic angle.

For a convex solid angle constructed with n equilateral triangles, n should be 3, 4, or 5. The angle of a vertex equal to or greater than 360° is flat or concave.

Theorem 5. If a polyhedron is constructed with 60 identical equilateral triangles, the polyhedron is not an isohedron or a regular polyhedron.

Proof. If a polyhedron is constructed with 60 identical equilateral triangles, as in Figure 1, then vertex Q in Figure 1 consists of 6 identical equilateral triangles.

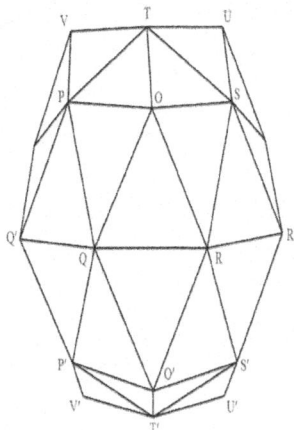

Figure 1: A convex polyhedron with 60 equilateral triangles.

Lemma 4 shows that vertex Q is not a convex solid angle. Instead, vertex Q may belong to the following:

1) Vertex Q is a concave solid angle.

    This hypothesis is not true because the polyhedron is convex.

2) Point Q is at the center of hexagon $PQ'P'O'RO$.

    Vertex O and pentagon PQRST form a regular pentagonal cone. A dihedral angle forms between triangles OPQ and OQR on line $\overline{OQ}$. Thus, polygon $PQ'P'O'RO$ is not a regular hexagon, and the hypothesis is not true.

3) Vertex Q locates on a dihedral angle.

Vertex Q' and pentagon $QP'T'S'R$ form a regular pentagonal cone. Line $\overline{O'Q}$ is on the dihedral angle between triangles $O'P'Q$ and $O'QR$. On the other hand, a dihedral angle forms between triangles OPQ and OQR on line $\overline{OQ}$, as in (2). Therefore, two dihedral angles appear at vertex Q, and the hypothesis is not true.

Theorem 6. A polyhedron becomes an icosahedron when the solid angles of the hexamers of the polyhedron spread to be flat and the connecting lines between the vertices of 2 nearby pentagonal cones become the ridges of dihedral angles.

Proof. As Figure 1 shows that the distances between the vertices of any 2 pentagonal cones inFigure 2 are the same. Then,

$$\overline{OO'} = \overline{OQ'} = \overline{OV} = \overline{OU} = \overline{OR'} \text{ and } \overline{OO'} = \overline{OQ'} = \overline{Q'V} = \overline{VU} = \overline{UR'} = \overline{R'O'}$$

Spreading and flattening the solid angles of the hexamers in the pentagonal cone $OO'Q'VUR'$ forms dihedral angles on lines $\overline{OO'}$, $\overline{OQ'}$, $\overline{OV}$, $\overline{OU}$, and $\overline{OR'}$. These lines become the edges of pentagonal cones, and vertex O has equilateral triangular faces. The vertices of 12 pentagonal cones have the same number of edges.

Therefore, a polyhedron becomes an icosahedron.

Theorem 7. A polyhedron becomes a dodecahedron when the solid angles of pentamers of the polyhedron spread out and become flat.

Proof. In Figure 3, for triangle OQR, triangle OQR becomes an isosceles triangle when lines $\overline{OQ}$ and $\overline{OR}$ decrease. Because

$$\overline{OR} < \overline{QR} \Rightarrow \angle OQR < \angle QOR$$

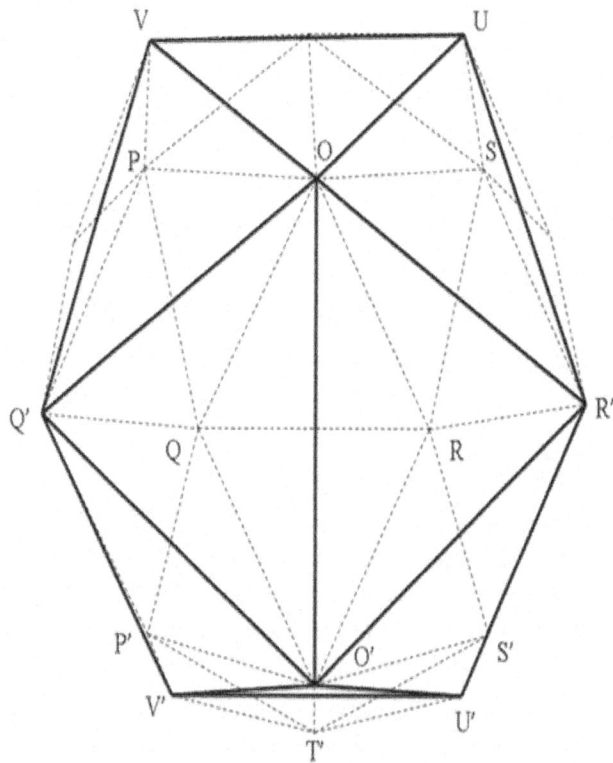

Figure 2: An icosahedron forms from a polyhedron after the solid angles of hexamers degenerate and the connecting lines (solid lines) between the vertices of 2 nearby pentagonal cones become the ridges of dihedral angles.

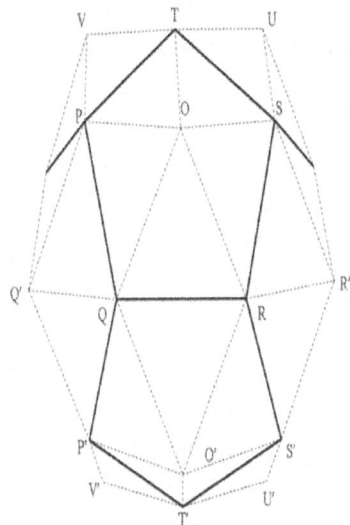

Figure 3. A dodecahedron forms from a polyhedron when the solid angles of pentamers (solid lines) spread out and become flat.

And $\angle RQO + \angle OQP + \angle PQQ' + \angle Q'QP' + \angle P'QO' + \angle O'QR < 360°$ Vertex Q becomes a convex solid angle and the polyhedron becomes a pentakis dodecahedron. When the edge lines at vertex O decrease to allow the solid angle O to disappear and become flat, the pentagonal cone OPQRST flattens to become a pentagon PQRST.

Finally, the polyhedron becomes a dodecahedron.

RESULTS AND DISCUSSION

Mathematical Theorems in Packing a Stable Particle

Theorems 1, 2, and 3 show that 45 quadrilaterals, 36 pentagons, or polygons with 6 or more edges cannot form an isohedron or a regular polyhedron with 180 subunits. Only 60 triangles can form a regular polyhedron for packing 180 subunits. In the MS2 particle, the way to pack 180 coat proteins is to form an isohedron or a regular polyhedron with 60 identical triangles (Lemma 4). The 3D polyhedron constructed with 60 identical equilateral triangles in Figure 1 is not an isohedron or a regular polyhedron (Theorem 5). That is, a polyhedron consisting of 60 identical equilateral triangles is not a convex isohedron. Therefore, a T = 3 MS2 phage with 180 coat protein subunits might become an icosahedron [3] only if the solid angles of hexamers in the polyhedron of 60 identical equilateral triangles spread to become flat and the connecting lines between the vertices of 2 nearby pentagonal cones become the ridges of dihedral angles (Theorem 6). Theorem 7 shows that the MS2 particle model may form a dodecahedron when the solid angles of pentamers spread out and become flat.

The 3D structure of MS2 [3] consists of symmetrical units that lie between 2 threefold axes and 1 fivefold axis. However, a particle with 180 coat protein subunits and T = 3 requires some geometrical adaptations to pack a stable regular polyhedron including an icosahedron or a dodecahedron (Theorem 6 and 7).

An array of hexamers is the basic unit for generating an icosahedron [4] . A hexagon consisting of 6 regular triangles cannot form a solid angle, but lies flat (Lemma 4). Removing a regular triangle from the triangles in the hexagon forms a pentagonal cone. The vertex of the pentagonal cone becomes one of the 12 solid angles of an icosahedron and the 20 triangular faces form part of hexagons [8] .

For MS2 capsids with the principal of quasi-equivalence [8] , the triangulation number $\left(T = h^2 + hk + k^2\right)$ in the hexagon net is 3, where h and k are nonnegative integers on the original hexagonal net, h and k cannot be zero simultaneously, and the capsid has 180 structure subunit $^{(S)}$ proteins $\left(S = 60T\right)$ and 32 morphological units $\left(M\right)$ $(M = 10T + 2)$ [8] . T values may be 1, 3, 4, 7, 9, 12, 13, 16, 19, 21, 25, etc [1] although T = 2 and 6 appear [9] [10] . When T > 1, the morphological unit appears to form pentamers or hexamers. In an assembly pathway, 5 dimers converge into a pentamer. Twelve pentamers are linked together with free dimers creating a complete particle [2] . According to Theorems 6 and 7, a T = 3 particle can form a regular polyhedron with geometrical degeneration. Although the numbers of subunit proteins vary, the particle morphology is quasi-equiva- lent.

This study hypothesizes that regular Platonic solids allow a single type of asymmetric subunit to assemble into a well-defined spherical structure [7] . The asymmetrical subunit contains 3 subunits, designated as A, B, and C, in an icosahedral particle of phage MS2. Pairwise interactions between the monomers form dimers. The capsid contains 2 types of dimers: one at the quasi-twofold axis composed of subunits A and B and the other at the icosahedral twofold axis consisting of 2 C subunits [7] . Therefore, the capsid is effectively constructed from 90 dimers [7] . The theorems above indicate that the T = 3 phage MS2 particles with 180 protein subunits may by icosahedral, dodecahedral, or pentakis dodecahedral (dual semiregular solid) particles.

Therefore, further studies with EM images are necessary to determine if the Platonic solid of MS2 is an icosahedron or a dodecahedron.

Mathematical Reason with Electron Micrographs of MS2 Particles

With electron micrographs [11] , it is difficult to distinguish between regular hexagons and hexagon-like dodecagons in the projections of icosahedron. Both regular hexagons and regular decagons appear in EM images, though regular decagons appear spherical (Figure 4). However, unequilateral hexagons with large obtuse angles instead of narrow obtuse angles appear in EM images of MS2 in twofold views (Figure 5).

The 5 regular Platonic solids are tetrahedron, octahedron, cube, dodecahedron, and icosahedron. The icosahedron has a common symmetry with the dodecahedron, and the octahedron is similar to the cube [7] . Most studies recognize the bacteriophage MS2 as an icosahedral particle [1] -[3] [6] [12] [13] , though the MS2 coat protein mutant corresponds to T = 3 octahedral particles [9] . The main difference in the subunit packing between the octahedral and icosahedral arrangements is close to the fourfold and fivefold symmetry axes [7] .

The mathematical reason from Theorem 6 and 7 shows that the T = 3 phage MS2 particles may be icosahedral, dodecahedral, or pentakis dodecahedral particles. The projections of icosahedral particles in phage MS2 EM images exhibit unequilateral hexagons in twofold views, regular hexagons in threefold views, and regular decagons in fivefold views (Figure 4). The projections of dodecahedron are unequilateral hexagons in twofold views, hexagon-like dodecagons in threefold views, and regular decagons in fivefold views and asymmetrical views. The projections of pentakis dodecahedra are regular decagons in twofold, threefold, and fivefold views (Table 1). In fivefold views, icosahedra, dodecahedra, and pentakis dodecahedra models show the same projection as a regular decagon.

In threefold views, 3 models have different projections: regular hexagons form from icosahedra, hexagon-like dodecagons form from dodecahedra, and regular decagons form from pentakis dodecahedra. In twofold views, the projections of icosahedra and dodecahedra are unequilateral hexagons (Figure 6). In the unequilateral hexagons, 4 vertices have the same angles, and are unlike the other 2 obtuse angles. The 2 big obtuse angles in the unequilateral hexagons of the projections from icosahedra and dodecahedra are 138° and 116°, respectively [14] . The obtuse angles of unequilateral hexagon projections from icosahedra are much larger than those from dodecahedra. EM images of the MS2 particles show the projections of icosahedral particles since the obtuse angles of unequilateral hexagons are 138°. The T = 3 phage MS2 particles with

180 coat protein subunits [3] form icosahedra. This suggests that the solid angles of hexamers in the polyhedral particles may spread to become flat, and the connecting lines between the vertices of 2 nearby pentagonal cones become the ridges of dihedral angles during packing the coat protein subunits in MS2 (Theorem 6 and 7). A few viral particles are dodecahedra [4] [15] . In nature, spreading in the solid angles of hexamers in a polyhedron seems to be easier and more common than spreading of pentamers. Most viruses and phages form icosahedral particles instead of dodecahedral or pentakis dodecahedral particles.

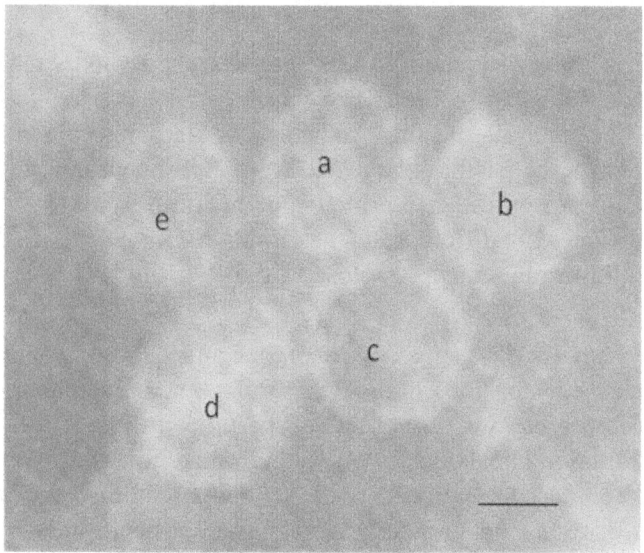

Figure 4: Micrograph of phage MS2 in twofold (a), threefold (b) and fivefold (c) views. Note particles attaching (d) and unattaching (e) on a pilus of E. coli. The bar represents 10 nm.

Mathematical Analysis in Particle Electron Micrographs for 3D Reconstruction

Although the 3D structure reconstructed from the EM images shows the Platonic solids of the particles, the wrong order of 3D reconstruction might yield a false solid figure [16] . It is easy to recognize an icosahedron or a dodecahedron in the Platonic solid of a particle using EM images and mathematical analysis. According to a primary survey of the Platonic solid of the particle, the 3D reconstruction can be performed confidently with a right symmetrical order [10] [14]

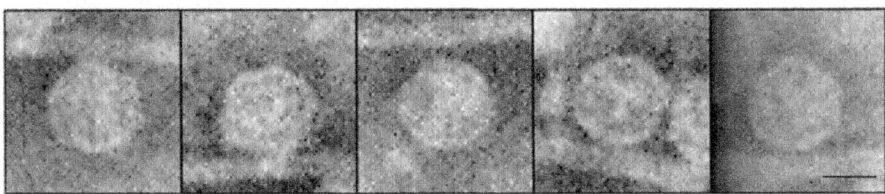

Figure 5: Micrographs of phage MS2 from twofold views. The bar represents 20 nm.

Table 1: Projections of polyhedra from various views.

	Symmetry			Asymmetry
	Twofold	Threefold	Fivefold	
Icosahedron	Unequilateral hexagon	Regular hexagon	Regular decagon	--
Dodecahedron	Unequilateral hexagon	Hexagon-like dodecagon	Regular decagon	(two forms)
Pentakis dodecahedron	regular decagon	Regular decagon	Regular decagon	--

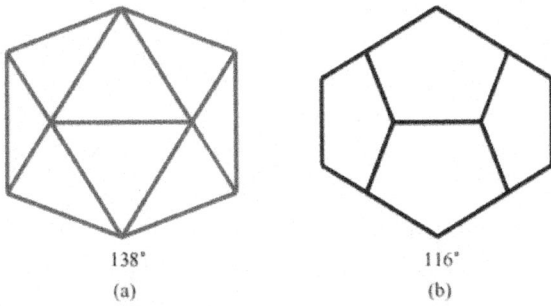

138°

(a)

116°

(b)

Figure 6: Projections of an icosahedron (a) and a dodecahedron (b) from a twofold view.

The mathematical analysis of EM images is a useful primary survey before performing the 3D reconstruction of a polyhedral particle.

CONCLUSION

In conclusion, a viral particle with 180 coat protein subunits and T = 3 requires some geometrical adaptation to form a stable regular polyhedron, such as an icosahedron or a dodecahedron. The mathematical analysis of MS2 particles reveals the EM projections of icosahedral particles. The MS2 particles are confirmed to be icosahedra.

ACKNOWLEDGEMENTS

We gratefully acknowledge Mr. Jaw-Fu Gau for his EM assistance.

REFERENCES

1.   Janner, A. (2006) Towards a Classification of Icosahedral Viruses in Terms of Indexed Polyhedral. Acta Crystallographica, Section A, 62, 319-330.

2.   Golmohammadi, R., Valegård, K., Fridborg, K. and Liljas, L. (1993) The Refined Structure of Bacteriophage MS2 at 2.8 Å Resolution. Journal of Molecular Biology, 234, 620-639.

3.   Valegård, K., Liljas, L., Fridborg, K. and Unge, T. (1990) The Three-Dimensional Structure of the Bacterial Virus MS2. Nature, 345, 36-41. http://dx.doi.org/10.1038/345036a0

4.   Baker, T.S., Olson, N.H. and Fuller, S.D. (1999) Adding the Third Dimension to Virus Life Cycles: Three-Dimen- sional Reconstruction of Icosahedral Viruses from Cryo-Electron Micrographs. Microbiology and Molecular Biology Reviews, 63, 862-922.

5.   Valegård, K., Murray, J.B., Stonehouse, N.J., van den Worm, S., Stockley, P.G. and Liljas, L. (1997) The Three-Di- mensional Structures of Two Complexes between Recombinant MS2 Capsids and RNA Operator Fragments Reveal Sequence-Specific Protein-RNA Interactions. Journal of Molecular Biology, 270, 724-738.

6.   Koning, R., van den Worm, S., Plaisier, J.R., van Duin, J., Abrahams, J.P. and Koerten, H. (2003) Visualization by Cryo- Electron Microscopy of Genomic RNA That Binds to the Protein Capsid inside Bacteriophage MS2. Journal of Molecular Biology, 332, 415-422.

7.   Plevka, P., Tars, K. and Liljas, L. (2008) Crystal Packing of a Bacteriophage MS2 Coat Protein Mutant Corresponds to Octahedral Particles. Protein Science, 17, 1731-1739.http://dx.doi.org/10.1110/ps.036905.108

8.   Caspar, D.L.D. and Klug, A. (1962) Physical Principles in the Construction of Regular Viruses. Cold Spring Harbor Symposia on Quantitative Biology, 27, 1-24.

9.   Flint, S.J., Enquist, L.W., Krug, R.M., Racaniello, V.R. and Skaika, A.M. (2000) Principles of Virology: Molecular Bi- ology, Pathogenesis, and Control. American Society for Microbiology Press, Washington DC, 58-98.

10. Reddy, V.S. and Johnson, J.E. (2005) Structure-Derived Insights into

Assembly. Advances in Virus Research, 64, 45- 68.

11. Gau, J.F. (2010) The Attachment of MS2 on E. coli. Master thesis, National Taiwan Ocean University, Keelung.

12. Plevka, P., Tars, K. and Liljas, L. (2009) Structure and Stability of Icosahedral Particles of a Covalent Coat Protein Dimer of Bacteriophage MS2. Protein Science, 18, 1653-1661.http://dx.doi.org/10.1002/pro.184

13. Schneemann, A. (2006) The Structural and Functional Role of RNA in Icosahedral Virus Assembly. Annual Review of Microbiology, 60, 51-67. http://dx.doi.org/10.1146/annurev.micro.60.080805.142304

14. Dr. Math, FAQ (1994-2009) (2009) Regular Polyhedral: Formulas. The Math Forum@Drexel. Drexel University, Phi- ladelphia. http://mathforum.org

15. Cann, A.J., (Au & Ed) (2005) The Principles of Molecular Virology. Chap. 2, Particles. Elsevier Academic Press, San Diego, 25-55.

16. Shih, N.C. (2009) Mathematical Studies on the 3D Reconstruction of Phage MS2. Master Thesis, Feng-Chia University, Taichung.

Chapter 2

A NEW APPROACH TO SOLVE INTUITIONISTIC FUZZY OPTIMIZATION PROBLEM USING POSSIBILITY, NECESSITY, AND CREDIBILITY MEASURES

Dipankar Chakraborty,[1] Dipak Kumar Jana,[2] and Tapan Kumar Roy[3]

[1]Department of Mathematics, Heritage Institute of Technology, Anandapur, Kolkata, West Bengal 700107, India

[2]Department of Applied Science, Haldia Institute of Technology, Haldia, Purba Midnapur, West Bengal 721657, India

[3]Department of Mathematics, Indian Institute of Engineering Science and Technology, Shibpur, Howrah, West Bengal 711103, India

ABSTRACT

Corresponding to chance constraints, real-life possibility, necessity, and credibility measures on intuitionistic fuzzy set are defined. For the first time the mathematical and graphical representations of different types of measures in trapezoidal intuitionistic fuzzy environment are defined in this paper. We have developed intuitionistic fuzzy chance constraints model (CCM) based on possibility and necessity measures. We have also proposed a new method for solving an intuitionistic fuzzy CCM using chance operators. To validate the proposed method, we have discussed three different approaches to solve the intuitionistic fuzzy linear programming (IFLPP) using possibility, necessity and credibility measures. Numerical and graphical representations of optimal solutions of the given example at different possibility and necessity, levels have been discussed.

INTRODUCTION

In the real world some data often provide imprecision and vagueness at certain level. Such vagueness has been represented through fuzzy sets. Zadeh [1] first introduced the fuzzy sets. The perception of intuitionistic fuzzy set (IFS) can be analysed as an unconventional approach to define a fuzzy set where available information is not adequate for the definition of an imprecise concept by means of a usual fuzzy set. This IFS was first introduced by Atanassov [2]. Many researchers have shown their interest in the study of intuitionistic fuzzy sets/numbers [3–7]. Fuzzy sets are defined by the membership function in all its entirety (c.f. Pramanik et al. [8, 9]), but IFS is characterized by a membership function and a nonmembership function so that the sum of both values lies between zero and one [10]. Esmailzadeh and Esmailzadeh [11] provided new distance between triangular intuitionistic fuzzy numbers.

Recently, the IFN has also found its application in fuzzy optimization. Angelov [12] proposed the optimization in an intuitionistic fuzzy environment. Dubey and Mehra [13] solved linear programming with triangular intuitionistic fuzzy number. Parvathi and Malathi [14] developed intuitionistic fuzzy simplex method. Hussain and Kumar [15] and Nagoor Gani and Abbas [16] proposed a method for solving intuitionistic fuzzy transportation problem. Ye [17] discussed expected value method for intuitionistic trapezoidal fuzzy multicriteria decision-making problems. Wan and Dong [18] used possibility degree method for interval-valued intuitionistic fuzzy for decision making.

Possibility, necessity, and credibility measures have a significant role in fuzzy and intuitionistic fuzzy optimization. Buckley [19] introduced possibility and necessity in optimization and Jamison and Lodwick [20] developed the construction of consistent possibility and necessity measures. Duality in fuzzy linear programming with possibility and necessity relations has been developed by Ramík [21]. Iskander [22] suggested an approach for possibility and necessity dominance indices in stochastic fuzzy linear programming. Sakawa et al. [23] used possibility and necessity to solve fuzzy random bilevel linear programming. Pathak et al. [24] discussed a possibility and necessity approach to solve fuzzy production inventory model for deteriorating items with shortages under the effect of time dependent learning and forgetting. Maity [25] established possibility and necessity representations of fuzzy inequality and its application to two warehouse production-inventory problem. Wu [26] presented possibility and necessity measures fuzzy optimization problems based on the embedding theorem. Xu and Zhou [27] discussed possibility, necessity, and credibility measures for fuzzy optimization. Maity and Maiti [28] developed the possibility and necessity constraints and their defuzzification for multiitem production-inventory scenario via optimal control theory. Das et al. [29] presented a two-warehouse supply-chain model under possibility, necessity, and credibility measures. Panda et al. [30] proposed a single period inventory model with imperfect production and stochastic demand under chance and imprecise constraints. Intuitionistic fuzzy-valued possibility and necessity measures have been devolved by Ban [31] using measure theory. With our best knowledge, however, none of them introduced chance constraints model based on possibility, necessity, and credibility measures on intuitionistic fuzzy set for membership and nonmembership functions. The rest of this paper is organized into different section as follows demonstrating the deduction of our theory and its application. In Section 2, we recall some preliminary knowledge about intuitionistic fuzzy and its arithmetic operation. Section 3 has provided possibility, necessity, and credibility measures in trapezoidal intuitionistic fuzzy number and its graphical representation. In Section 4, we have proposed intuitionistic fuzzy chance constraint models based on possibility, necessity, and credibility measures. The solution methodology of the proposed models using chance operator has been discussed in Section 5. In Section 6, a numerical example is presented to validate the proposed method. The numerical and graphical results at different possibility and necessity levels of the given problems have also been discussed here. Section 7 summarizes the paper and also discusses about the scope of future work.

PRELIMINARIES

Definition 1 (intuitionistic fuzzy set [2, 10]). Let E be a given set and let $A \subset E$ be a set. An IFS A^* in E is given by $A^* = \{\langle x, \mu(x), v_A(x)\rangle; x \in E\}$, where $\mu_A : E \to [0, 1]$ and $v_A : E \to [0, 1]$ define the degree of membership and the degree of nonmembership of the element $x \in E$ to $A \subset E$ satisfying the condition $0 \leq \mu_A(x) + v_A(x) \leq 1$.

Definition 2 (intuitionistic fuzzy number [7]). An IFN $\widetilde{A}^I$ is

(i)   an intuitionistic fuzzy subset on real line,

(ii)   there exist $m \in \mathfrak{R}$, such that $\mu_{\widetilde{A}^I}(m) = 1$, and $v_{\widetilde{A}^I}(m) = 0$.

(iii) convex   for   the   membership   function $v_{\widetilde{A}^I}$;   that   is, $\mu_{\widetilde{A}^I}(\lambda x_1 + (1 - \lambda)x_2) \geq \min(\mu_{\widetilde{A}^I}(x_1), \mu_{\widetilde{A}^I}(x_2)), x_1, x_2 \in R, \lambda \in [0, 1]$.

(iv)   concave for the nonmembership function $v_{\widetilde{A}^I}$; that is, $v_{\widetilde{A}^I}(\lambda x_1 + (1 - \lambda)x_2) \leq \max(v_{\widetilde{A}^I}(x_1), v_{\widetilde{A}^I}(x_2)), x_1, x_2 \in R, \lambda \in [0, 1]$.

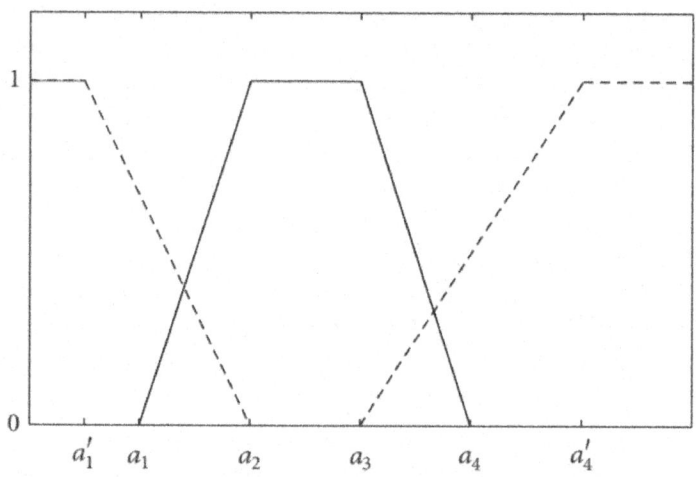

Figure 1: Membership and nonmembership functions of TIFN.

Definition 3 (trapezoidal intuitionistic fuzzy number (TIFN)). Let $a_1' \leq a_1 \leq a_2 \leq a_3 \leq a_4 \leq a_4'$. A TIFN $\widetilde{A}^I$ in $\mathfrak{R}$ written as $(a_1, a_2, a_3, a_4)(a_1', a_2, a_3, a_4')$ has membership function (c.f. Figure 1)

$$\mu_{\widetilde{A}^I}(x) = \begin{cases} \dfrac{x - a_1}{a_2 - a_1}, & a_1 \leq x \leq a_2; \\ 1, & a_2 \leq x \leq a_3; \\ \dfrac{a_4 - x}{a_4 - a_3}, & a_3 \leq x \leq a_4; \\ 0, & \text{otherwise}, \end{cases} \tag{1}$$

and nonmembership function

$$\nu_{\widetilde{A}^I}(x) = \begin{cases} \dfrac{a_2 - x}{a_2 - a_1'}, & a_1' \leq x \leq a_2; \\ 0, & a_2 \leq x \leq a_3; \\ \dfrac{x - a_3}{a_4' - a_3}, & a_3 \leq x \leq a_4'; \\ 1, & \text{otherwise}. \end{cases} \tag{2}$$

Definition 4 (triangular intuitionistic fuzzy number (TrIFN)). Let $a_1' \leq a_1 \leq a_2 \leq a_3 \leq a_3'$. A TrIFN $\widetilde{A}^I$ in $\mathfrak{R}$ written as $(a_1, a_2, a_3)(a_1', a_2, a_3')$ has membership function (c.f. Figure 2)

$$\mu_{\widetilde{A}^I}(x) = \begin{cases} \dfrac{x - a_1}{a_2 - a_1}, & a_1 \le x \le a_2; \\ 1, & x = a_2; \\ \dfrac{a_3 - x}{a_3 - a_2}, & a_2 \le x \le a_3; \\ 0, & \text{otherwise}, \end{cases}$$

(3)

and nonmembership function

$$\nu_{\widetilde{A}^I}(x) = \begin{cases} \dfrac{a_2 - x}{a_2 - a_1'}, & a_1' \le x \le a_2; \\ 0, & x = a_2; \\ \dfrac{x - a_2}{a_3' - a_2}, & a_2 \le x \le a_3'; \\ 1, & \text{otherwise}. \end{cases}$$

(4)

Definition 5. A positive TIFN $\widetilde{A}^I$ is denoted by $\widetilde{A}^I = (a_1, a_2, a_3, a_4)(a_1', a_2, a_3, a_4')$, where all $a_i > 0$ for all $i=1, 2, 3, 4$, and $a_i' > 0$ for $i = 1, 4$.

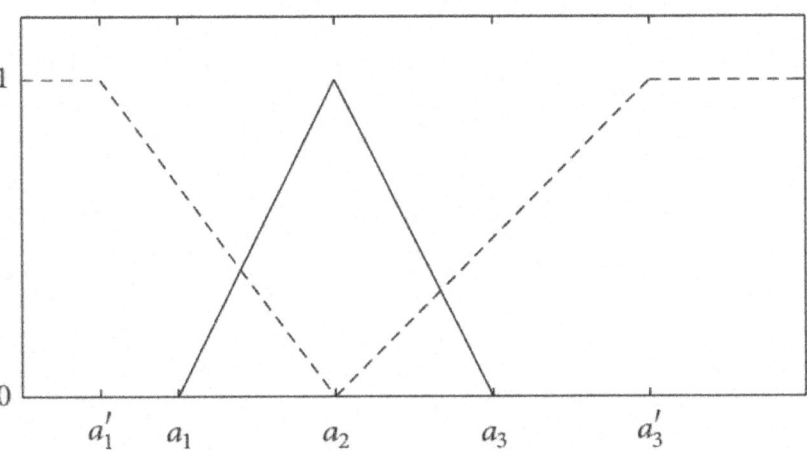

Figure 2: Membership and nonmembership functions of TrIFN.

Definition 6. Two TIFN $\widetilde{A}^I = (a_1, a_2, a_3, a_4)(a_1', a_2, a_3, a_4')$ and $\widetilde{B}^I = (b_1, b_2, b_3, b_4)(b_1', b_2, b_3, b_4')$ are said to be equal if and only if $a_1 = b_1, a_2 = b_2, a_3 = b_3, a_4 = b_4, a_1' = b_1'$ and $a_4' = b_4'$.

Definition 7. Let $\widetilde{A}^I = (a_1, a_2, a_3, a_4)(a_1', a_2', a_3', a_4')$ and $\widetilde{B}^I = (b_1, b_2, b_3, b_4)(b_1', b_2', b_3', b_4')$ be two TIFN; then

(i) $\widetilde{A}^I \oplus \widetilde{B}^I = (a_1 + b_1, a_2 + b_2, a_3 + b_3, a_4 + b_4)(a_1' + b_1', a_2' + b_2', a_3' + b_3', a_4' + b_4')$;

(ii) $k\widetilde{A}^I = k(a_1, a_2, a_3, a_4)(a_1', a_2', a_3', a_4') = (ka_1, ka_2, ka_3, ka_4)(ka_1', ka_2', ka_3', ka_4')$ if $k \geq 0$;

(iii) $k\widetilde{A}^I = k(a_1, a_2, a_3, a_4)(a_1', a_2', a_3', a_4') = (ka_4, ka_3, ka_2, ka_1)(ka_4', ka_3', ka_2', ka_1')$ if $k < 0$;

(iv) $\widetilde{A}^I \ominus \widetilde{B}^I = (a_1 - b_4, a_2 - b_3, a_3 - b_2, a_4 - b_1)(a_1' - b_4', a_2' - b_3', a_3' - b_2', a_4' - b_1')$;

(v) $\widetilde{A}^I \otimes \widetilde{B}^I = (a_1 b_1, a_2 b_2, a_3 b_3, a_4 b_4)(a_1' b_1', a_2' b_2', a_3' b_3', a_4' b_4')$.

POSSIBILITY, NECESSITY, AND CREDIBILITY MEASURES OF INTUITIONISTIC FUZZY NUMBER

Definition 8. Let $\widetilde{A}^I$ and $\widetilde{B}^I$ be two IFN with membership function $\mu_{\widetilde{A}^I}$, $\mu_{\widetilde{B}^I}$ and nonmembership function $\nu_{\widetilde{A}^I}$, $\nu_{\widetilde{B}^I}$, respectively, and R is the set of real numbers. Then

$$\text{Pos}_\mu \left(\widetilde{A}^I * \widetilde{B}^I \right) = \sup \left\{ \min \left(\mu_{\widetilde{A}^I}, \mu_{\widetilde{B}^I} \right), x, y \in R, x * y \right\}$$

$$\text{Pos}_\nu \left(\widetilde{A}^I * \widetilde{B}^I \right) = \sup \left\{ \min \left(\nu_{\widetilde{A}^I}, \nu_{\widetilde{B}^I} \right), x, y \in R, x * y \right\},$$

$$\text{Nes}_\mu \left(\widetilde{A}^I * \widetilde{B}^I \right) = \inf \left\{ \max \left(\mu_{\widetilde{A}^I}, \mu_{\widetilde{B}^I} \right), x, y \in R, x * y \right\}$$

$$\text{Nes}_\nu \left(\widetilde{A}^I * \widetilde{B}^I \right) = \inf \left\{ \max \left(\nu_{\widetilde{A}^I}, \nu_{\widetilde{B}^I} \right), x, y \in R, x * y \right\},$$

(5)

where the abbreviations Pos_μ and Pos_ν represent possibility of membership and nonmembership function, and Nes_μ and Nes_ν represent necessity of membership and nonmembership function. $*$ is any of the relations , $\leq, \geq, =$.

The dual relationship of possibility and necessity gives

$$\text{Nes}_\mu\left(\widetilde{A}^I * \widetilde{B}^I\right) = 1 - \text{Pos}_\mu\left(\overline{\widetilde{A}^I * \widetilde{B}^I}\right)$$

$$\text{Nes}_\nu\left(\widetilde{A}^I * \widetilde{B}^I\right) = 1 - \text{Pos}_\nu\left(\overline{\widetilde{A}^I * \widetilde{B}^I}\right),$$

(6)

where $\overline{\widetilde{A}^I * \widetilde{B}^I}$ represents complement of the event $\widetilde{A}^I * \widetilde{B}^I$.

Definition 9. Let $\widetilde{A}^I$ be a IFN. Then the intuitionistic fuzzy measures of $\widetilde{A}^I$ for membership and nonmembership function are

$$\text{Me}_\mu\left\{\widetilde{A}^I\right\} = \lambda\text{Pos}_\mu\left\{\widetilde{A}^I\right\} + (1 - \lambda)\,\text{Nec}_\mu\left\{\widetilde{A}^I\right\}$$

$$\text{Me}_\nu\left\{\widetilde{A}^I\right\} = \lambda\text{Pos}_\nu\left\{\widetilde{A}^I\right\} + (1 - \lambda)\,\text{Nec}_\nu\left\{\widetilde{A}^I\right\},$$

(7)

where the abbreviation Me_μ and Me_ν represent measures of membership and nonmembership functions and λ $(0 \le \lambda \le 1)$ is the optimistic-pessimistic parameter to determine the combined attitude of a decision maker.

If $\lambda = 1$ then $\text{Me}_\mu = \text{Pos}_\mu$, $\text{Me}_\nu = \text{Pos}_\nu$; it means

the decision maker is optimistic and maximum chance of $\widetilde{A}^I$ holds

If $\lambda = 0$, then $\text{Me}_\mu = \text{Nes}_\mu$, $\text{Me}_\nu = \text{Nes}_\nu$; it means

the decision maker is pessimistic and minimal chance of $\widetilde{A}^I$ holds

If $\lambda = 0.5$, then $\text{Me}_\mu = \text{Cr}_\mu$, $\text{Me}_\nu = \text{Cr}_\nu$, where Cr is the credibility measure; it means the decision maker takes compromise attitude.

Measures of Trapezoidal Intuitionistic Fuzzy Number.

Let $\widetilde{A}^I = (a_1, a_2, a_3, a_4)(a_1', a_2, a_3, a_4')$ and $\widetilde{B}^I = (b_1, b_2, b_3, b_4)(b_1', b_2, b_3, b_4')$ be two TIFN. From Definition 8 the possibilities of $\widetilde{A}^I \le \widetilde{B}^I$ for membership and nonmembership functions (c.f. Figures 3 and 4) are as follows:

$$\text{Pos}_\mu\left(\widetilde{A}^I \le \widetilde{B}^I\right) = \begin{cases} 1, & a_2 \le b_3; \\ \dfrac{b_4 - a_1}{b_4 - b_3 + a_2 - a_1}, & a_1 < b_4,\ a_2 > b_3; \\ 0, & b_4 \le a_1, \end{cases}$$

$$\text{Pos}_\nu\left(\widetilde{A}^I \le \widetilde{B}^I\right) = \begin{cases} 0, & a_2 \le b_3; \\ \dfrac{a_2 - b_3}{a_2 - a_1' + b_4' - b_3}, & a_2 > b_3,\ b_4' > a_1'; \\ 1, & b_4' \le a_1'. \end{cases}$$

(8)

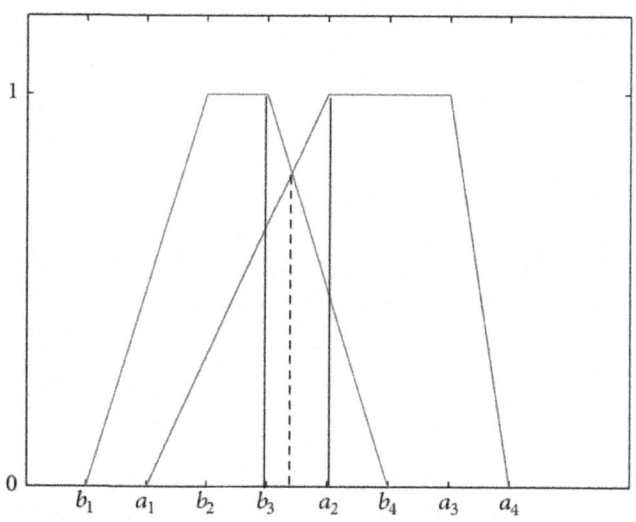

Figure 3: Membership function of TIFN $\widetilde{A}^I$ and $\widetilde{B}^I$ and $\text{Pos}_\mu(\widetilde{A}^I \leq \widetilde{B}^I)$

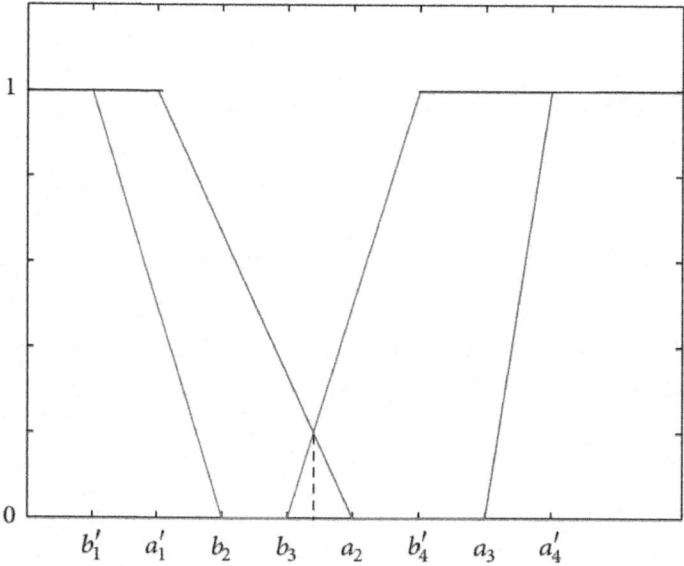

Figure 4: Nonmembership function of TIFN $\widetilde{A}^I$ and $\widetilde{B}^I$ and $\text{Pos}_\mu(\widetilde{A}^I \leq \widetilde{B}^I)$

From the Definition 8 the possibilities of $\widetilde{A}^I \geq \widetilde{B}^I$ for membership and nonmembership function (c.f. Figures 5 and 6) are as follows:

$$\text{Pos}_\mu\left(\widetilde{A}^I \geq \widetilde{B}^I\right) = \begin{cases} 1, & a_3 \geq b_2; \\ \dfrac{a_4 - b_1}{a_4 - a_3 + b_2 - b_1}, & a_3 < b_2,\ a_4 > b_1; \\ 0, & a_4 \leq b_1, \end{cases}$$

$$\text{Pos}_\nu\left(\widetilde{A}^I \geq \widetilde{B}^I\right) = \begin{cases} 0, & b_2 \leq a_3; \\ \dfrac{b_2 - a_3}{b_2 - b_1' + a_4' - a_3}, & b_2 > a_3,\ a_4' > b_1'; \\ 1, & b_1' \geq a_4'. \end{cases} \qquad (9)$$

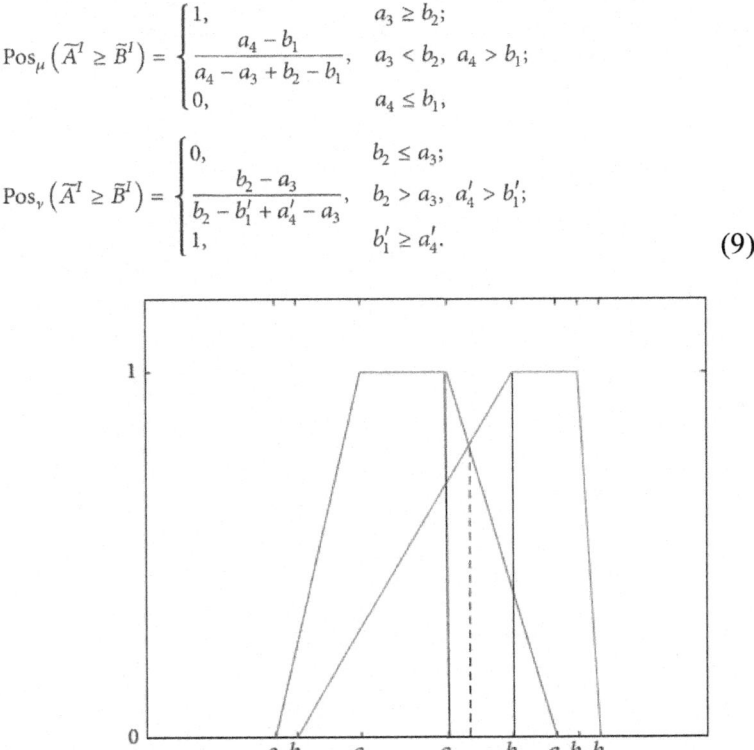

Figure 5: Membership function of TIFN $\widetilde{A}^I$ and $\widetilde{B}^I$ and $\text{Pos}_\nu(\widetilde{A}^I \geq \widetilde{B}^I)$

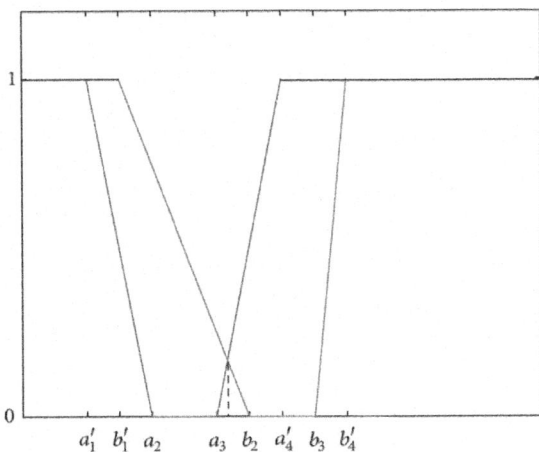

Figure 6: Nonmembership function of TIFN $\widetilde{A}^I$ and $\widetilde{B}^I$ and $\text{Pos}_\nu(\widetilde{A}^I \geq \widetilde{B}^I)$

Now, by Definition 8, necessity of the event $\widetilde{A}^I \le \widetilde{B}^I$ are as follows:

$$\text{Nes}_\mu\left(\widetilde{A}^I \le \widetilde{B}^I\right) = 1 - \text{Pos}_\mu\left(\widetilde{A}^I > \widetilde{B}^I\right)$$

$$= \begin{cases} 0, & b_3 \ge a_1; \\ \dfrac{b_1 - a_3}{a_4 - a_3 - b_2 + b_1}, & b_1 > a_3,\ a_4 > b_2; \\ 1, & b_2 \ge a_4, \end{cases}$$

$$\text{Nes}_y\left(\widetilde{A}^I \le \widetilde{B}^I\right) = 1 - \text{Pos}_y\left(\widetilde{A}^I > \widetilde{B}^I\right)$$

$$= \begin{cases} 0, & b_2 \ge a_4'; \\ \dfrac{a_4' - b_2}{a_4' - a_3 - b_2 + b_1'}, & b_2 < a_4',\ a_3 < b_1; \\ 1, & a_3 \ge b_1'. \end{cases} \tag{10}$$

By Definition 8 necessity of the event $\widetilde{A}^I \ge \widetilde{B}^I$ are as follows:

$$\text{Nes}_\mu\left(\widetilde{A}^I \ge \widetilde{B}^I\right) = 1 - \text{Pos}_\mu\left(\widetilde{A}^I < \widetilde{B}^I\right)$$

$$= \begin{cases} 0, & a_2 \le b_4; \\ \dfrac{a_2 - b_4}{a_2 - a_1 - b_4 + b_3}, & a_2 > b_4,\ a_1 < b_3; \\ 1, & a_1 \ge b_3, \end{cases}$$

$$\text{Nes}_y\left(\widetilde{A}^I \ge \widetilde{B}^I\right) = 1 - \text{Pos}_y\left(\widetilde{A}^I < \widetilde{B}^I\right)$$

$$= \begin{cases} 0, & b_3 \le a_1'; \\ \dfrac{b_3 - a_1'}{a_2 - a_1' - b_4' + b_3}, & b_3 > a_1',\ a_2 > b_4'; \\ 1, & a_2 \le b_4'. \end{cases} \tag{11}$$

By Definition 9 measures of the event $\widetilde{A}^I \le \widetilde{B}^I$ are as follows:

$$\text{Me}_\mu\left(\widetilde{A}^I \le \widetilde{B}^I\right)$$

$$= \lambda\text{Pos}_\mu\left(\widetilde{A}^I \le \widetilde{B}^I\right) + (1 - \lambda)\,\text{Nes}_\mu\left(\widetilde{A}^I \le \widetilde{B}^I\right)$$

$$= \begin{cases} 0, & b_4 \le a_1; \\ \lambda\dfrac{b_4 - a_1}{b_4 - b_3 + a_2 - a_1}, & b_3 > a_2,\ b_1 < a_2; \\ \lambda, & b_3 > a_2,\ b_1 < a_3; \\ \lambda + (1 - \lambda)\dfrac{b_1 - a_3}{a_4 - a_3 - b_2 + b_1}, & b_1 > a_3,\ a_4 > b_2; \\ 1, & b_2 \ge a_4, \end{cases}$$

$$\text{Me}_y\left(\widetilde{A}^I \le \widetilde{B}^I\right)$$

$$= \lambda\text{Pos}_y\left(\widetilde{A}^I \le \widetilde{B}^I\right) + (1 - \lambda)\,\text{Nes}_y\left(\widetilde{A}^I \le \widetilde{B}^I\right)$$

$$= \begin{cases} 1, & b_4' \le a_1'; \\ \lambda\dfrac{a_2 - b_3}{a_2 - a_1' + b_4' - b_3} + (1 - \lambda), & b_4' > a_1',\ a_2 > b_3; \\ (1 - \lambda), & a_2 < b_3,\ b_1' < a_3; \\ (1 - \lambda)\dfrac{a_4' - b_2}{a_4' - a_3 - b_2 + b_1'}, & a_3 < b_1',\ a_4' > b_2; \\ 0, & b_2 \ge a_4'. \end{cases} \tag{12}$$

By Definition 9 measures of the event $\tilde{A}^I \geq \tilde{B}^I$ are as follows:

$$\text{Me}_\mu\left(\tilde{A}^I \geq \tilde{B}^I\right)$$

$$= \lambda \text{Pos}_\mu\left(\tilde{A}^I \geq \tilde{B}^I\right) + (1-\lambda)\text{Nes}_\mu\left(\tilde{A}^I \geq \tilde{B}^I\right)$$

$$= \begin{cases} 1, & b_3 \leq a_2; \\ (1-\lambda)\dfrac{a_2 - b_4}{a_2 - a_1 - b_4 + b_3} + \lambda, & b_3 > a_1,\ b_4 < a_2; \\ \lambda, & b_4 > a_2,\ b_2 < a_3; \\ \lambda\dfrac{a_4 - b_1}{a_4 - a_3 + b_2 - b_1}, & a_3 < a_2,\ a_4 > b_1; \\ 0, & b_1 \geq a_4, \end{cases}$$

$$\text{Me}_\nu\left(\tilde{A}^I \geq \tilde{B}^I\right)$$

$$= \lambda \text{Pos}_\nu\left(\tilde{A}^I \geq \tilde{B}^I\right) + (1-\lambda)\text{Nes}_\nu\left(\tilde{A}^I \geq \tilde{B}^I\right)$$

$$= \begin{cases} 0, & b_3 \leq a_1'; \\ (1-\lambda)\dfrac{b_3 - a_1'}{a_2 - a_1' - b_4' + b_3}, & b_3 > a_1',\ a_2 > b_4'; \\ (1-\lambda), & a_2 < b_4',\ b_2 < a_3; \\ (1-\lambda) + \lambda\dfrac{b_2 - a_3}{b_2 - b_1' + a_4' - a_3}, & a_3 < b_2,\ a_4' > b_1'; \\ 1, & b_1' \geq a_4'. \end{cases} \qquad (13)$$

For $\lambda = 0.5$,

$$\text{Cr}_\mu\left(\tilde{A}^I \leq \tilde{B}^I\right) = \begin{cases} 0, & b_1 \leq a_1; \\ \dfrac{b_4 - a_1}{2\left(b_4 - b_3 + a_2 - a_1\right)}, & b_4 > a_1,\ a_2 > b_3; \\ \dfrac{1}{2}, & b_3 > a_2,\ b_1 < a_3; \\ \dfrac{a_4 - 2a_3 + 2b_1 - b_2}{2\left(a_4 - a_3 - b_2 + b_1\right)}, & b_1 > a_3,\ a_4 > b_2; \\ 1, & b_2 \geq a_4, \end{cases}$$

$$\text{Cr}_\nu\left(\tilde{A}^I \leq \tilde{B}^I\right) = \begin{cases} 1, & b_4' \leq a_1'; \\ \dfrac{2a_2 - 2b_3 + b_4' - a_1'}{2\left(a_2 - a_1' + b_4' - b_3\right)}, & b_4' > a_1',\ a_2 > b_3; \\ \dfrac{1}{2}, & a_2 < b_3,\ b_1' > a_3; \\ \dfrac{a_4' - b_2}{2\left(a_4' - a_3 - b_2 + b_1'\right)}, & a_3 < b_1',\ b_2 < a_4'; \\ 0, & b_2 \geq a_4', \end{cases}$$

$$Cr_\mu \left(\widetilde{A}^I \geq \widetilde{B}^I \right) = \begin{cases} 1, & b_3 \leq a_2; \\ \dfrac{2a_2 - 2b_4 + b_3 - a_1}{2\left(a_2 - a_1 - b_4 + b_3 \right)}, & b_3 > a_1, \ a_2 > b_4; \\ \dfrac{1}{2}, & b_4 > a_2, \ b_2 < a_3; \\ \dfrac{a_4 - b_1}{2\left(a_4 - a_3 + b_2 - b_1 \right)}, & b_2 > a_3, \ a_4 > b_1; \\ 0, & b_1 \geq a_4, \end{cases}$$

$$Cr_\nu \left(\widetilde{A}^I \geq \widetilde{B}^I \right) = \begin{cases} 0, & b_3' \leq a_1'; \\ \dfrac{b_3 - a_1'}{2\left(a_2 - a_1' - b_4' + b_3 \right)}, & b_3 > a_1', \ a_2 > b_4'; \\ \dfrac{1}{2}, & a_2 < b_4', \ b_2 < a_3; \\ \dfrac{2b_2 - 2a_3 + a_4' - b_1'}{2\left(b_2 - b_1' + a_4' - a_3 \right)}, & b_2 > a_3, \ b_1' < a_4'; \\ 1, & b_1' \geq a_4'. \end{cases}$$

$$(14)$$

Lemma 10. If $\widetilde{A}^I = (a_1, a_2, a_3, a_4)(a_1', a_2, a_3, a_4')$ and $\widetilde{B}^I = (b_1, b_2, b_3, b_4)(b_1', b_2, b_3, b_4')$,

$$Pos_\mu \left(\widetilde{A}^I \leq \widetilde{B}^I \right) \geq \alpha, \qquad Pos_\nu \left(\widetilde{A}^I \leq \widetilde{B}^I \right) \leq \beta$$

$$\Longleftrightarrow \frac{b_4 - a_1}{b_4 - b_3 + a_2 - a_1} \geq \alpha,$$

$$\frac{a_2 - b_3}{a_2 - a_1' + b_4' - b_3} \leq \beta.$$

$$(15)$$

Proof. Let us consider

$$Pos_\mu \left(\widetilde{A}^I \leq \widetilde{B}^I \right) \geq \alpha, \qquad Pos_\nu \left(\widetilde{A}^I \geq \widetilde{B}^I \right) \leq \beta.$$

$$(16)$$

Now from (8)

$$Pos_\mu \left(\widetilde{A}^I \leq \widetilde{B}^I \right) \geq \alpha \Longleftrightarrow \frac{b_4 - a_1}{b_4 - b_3 + a_2 - a_1} \geq \alpha,$$

$$Pos_\nu \left(\widetilde{A}^I \leq \widetilde{B}^I \right) \leq \beta \Longleftrightarrow \frac{a_2 - b_3}{a_2 - a_1' + b_4' - b_3} \leq \beta.$$

$$(17)$$

Note. $Pos(\tilde{A}I \leq x) \geq \alpha$ and $Pos_\nu(\tilde{A}I \leq x) \leq \beta \Leftrightarrow (x - a_1)/(a_2 - a_1) \geq \alpha$ and $(a_2 - x)/(a_2 - a_1') \leq \beta$.

$$Nes_\mu\left(\widetilde{A}^I \le \widetilde{B}^I\right) \ge \alpha, \qquad Nes_\nu\left(\widetilde{A}^I \le \widetilde{B}^I\right) \le \beta$$

$$\Longleftrightarrow \frac{b_1 - a_3}{a_4 - a_3 - b_2 + b_1} \ge \alpha,$$

$$\frac{a_4' - b_2}{a_4' - a_3 - b_2 + b_1'} \le \beta.$$

(18)

Proof. Let us consider

$$Nes_\mu\left(\widetilde{A}^I \le \widetilde{B}^I\right) \ge \alpha, \qquad Nes_\nu\left(\widetilde{A}^I \le \widetilde{B}^I\right) \le \beta.$$

(19)

Now from (10),

$$Nes_\mu\left(\widetilde{A}^I \le \widetilde{B}^I\right) \ge \alpha \Longleftrightarrow \frac{b_1 - a_3}{a_4 - a_3 - b_2 + b_1} \ge \alpha,$$

$$Nes_\nu\left(\widetilde{A}^I \le \widetilde{B}^I\right) \le \beta \Longleftrightarrow \frac{a_4' - b_2}{a_4' - a_3 - b_2 + b_1'} \le \beta.$$

(20)

Note. $Nes(\tilde{A}I \le x) \ge \alpha$ and $Nes_\nu(\tilde{A}I \le x) \le \beta \Leftrightarrow (x - a_3)/(a_4 - a_3) \ge \alpha$ and $(a_4' - x)/(a_4' - a_3) \le \beta$.

Lemma 12. If $\widetilde{A}^I = (a_1, a_2, a_3, a_4)(a_1', a_2, a_3, a_4')$ and $\widetilde{B}^I = (b_1, b_2, b_3, b_4)(b_1', b_2, b_3, b_4')$, them

$$Cr_\mu\left(\widetilde{A}^I \le \widetilde{B}^I\right) \ge \alpha, \qquad Cr_\nu\left(\widetilde{A}^I \le \widetilde{B}^I\right) \le \beta$$

$$\Longleftrightarrow \frac{b_4 - a_1}{2\left(b_4 - b_3 + a_2 - a_1\right)} \ge \alpha,$$

$$\frac{a_4 - 2a_3 + 2b_1 - b_2}{2\left(a_4 - a_3 - b_2 + b_1\right)} \ge \alpha,$$

$$\frac{2a_2 - 2b_3' + b_4' - a_1'}{2\left(a_2 - a_1' + b_4' - b_3\right)} \le \beta,$$

$$\frac{a_4' - b_2}{2\left(a_4' - a_3 - b_2 + b_1'\right)} \le \beta.$$

(21)

Proof. Let us consider

$$Cr_\mu\left(\widetilde{A}^I \le \widetilde{B}^I\right) \ge \alpha, \qquad Cr_\nu\left(\widetilde{A}^I \le \widetilde{B}^I\right) \le \beta.$$

(22)

Now, from (14),

$$\mathrm{Cr}_\mu\left(\widetilde{A}^I \le \widetilde{B}^I\right) \ge \alpha \Longleftrightarrow \frac{b_4 - a_1}{2\left(b_4 - b_3 + a_2 - a_1\right)} \ge \alpha,$$

$$\frac{a_4 - 2a_3 + 2b_1 - b_2}{2\left(a_4 - a_3 - b_2 + b_1\right)} \ge \alpha,$$

$$\mathrm{Cr}_\nu\left(\widetilde{A}^I \le \widetilde{B}^I\right) \le \beta \Longleftrightarrow \frac{2a_2 - 2b_3' + b_4' - a_1'}{2\left(a_2 - a_1' + b_4' - b_3\right)} \le \beta,$$

$$\frac{b_2 - a_4'}{2\left(b_2 - b_1' - a_4' + a_3\right)} \le \beta.$$

$$(23)$$

Note. $\mathrm{Cr}_\mu(\widetilde{A}^I \le x) \ge \alpha$, $\mathrm{Cr}_\nu(\widetilde{A}^I \le x) \ge \beta \Leftrightarrow (x - a_1)/2(a_2 - a_1) \ge \alpha$, $(a_4 - 2a_3 + x)/2(a_4 - a_3) \ge \alpha$ and $(2a_2 - x - a_1')/2(a_2 - a_1') \le \beta$, $(a_4' - x)/2(a_4' - a_3) \le \beta$.

Intuitionistic Fuzzy CCM

The chance operator is actually taken as possibility or necessity or credibility measures. We can use chance operator to transform the intuitionistic fuzzy problem into crisp problem, which is called as CCM [27]. A general singleobjective mathematical programming problem with intuitionistic fuzzy parameter should have the following form:

$$\text{Max} \qquad f\left(x, \xi^I\right)$$

$$\text{subject to} \quad g_i\left(x, \xi^I\right) \le \widetilde{b}_i^I, \quad i = 1, 2, \ldots, n,$$

$$x \ge 0, \qquad (24)$$

where x is the decision vector, ξ^I and $\widetilde{b}_i^I$ are intuitionistic fuzzy parameters, $f(x, \xi^I)$ is an imprecise objective function, and $g_i(x, \xi^I)$ are constraints function for $i = 1, 2, \ldots, n$.

The general chance-constraints model for problem (24) is as follows:

$$\text{Max} \qquad f_1 + f_2$$

$$\text{subject to} \quad \mathrm{Ch}_\mu\left\{f\left(x, \xi^I\right) \ge f_1\right\} \ge \alpha$$

$$\mathrm{Ch}_\nu\left\{f\left(x, \xi^I\right) \ge f_2\right\} \le \beta$$

$$\mathrm{Ch}_\mu\left\{g_i\left(x, \xi^I\right) \le \widetilde{b}_i^I\right\} \ge \lambda_i$$

$$\mathrm{Ch}_\nu\left\{g_i\left(x, \xi^I\right) \le \widetilde{b}_i^I\right\} \le \psi_i$$

$$x \ge 0, \quad i = 1, 2, \ldots, n. \qquad (25)$$

The abbreviations Ch_μ and Ch_ν represent chance operator (i.e., Pos or Nec measure) for membership and nonmembership functions. α, β, λ_i, and ψ_i are the predetermined confidence levels such that $0 \leq \lambda_i + \psi_i \leq 1$ and $0 \leq \alpha + \beta \leq 1$ for $i = 1, 2 \ldots n$.

Intuitionistic Fuzzy CCM Based on Possibility Measure.

The CCM based on possibility measure is as follows:

Max $\qquad f_1 + f_2$

Subject to $\quad \text{Pos}_\mu \left\{ f\left(x, \xi^I\right) \geq f_1 \right\} \geq \alpha$

$\qquad \text{Pos}_v \left\{ f\left(x, \xi^I\right) \geq f_2 \right\} \leq \beta$

$\qquad \text{Pos}_\mu \left\{ g_i\left(x, \xi^I\right) \leq \overline{b}_i^I \right\} \geq \lambda_i$

$\qquad \text{Pos}_v \left\{ g_i\left(x, \xi^I\right) \leq \overline{b}_i^I \right\} \leq \psi_i$

$\qquad x \geq 0, \quad i = 1, 2, \ldots, n,$ $\qquad\qquad\qquad\qquad$ (26)

Where α, β, λ_i, and ψ_i are the predetermined confidence levels such that $0 \leq \lambda_i + \psi_i \leq 1$ and $0 \leq \alpha + \beta \leq 1$ for $i = 1, 2 \ldots n$.

Definition 13. A solution x^* of the problem (26) satisfies $\text{Pos}_\mu(g_i(x, \xi^I) \leq \overline{b}_i^I) \geq \lambda_i$ and $\text{Pos}_v(g_i(x, \xi^I) \leq \overline{b}_i^I) \leq \psi_i$ for $i = 1, 2 \ldots\ldots\ldots n$ is called a feasible solution at (λ_i, ψ_i) possibility levels, $i = 1, 2 \ldots, n$

Definition 14. A feasible solution at (λ_i, ψ_i) possibility levels, x^*, is said to be (α, β) efficient solution for problem (26) if and only if there exists no other feasible solution at (λ_i, ψ_i) possibility levels, such that $\text{Pos}_\mu\{f(x, \xi^I)\} \geq \alpha$ and $\text{Pos}_v\{f(x, \xi^I)\} \leq \beta$ with $f(x) \geq f_1(x^*) + f_2(x^*)$.

Intuitionistic Fuzzy CCM Based on Necessity Measure.

The CCM based on necessity measure is as follows:

Max $\qquad f_1 + f_2$

Subject to $\quad \text{Nes}_\mu \left\{ f\left(x, \xi^I\right) \geq f_1 \right\} \geq \alpha$

$\qquad \text{Nes}_v \left\{ f\left(x, \xi^I\right) \geq f_2 \right\} \leq \beta$

$\qquad \text{Nes}_\mu \left\{ g_i\left(x, \xi^I\right) \leq \overline{b}_i^I \right\} \geq \lambda_i$

$\qquad \text{Nes}_v \left\{ g_i\left(x, \xi^I\right) \leq \overline{b}_i^I \right\} \leq \psi_i$

$\qquad x \geq 0, \quad i = 1, 2, \ldots, n,$ $\qquad\qquad\qquad\qquad$ (27)

where α, β, λ_i, and ψ_i are the predetermined confidence levels such that $0 \leq \lambda_i + \psi_i \leq 1$ and $0 \leq \alpha + \beta \leq 1$ for $i = 1, 2, \ldots, n$.

Definition 15. A solution $x*$ of the problem (27) satisfies $\text{Nes}_\mu(g_i(x, \xi^I) \leq \tilde{b}_i^I) \geq \lambda_i$ and $\text{Nes}_\nu(g_i(x, \xi^I) \leq \tilde{b}_i^I) \leq \psi_i$ for $i = 1, 2, \ldots,$ n is called a feasible solution (λ_i, ψ_i) necessity levels, $i = 1, 2, \ldots, n$.

Definition 16. A feasible solution at (λ^i, ψ^i) necessity levels, x^*, is said to be (α, β) efficient solution for problem (27) if and only if there exists no other feasible solution at (λ^i, ψ^i) necessity levels, such that $\text{Nes}_\mu\{f(x, \xi^I)\} \geq \alpha$ and $\text{Nes}_\nu\{f(x, \xi^I)\} \leq \beta$ with $f(x) \geq f_1(x^*) + f_2(x^*)$.

Intuitionistic Fuzzy CCM Based on Credibility Measure.

The CCM based on credibility measure is as follows:

$$\text{Max} \quad f_1 + f_2$$

$$\text{Subject to} \quad \text{Cr}_\mu\left\{ f\left(x, \xi^I\right) \geq f_1 \right\} \geq \alpha$$

$$\text{Cr}_\nu\left\{ f\left(x, \xi^I\right) \geq f_2 \right\} \leq \beta$$

$$\text{Cr}_\mu\left\{ g_i\left(x, \xi^I\right) \leq \tilde{b}_i^I \right\} \geq \lambda_i$$

$$\text{Cr}_\nu\left\{ g_i\left(x, \xi^I\right) \leq \tilde{b}_i^I \right\} \leq \psi_i$$

$$x \geq 0, \quad i = 1, 2, \ldots, n, \qquad (28)$$

where α, β, λ_i, and ψ_i are the predetermined confidence levels such that $0 \leq \lambda_i + \psi_i \leq 1$ and $0 \leq \alpha + \beta \leq 1$ for $i = 1, 2, \ldots, n$.

Definition 17. A solution $x*$ of the problem (28) satisfies $\text{Cr}_\mu(g_i(x, \xi^I) \leq \tilde{b}_i^I) \geq \lambda_i$ and $\text{Cr}_\nu(g_i(x, \xi^I) \leq \tilde{b}_i^I) \leq \psi_i$ for $i = 1, 2, \ldots, n$ is called a feasible solution (λ_i, ψ_i) credibility levels, $i = 1, 2, \ldots, n$

Definition 18. A feasible solution at (λ_i, ψ_i) credibility levels, x^*, is said to be (α, β) efficient solution for problem (28) if and only if there exists no other feasible solution at (λ_i, ψ_i) credibility levels, such that $\text{Cr}_\mu\{f(x, \xi^I)\} \geq \alpha$ and C $\text{Cr}_\nu\{f(x, \xi^I)\} \leq \beta$ with $f(x) \geq f_1(x^*) + f_2(x^*)$.

PROPOSED METHOD TO SOLVE IFLPP USING CHANCE OPERATOR

To solve intuitionistic fuzzy CCM based on possibility or necessity or credibility measures we propose the following method.

Step 1. Apply chance operator possibility/necessity/credibility in intuitionistic fuzzy programming (24). Problem (24) can be converted into following problem:

Max $\qquad f_1 + f_2$

Subject to $\quad \text{Pos}_\mu \left\{ f\left(x, \Im^I\right) \geq f_1 \right\} \geq \alpha$

$\qquad$ or $\text{Nes}_\mu \left\{ f\left(x, \Im^I\right) \geq f_1 \right\} \geq \alpha$

$\qquad$ or $\text{Cr}_\mu \left\{ f\left(x, \Im^I\right) \geq f_1 \right\} \geq \alpha$

$\qquad$ $\text{Pos}_\nu \left\{ f\left(x, \Im^I\right) \geq f_2 \right\} \leq \beta$ $\hspace{3cm}$ (29)

or $\text{Nes}_\nu \left\{ f\left(x, \Im^I\right) \geq f_2 \right\} \leq \beta$

or $\text{Cr}_\nu \left\{ f\left(x, \Im^I\right) \geq f_2 \right\} \leq \beta$ $\hspace{3cm}$ (30)

$\text{Pos}_\mu \left\{ g_i\left(x, \Im^I\right) \leq \tilde{b}_i^I \right\} \geq \lambda_i$

or $\text{Nes}_\mu \left\{ g_i\left(x, \Im^I\right) \leq \tilde{b}_i^I \right\} \geq \lambda_i$

or $\text{Cr}_\mu \left\{ g_i\left(x, \Im^I\right) \leq \tilde{b}_i^I \right\} \geq \lambda_i$

$\text{Pos}_\nu \left\{ g_i\left(x, \Im^I\right) \leq \tilde{b}_i^I \right\} \leq \psi_i$ $\hspace{3cm}$ (31)

or $\text{Nes}_\nu \left\{ g_i\left(x, \Im^I\right) \leq \tilde{b}_i^I \right\} \leq \psi_i$

or $\text{Cr}_\nu \left\{ g_i\left(x, \Im^I\right) \leq \tilde{b}_i^I \right\} \leq \psi_i$ $\hspace{3cm}$ (32)

$x \geq 0, \quad$ for $i = 1, 2, \ldots, n$ $\hspace{3cm}$ (33)

$0 \leq \lambda_i + \psi_i \leq 1, \quad 0 \leq \alpha + \beta \leq 1$

for $i = 1, 2, \ldots, n,$ $\hspace{3cm}$ (34)

where (λ_i, ψ_i) and (α, β) are the predefined confidence levels.

Step 2. Using Lemmas 10, 11, and/or Lemma 12, the above problem in Step 1 can also be written as

Max $f_1 + f_2$

Subject to $f_1 + f_2 \geq Z$

$\qquad$ (31) – (34),   (35)

where Z is obtained by applying Lemmas 10, 11, and/or Lemma 12 in (30) and (29).

Step 3. The above problem is equivalent to

Max Z

Subject to   (31) – (34).   (36)

Step 4. Crisp programming problem obtained in Step 2 can be solved using any well-known method to get the optimal solution.

NUMERICAL EXAMPLE

Let us consider the following intuitionistic fuzzy mathematical programming problem as:

Maximize $\xi_1^I x_1 \oplus \xi_2^I x_2$

subject to $\xi_3^I x_1 \oplus \xi_4^I x_2 \leq \xi_5^I$

$\qquad$ $\xi_6^I x_1 \oplus \xi_7^I x_2 \leq \xi_8^I$

$\qquad$ $x_1, x_2 \geq 0,$ (37)

where $\xi_1^I = (5,6,7,8)(4,6,7,9)$, $\xi_2^I = (4,5,6,7)(3,5,6,8)$, $\xi_3^I = (1,2,3,4)(0.5,2,3,6)$, $\xi_4^I = (2,3,4,5)(1,3,4,6)$, $\xi_5^I = (6,7,8,9)(5,7,8,10)$, $\xi_6^I = (3,4,5,6)(2,4,5,7)$, $\xi_7^I = (1,2,3,4)(0,2,3,4)$, and $\xi_8^I = (10,11,12,14)(9,11,12,16)$.

Intuitionistic Fuzzy CCM Based on Possibility Measure.

Now by using Step 2 of the method explained in Section 4 and

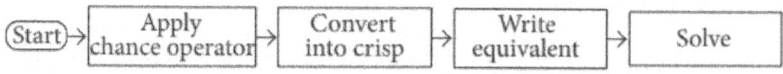

Figure 7: Flow chart of the proposed algorithm

Lemma 10, if we apply the possibility measure in intuitionistic fuzzy mathematical programming (37), problem (37) is converted into the following crisp programming problem:

$$
\begin{aligned}
\text{Maximize} \quad & (15 - \alpha + 2\beta)\, x_1 + (13 - \alpha + 2\beta)\, x_2 \\
\text{subject to} \quad & (1 + \lambda_1)\, x_1 + (2 + \lambda_1)\, x_2 \le 9 - \lambda_1 \\
& (2 - 1.5\psi_1)\, x_1 + (3 - 2\psi_1)\, x_2 \le 8 + 2\psi_1 \\
& (3 + \lambda_2)\, x_1 + (1 + \lambda_2)\, x_2 \le 14 - 2\lambda_2 \\
& (4 - 2\psi_2)\, x_1 + (2 - 2\psi_2)\, x_2 \le 12 + 4\psi_2 \\
& x_1, x_2 \ge 0, \quad 0 \le \alpha + \beta \le 1, \\
& 0 \le \lambda_i + \psi_i \le 1, \quad \text{for } i = 1, 2.
\end{aligned}
\tag{38}
$$

Solving the above crisp problem for efficient levels ($\alpha = 0.6$, $\beta = 0.4$) and different possibility levels, we get different optimal solutions. Optimal solution of (38) at different possibility levels (in Figure 7) are presented in Table 1. From Table 1, we can observe that maximum value (= 73.09) can be obtained at ($\lambda_1 = 0.40$, $\psi_1 = 0.35$) and ($\lambda_2 = 0.30$, $\psi_2 = 0.40$) possibility levels.

Intuitionistic Fuzzy CCM Based on Necessity Measure

Now by using Step 2 of the method explained in Section 4 and Lemma 11, if we apply the necessity measure in (37), problem (37) is converted into following crisp programming problem:

$$
\begin{aligned}
\text{Maximize} \quad & (10 - \alpha + 2\beta)\, x_1 + (8 - \alpha + 2\beta)\, x_2 \\
\text{subject to} \quad & (3 + 2\lambda_1)\, x_1 + (4 + 2\lambda_1)\, x_2 \le 6 + \lambda_1 \\
& (5 - 2\psi_1)\, x_1 + (6 - 2\psi_1)\, x_2 \le 7 - 2\psi_1 \\
& (5 + \lambda_2)\, x_1 + (3 + \lambda_2)\, x_2 \le 6 + \lambda_2 \\
& (7 - 2\psi_2)\, x_1 + (4 - \psi_2)\, x_2 \le 11 - 2\psi_2 \\
& x_1, x_2 \ge 0, \quad 0 \le \alpha + \beta \le 1, \\
& 0 \le \lambda_i + \psi_i \le 1, \quad \text{for } i = 1, 2.
\end{aligned}
\tag{39}
$$

Solving the above crisp linear programming problem for efficient levels ($\alpha = 0.6$, $\beta = 0.4$) and different necessity levels, we get different optimal solutions. Optimal solutions of (39) at different necessity levels (in Figures 8 and 9) are presented in Table 2. From Table 2, we can observed that at ($\lambda_1 = 0.35$, $\psi_1 = 0.45$) and ($\lambda_2 = 0.35$, $\psi_2 = 0.45$) the decision maker will get the maximum value = 12.98.

Table 1: Optimal solution of (38) at different possibility levels

λ_1	ψ_1	λ_2	ψ_2	Optimal solution	Optimal value (f^*)
0.30	0.30	0.35	0.35	$x_1 = 3.41, x_2 = 1.37$	70.09
0.30	0.35	0.35	0.40	$x_1 = 3.29, x_2 = 1.66$	72.14
0.40	0.35	0.30	0.40	$x_1 = 3.43, x_2 = 1.57$	73.09
0.35	0.40	0.40	0.30	$x_1 = 3.10, x_2 = 1.9$	72.2
0.40	0.45	0.40	0.45	$x_1 = 3.16, x_2 = 1.73$	71.05
0.50	0.45	0.45	0.5	$x_1 = 3.16, x_2 = 1.50$	67.93
0.45	0.50	0.50	0.40	$x_1 = 2.97, x_2 = 1.73$	68.02
0.60	0.45	0.45	0.6	$x_1 = 3.29, x_2 = 1.20$	65.93
0.50	0.50	0.50	0.50	$x_1 = 3.03, x_2 = 1.57$	67.00
0.55	0.40	0.55	0.40	$x_1 = 2.97, x_2 = 1.50$	65.10

Table 2: Optimal solution of (39) at different necessity levels

λ_1	ψ_1	λ_2	ψ_2	Optimal solution	Optimal value (f^*)
0.35	0.30	0.30	0.35	$x_1 = 0.91, x_2 = 0.43$	12.93
0.45	0.30	0.40	0.35	$x_1 = 0.90, x_2 = 0.45$	12.89
0.50	0.35	0.50	0.35	$x_1 = 0.87, x_2 = 0.47$	12.86
0.60	0.40	0.60	0.40	$x_1 = 0.85, x_2 = 0.50$	12.84
0.70	0.20	0.70	0.20	$x_1 = 0.87, x_2 = 0.45$	12.71
0.60	0.30	0.60	0.20	$x_1 = 0.87, x_2 = 0.47$	12.79
0.65	0.35	0.75	0.20	$x_1 = 0.84, x_2 = 0.50$	12.75
0.70	0.10	0.70	0.10	$x_1 = 0.89, x_2 = 0.43$	12.67
0.50	0.50	0.50	0.50	$x_1 = 0.85, x_2 = 0.51$	12.94
0.35	0.45	0.35	0.45	$x_1 = 0.88, x_2 = 0.48$	12.98

Table 3: Input data for IFTP

	$\overline{D}_1^I$	$\overline{D}_2^I$	$\overline{D}_3^I$	Availability ($\overline{a}_i^I$)
$\overline{S}_1^I$	(2, 4, 6, 7) (1, 4, 6, 9)	(4, 6, 7, 8) (3, 6, 7, 9)	(3, 7, 9, 12) (2, 7, 9, 13)	(4, 6, 8, 9) (2, 6, 8, 10)
$\overline{S}_2^I$	(1, 3, 4, 6) (0.5, 3, 4, 7)	(3, 5, 6, 7) (2, 5, 6, 9)	(2, 6, 7, 11) (1, 6, 7, 12)	(0, 0.5, 1, 2) (0, 0.5, 1, 5)
$\overline{S}_3^I$	(3, 4, 5, 8) (2, 4, 5, 10)	(1, 2, 3, 4) (0.5, 2, 3, 5)	(2, 4, 5, 10) (1, 4, 5, 11)	(8, 9.5, 10, 11) (6.5, 9.5, 10, 11)
Demand ($\overline{b}_j^I$)	(6, 7, 8, 10) (5, 7, 8, 12)	(4, 5, 6, 9) (3, 5, 6, 11)	(2, 4, 5, 7) (0.5, 4, 5, 8)	

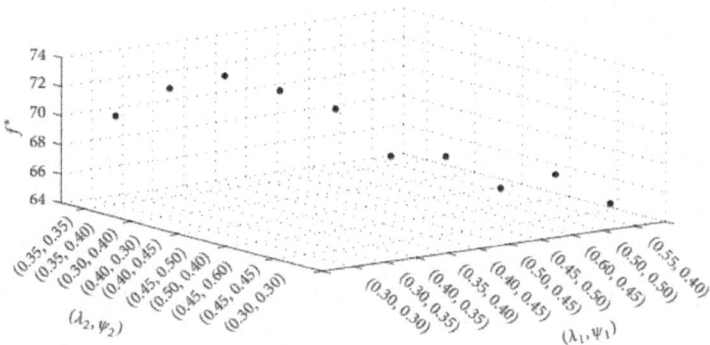

Figure 8: Optimal solution at different possibility levels.

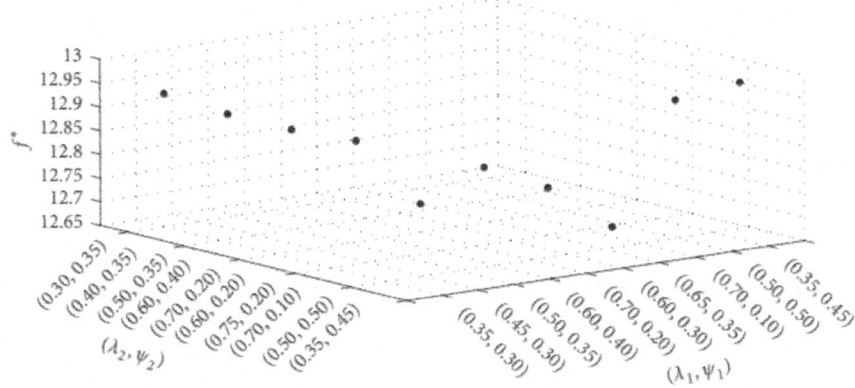

Figure 9: Optimal solution at different necessity levels.

INTUITIONISTIC FUZZY TRANSPORTATION PROBLEM BASED ON POSSIBILITY MEASURE

Let us consider the following intuitionistic fuzzy transportation problem (IFTP) (in Table 3).

Above transportation problem is a balanced transportation problem as $\oplus_{i=1}^{3} \bar{a}_i^I = \oplus_{j=1}^{3} \bar{b}_j^I$. The above IFTP can be written as

Minimize $(2, 4, 6, 7)(1, 4, 6, 9)\, x_{11}$

$\oplus (4, 6, 7, 8)(3, 6, 7, 9)\, x_{12}$

$\oplus (3, 7, 9, 12)(2, 7, 9, 13)\, x_{13}$

$\oplus (1, 3, 4, 6)(0.5, 3, 4, 7)\, x_{21}$

$\oplus (3, 5, 6, 7)(2, 5, 6, 9)\, x_{22}$

$\oplus (2, 6, 7, 11)(1, 6, 7, 12)\, x_{23}$

$\oplus (3, 4, 5, 8)(2, 4, 5, 10)\, x_{31}$

$\oplus (1, 2, 3, 4)(0.5, 2, 3, 5)\, x_{32}$

$\oplus (2, 4, 5, 10)(1, 4, 5, 11)\, x_{33}$

subject to $x_{11} + x_{12} + x_{13}$

$$\leq (4, 6, 8, 9)\,(2, 6, 8, 10)$$

$x_{21} + x_{22} + x_{23}$

$$\leq (0, 0.5, 1, 2)\,(0, 0.5, 1, 5)$$

$x_{31} + x_{32} + x_{33}$

$$\leq (8, 9.5, 10, 11)\,(6.5, 9.5, 10, 11)$$

$x_{11} + x_{21} + x_{31}$

$$\geq (6, 7, 8, 10)\,(5, 7, 8, 12)$$

$x_{12} + x_{22} + x_{32}$

$$\geq (4, 5, 6, 9)\,(3, 5, 6, 11)$$

$x_{13} + x_{23} + x_{33}$

$$\geq (2, 4, 5, 7)\,(0.5, 4, 5, 8)$$

$x_{ij} \geq 0, \quad \forall i, j.$

$$(40)$$

Now by using Step 2 of the method explained in Section 4 and Lemma 10, if we apply the possibility measure in intuitionistic fuzzy mathematical programming (40), problem (40) is converted into following crisp programming problem:

Minimize $(6 + 2\alpha - 3\beta)\,x_{11} + (10 + 2\alpha - 2\beta)\,x_{12}$

$$+ (10 + 4\alpha - 5\beta)\,x_{13} + (4 + 2\alpha - 2.5\beta)\,x_{21}$$

$$+ (8 + 2\alpha - 3\beta)\,x_{22} + (8 + 4\alpha - 5\beta)\,x_{23}$$

$$+ (7 + \alpha - 5\beta)\,x_{31} + (3 + \alpha - 1.5\beta)\,x_{32}$$

$$+ (6 + 2\alpha - 3\beta)\,x_{33}$$

subject to $x_{11} + x_{12} + x_{13} \leq (17 - \lambda_1 + 2\psi_1)$

$$x_{21} + x_{22} + x_{23} \leq (3 - \lambda_2 + 4\psi_2)$$

$$x_{31} + x_{32} + x_{33} \leq (21 - \lambda_3 + \psi_3)$$

$$x_{11} + x_{21} + x_{31} \geq (13 + \lambda_4 - 2\psi_4)$$

$$x_{12} + x_{22} + x_{32} \geq (9 + \lambda_5 - 2\psi_5)$$

$$x_{13} + x_{23} + x_{33} \geq (6 + 2\lambda_6 - 3.5\psi_6)$$

$$x_{ij} \geq 0, \quad \text{for } i, j = 1, 2, 3, \quad 0 \leq \alpha + \beta \leq 1,$$

$$0 \leq \lambda_k + \psi_k \leq 1, \quad \text{for } k = 1, 2, \ldots, 6.$$

$$(41)$$

Solving the above crisp problem for efficient levels ($\alpha = 0.6$, $\beta = 0.4$) and possibility levels ($\lambda_i = 0.5$, $\psi_i = 0.5$) for $i = 1, 2, \ldots\ldots, 6$, using Lingo-11.0, we get $x_{11} = 0.75$, $x_{12} = 0$, $x_{13} = 0$, $x_{21} = 4.5$, $x_{22} = 0$, $x_{23} = 0$, $x_{31} = 7.25$, $x_{32} = 8.5$, and $x_{33} = 5.25$. Now the minimum intuitionstic fuzzy optimal cost is

$$\tilde{c}^I = (72.25, 109, 136, 203.5)\,(45.62, 109, 136, 246.25)\,. \tag{42}$$

DISCUSSION

Intuitionistic fuzzy sets being a generalization of fuzzy sets give us an additional possibility to represent imperfect knowledge, making it possible to describe many real problems in a more adequate way. So in this paper, we have developed the possibility and necessity measures on intuitionistic fuzzy set. Here we have presented first time the mathematical representation of different types of measures in intuitionistic fuzzy environments and some graphical representations of them are depicted. We have also developed the theoretical calculation on possibility, necessity, and credibility measures for defuzzify intuitionistic fuzzy linear programming problem using chance operators. To validate the proposed method, we have discussed three different approaches to defuzzify the intuitionistic fuzzy relations using possibility, necessity, and credibility measures. Using chance operator we can convert a problem under imprecise models to corresponding crisp models. At different levels of possibility, necessity, and credibility, we have achieved different optimal solution. A numerical example is presented and solved using LINGO-11.0 to illustrate the proposed approaches. The proposed method can be applied for multiobjective, multiitem transportation problem. This method can be also extended to be applied into different types of optimization problem, namely, optimal control and solid transportation problems.

CONFLICT OF INTERESTS

The authors declare that there is no conflict of interests regarding the publication of this paper.

REFERENCES

1. L. A. Zadeh, "Fuzzy sets," Information and Computation, vol. 8, pp. 338–353, 1965.

2. K. T. Atanassov, Intuitionistic Fuzzy Sets, VII ITKR's Session, Sofia, Bulgarian, 1983.

3. S. K. De and S. S. Sana, "A multi-periods production-inventory model

with capacity constraints for multi-manufacturers—a global optimality in intuitionistic fuzzy environment," Applied Mathematics and Computation, vol. 242, no. 1, pp. 825–841, 2014.

4.   H. Garg, M. Rani, S. P. Sharma, and Y. Vishwakarma, "Intuitionistic fuzzy optimization technique for solving multi-objective reliability optimization problems in interval environment," Expert Systems with Applications, vol. 41, no. 7, pp. 3157–3167, 2014.

5.   J. Wu and Y. Liu, "An approach for multiple attribute group decision making problems with interval-valued intuitionistic trapezoidal fuzzy numbers," Computers and Industrial Engineering, vol. 66, no. 2, pp. 311–324, 2013.

6.   A. Nagoorgani and K. Ponnalagu, "A new approach on solving intuitionistic fuzzy linear programming problem," Applied Mathematical Sciences, vol. 6, no. 70, pp. 3467–3474, 2012.

7.   G. S. Mahapatra and T. K. Roy, "Intuitionistic fuzzy number and its arithmetic operation with application on system failure," Journal of Uncertain Systems, vol. 7, no. 2, pp. 92–107, 2013.

8.   S. Pramanik, D. K. Jana, and M. Maiti, "Multi-objective solid transportation problem in imprecise environments," Journal of Transportation Security, vol. 6, no. 2, pp. 131–150, 2013.

9.   S. Pramanik, D. K. Jana, and M. Maiti, "A multi objective solid transportation problem in fuzzy, Bi-fuzzy environment via genetic algorithm," International Journal of Advanced Operations Management, vol. 6, no. 1, pp. 4–26, 2014.

10. K. T. Atanassov, "Intuitionistic fuzzy sets," Fuzzy Sets and Systems, vol. 20, no. 1, pp. 87–96, 1986.

11. M. Esmailzadeh and M. Esmailzadeh, "New distance between triangular intuitionistic fuzzy numbers,"Advances in Computational Mathematics and Its Applications, vol. 2, no. 3, pp. 3–10, 2013.

12. P. P. Angelov, "Optimization in an intuitionistic fuzzy environment," Fuzzy Sets and Systems, vol. 86, no. 3, pp. 299–306, 1997.

13. D. Dubey and A. Mehra, "Linear programming with triangular intuitionistic fuzzy number," Advances in Intelligent Systems Research, vol. 1, no. 1, pp. 563–569, 2011.

14. R. Parvathi and C. Malathi, "Intuitionistic fuzzy simplex method," International Journal of Computer Applications, vol. 48, no. 6, pp. 39–48, 2012.

15. R. J. Hussain and S. P. Kumar, "Algorithmic approach for solving

intuitionistic fuzzy transportation problem," Applied Mathematical Sciences, vol. 6, no. 77-80, pp. 3981–3989, 2012.

16. A. Nagoor Gani and S. Abbas, "Solving intuitionstic fuzzy transportation problem using zero suffix algorithm," International Journal of Mathematics Sciences & Enggineering Applications, vol. 6, pp. 73–82, 2012.

17. J. Ye, "Expected value method for intuitionistic trapezoidal fuzzy multicriteria decision-making problems," Expert Systems with Applications, vol. 38, no. 9, pp. 11730–11734, 2011.

18. S. Wan and J. Dong, "A possibility degree method for interval-valued intuitionistic fuzzy multi-attribute group decision making," Journal of Computer and System Sciences, vol. 80, no. 1, pp. 237–256, 2013.

19. J. J. Buckley, "Possibility and necessity in optimization," Fuzzy Sets and Systems, vol. 25, no. 1, pp. 1–13, 1988.

20. K. D. Jamison and W. A. Lodwick, "The construction of consistent possibility and necessity measures,"Fuzzy Sets and Systems, vol. 132, no. 1, pp. 1–10, 2002.

21. J. Ramík, "Duality in fuzzy linear programming with possibility and necessity relations," Fuzzy Sets and Systems, vol. 157, no. 10, pp. 1283–1302, 2006.

22. M. G. Iskander, "A suggested approach for possibility and necessity dominance indices in stochastic fuzzy linear programming," Applied Mathematics Letters, vol. 18, no. 4, pp. 395–399, 2005.

23. M. Sakawa, H. Katagiri, and T. Matsui, "Fuzzy random bilevel linear programming through expectation optimization using possibility and necessity," International Journal of Machine Learning and Cybernetics, vol. 3, no. 3, pp. 183–192, 2012.

24. S. Pathak, S. Kar, and S. Sarkar, "Fuzzy production inventory model for deteriorating items with shortages under the effect of time dependent learning and forgetting: a possibility/necessity approach,"OPSEARCH, vol. 50, no. 2, pp. 149–181, 2013.

25. K. Maity, "Possibility and necessity representations of fuzzy inequality and its application to two warehouse production-inventory problem," Applied Mathematical Modelling, vol. 35, no. 3, pp. 1252–1263, 2011.

26. H.-C. Wu, "Fuzzy optimization problems based on the embedding theorem and possibility and necessity measures," Mathematical and Computer Modelling, vol. 40, no. 3-4, pp. 329–336, 2004.

27. J. Xu and X. Zhou, Fuzzy-Like Multiple Objective Decesion Making,

Springer, 2010.

28. K. Maity and M. Maiti, "Possibility and necessity constraints and their defuzzification—a multi-item production-inventory scenario via optimal control theory," European Journal of Operational Research, vol. 177, no. 2, pp. 882–896, 2007.

29. B. Das, K. Maity, and M. Maiti, "A two warehouse supply-chain model under possibility/necessity/credibility measures," Mathematical and Computer Modelling, vol. 46, no. 3-4, pp. 398–409, 2007.

30. D. Panda, S. Kar, K. Maity, and M. Maiti, "A single period inventory model with imperfect production and stochastic demand under chance and imprecise constraints," European Journal of Operational Research, vol. 188, no. 1, pp. 121–139, 2008.

31. A. I. Ban, "Intuitionistic fuzzy-valued possibility and necessity measures," in Proceedings of the 8th International Conference on Intuitionistic Fuzzy Sets, vol. 10, pp. 1–7, Varna, Bulgaria, 2004.

Chapter 3

NUMERICAL ANALYSIS OF VIBRATION ISOLATION USING PILE ROWS AGAINST THE VIBRATION DUE TO MOVING LOADS IN A VISCOELASTIC MEDIUM

Bin Xu and Man-Qing Xu

Department of Civil Engineering, Nanchang Institute of Technology, Nanchang, Jiangxi 330029, China

ABSTRACT

A numerical method for evaluating the vertical vibration isolation effect of pile rows embedded in a viscoelastic half space subjected to a moving load is developed in this paper on the basis of the Cole-Cole model and Muki's method. Based on the proposed method, the influence of various parameters on the vibration isolation effect of pile rows embedded in the viscoelastic half space is investigated numerically.

INTRODUCTION

Vibration induced by railway traffic is a major concern for civil engineers as it causes annoyance to residents or even damage to adjacent structures. Generally, the effects of ground vibrations can be mitigated by two kinds of vibration isolation methods: the active and the passive vibration isolation methods. The active isolation system is often used to reduce the ground vibration near the source. It is usually installed either around the vibration source or at a close distance to the source. The passive isolation system, on the other hand, usually is far away from the source and surrounds the protected structure. Normally, there are two passive vibration isolation methods: the trench (open or infilled) isolation method and the pile (pile rows or sheet piles) isolation method. To date, many studies concerning vibration isolation using trenches or piles have been conducted. For example, Emad and Manolis [1] utilized the boundary element method (BEM) with constant elements to examine the efficiency of vibration reduction by open trench with a rectangular or a circular cross-section. Considering the coupling effects between the soil skeleton and underground water, Cao et al. [2] proposed an analytical model to investigate the screening efficiency of trenches to moving-load induced ground vibrations based on Biot's dynamic poroelastic theory. Cai et al. [3] investigated the vibration isolation effect of pile rows embedded in a poroelastic medium by using the wave function expansion method. Kattis et al. [4, 5] used 3D BEM to calculate the screening effectiveness of a pile row in the frequency domain. Also, by means of the frequency domain BEM, the screening effectiveness of four types of circular piles in a row against the vibration due to a massless square foundation subjected to a harmonic vertical loading is studied by Tsai et al. [6]. Besides, by means of the fictitious pile method developed by Muki and Sternberg [7] and the direct superposition method, Lu et al. [8, 9] analyzed the vibration isolation effect of pile rows.

It is noticed that previous studies concern the vibration isolation modelling in a generalized standard linear viscoelastic solid. For the generalized standard linear viscoelastic solid, the complex modulus belongs to the rational function. However, both creep tests [10] and vibration tests [11] suggest that the derivatives of stress relaxation functions have an asymptotic behaviour const. As a result, the convolution in the constitutive relation of the heterogeneous viscoelastic medium cannot be eliminated by the method used in the generalized standard linear viscoelastic solid. Unlike the standard linear solid, the complex modulus of the Cole-Cole viscoelastic medium is not the rational function, a viscoelastic kernel having a singularity const $\times t^{-\alpha}$ at $t \to 0$, where $0 < \alpha$ in both quasi-static and vibrations experiments [10]. It has been pointed out in several studies, for example Bagley and Torvik [13], Lu and Hanyga [14], and Soula et al. [15], that the Cole-Cole model fits experimental data over several decades of frequency. Experimental evidences in rock physics [16, 17] also point out to the Cole-Cole type behavior. Furthermore, the Cole-Cole relaxation model has many advantages over other models. In particular, the Cole-Cole relaxation model is compatible with a finite speed of wave propagation. It is very important to conduct accurate analysis of the wave attenuation and dispersion when moving loads are of high velocities.

In this study, a numerical method for evaluating the vertical vibration isolation effect of pile rows embedded in a viscoelastic half space subjected to a moving load is developed on the basis of the Cole-Cole model [12] and Muki's method [7]. Based on the proposed method, the influence of various parameters on the vibration isolation effect of pile rows embedded in the viscoelastic half space is investigated numerically. It is noted that the proposed method in this study belongs to the semianalytical category. Thus, compared with conventional domain discretization methods such as the finite element method (FEM) and the boundary element method (BEM), it significantly reduces the computational time.

THE FREE WAVE FIELD SOLUTION AND THE FUNDAMENTAL SOLUTION FOR A CIRCULAR UNIFORM PATCH LOAD

Usually, for a linear isotropic viscoelastic medium described by the Cole-Cole model, the P and the S wave modes satisfy the Cole-Cole relaxation law with different parameters. Consequently, in terms of the Cole-Cole models, the complex moduli for the P and the S waves in the frequency domain have the following forms:

$$M_p(\omega) = \bar{\lambda}(\omega) + 2\bar{\mu}(\omega)$$

$$= M_{\infty p} \frac{1 + a_p(i\eta_p\omega)^{-\alpha_p}}{1 + (i\eta_p\omega)^{-\alpha_p}} = M_{\infty p} + \frac{M_{\infty p}(a_p - 1)}{1 + (i\eta_p\omega)^{-\alpha_p}}, \tag{1a}$$

$$M_s(\omega) = \bar{\mu}(\omega)$$

$$= M_{\infty s} \frac{1 + a_s(i\eta_s\omega)^{-\alpha_s}}{1 + (i\eta_s\omega)^{-\alpha_s}} = M_{\infty s} + \frac{M_{\infty s}(a_s - 1)}{1 + (i\eta_s\omega)^{-\alpha_s}}, \tag{1b}$$

Where a bar over the function denotes the Fourier transform for time $t \rightarrow$ frequency ω. The subscripts p and s denote the P and S waves in a viscoelastic medium and $\bar{\lambda}(\omega)$ and $\bar{\mu}(\omega)$ are the complex modulus corresponding to the two Lame constants. M_∞ is the limit of the complex modulus for $\omega \rightarrow {}_\infty$, $M0$ is the value of the complex modulus for $\omega=0$, η is a characteristic relaxation time, and α controls the width of the transition zone between M_∞ and M_0. Besides, the conditions $0 < \alpha < 1$ and $a \leq 1$ follow from thermodynamics argument [12].

In this study, the free wave field solution is defined as the solution of the moving load in the absence of the pile rows. For a moving load, axisymmetry is lost due to the orientation of the load speed; thus, it is more convenient to consider the moving load problem in the Cartesian coordinate system (Figure 1). A moving load with a constant speed c and an oscillating frequency ω_0 is applied on the surface of the viscoelastic half space. The load moves along the negative direction of the y-axis and the distance between the load and y-axis is

ds (Figure 1). For the moving load applied over a rectangular area $2a \times 2_b$, the boundary conditions in the time-space domain are as follows:

$$\sigma_{zx}(x, y, z, t)\big|_{z=0} = 0, \tag{2a}$$

$$\sigma_{zy}(x, y, z, t)\big|_{z=0} = 0, \tag{2b}$$

$$\sigma_{zz}(x, y, z, t)\big|_{z=0}$$
$$= -q_z \left[H(x + d_s + a) - H(x + d_s - a) \right]$$
$$\times \left[H(y - y_0 + b + ct) - H(y - y_0 - b + ct) \right] e^{i\omega_0 t}, \tag{2c}$$

where q_z is the intensity of the distributed load, ω_0 is the frequency of moving load, $H(*)$ is the Heaviside step function, and $y0$ is the y coordinate of the center of the distributed load at $t=0$.

Performing a triple Fourier transform with respect to time and the two horizontal coordinates on ((2a), (2b), and (2c)), respectively, the following boundary conditions are derived in the frequency wavenumber domain:

$$\tilde{\tilde{\tilde{\sigma}}}_{zx}\left(k_x, k_y, 0, \omega\right) = 0, \tag{3a}$$

$$\tilde{\tilde{\tilde{\sigma}}}_{zy}\left(k_x, k_y, 0, \omega\right) = 0, \tag{3b}$$

$$\tilde{\tilde{\tilde{\sigma}}}_{zz}\left(k_x, k_y, 0, \omega\right)$$
$$= -8\pi q_z \frac{\sin(k_x a)}{k_x} \frac{\sin(k_y b)}{k_y} e^{i(k_x d_s - k_y y_0)} \delta\left(\omega - \omega_0 - k_y c\right) \tag{3c}$$

in which k_x and k_y represent the two horizontal wave numbers corresponding to x- and y-coordinates, respectively, and $\delta(*)$ is the Dirac delta function.

For a moving point load, the boundary conditions for σ_{zx}, σ_{zy} are the same as those for the moving rectangular distributed load, while the boundary condition for σ_{zz} is as follows

$$\sigma_{zz}(x, y, z, t)\big|_{z=0} = -F_z \delta(x + d_s) \delta(y - y_0 + ct) e^{i\omega_0 t}. \tag{4}$$

Likewise, the boundary condition for $\tilde{\tilde{\tilde{\sigma}}}_{zz}$ in the frequency wavenumber domain is given by

$$\bar{\bar{\sigma}}_{zz}\left(k_x, k_y, 0, \omega\right) = -2\pi F_z e^{i\left(k_x d_s - k_y y_0\right)} \delta\left(\omega - \omega_0 - c k_y\right).$$

(5)

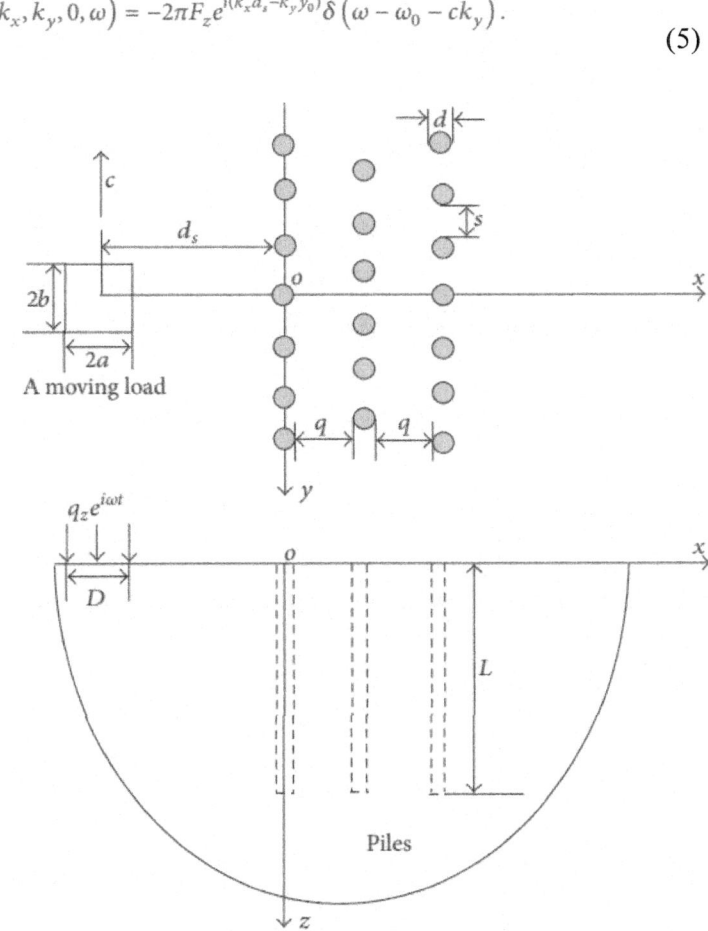

Figure 1: Piles rows embedded in a viscoelastic half space as the vibration isolation system against the vibration due to moving loads.

Using ((3a), (3b), and (3c)) or (5) as well as the expressions for the stress in the frequency wavenumber domain, the free wave field due to the surface moving load can be determined. In view of (3a), (3b), and (3c), all the variables in the frequency wavenumber domain due to the moving distributed rectangular load can be expressed in the following form:

$$\overline{\overline{\tilde{\Omega}}}\left(k_x, k_y, z, \omega\right)$$

$$= \left[-8\pi q_z \frac{\sin\left(k_x a\right)}{k_x} \frac{\sin\left(k_y b\right)}{k_y} e^{i\left(k_x d_s - k_y y_0\right)} \delta\left(\omega - \omega_0 - k_y c\right)\right]$$

$$\times \overline{\overline{\tilde{\Omega}}}^{*}\left(k_x, k_y, z, \omega\right),$$ (6)

where $\overline{\overline{\tilde{\Omega}}}^{*}(k_x, k_y, z, \omega)$ is the solution of a variable corresponding to a unit boundary value $\overline{\overline{\tilde{\sigma}}}_{zz}$ in (3c). The moving point load has similar expression

$$\overline{\overline{\tilde{\Omega}}}\left(k_x, k_y, z, \omega\right)$$

$$= \left[-2\pi F_z e^{i\left(k_x d_s - k_y y_0\right)} \delta\left(\omega - \omega_0 - c k_y\right)\right] \times \overline{\overline{\tilde{\Omega}}}^{*}\left(k_x, k_y, z, \omega\right).$$ (7)

Performing the inverse Fourier transform with respect to the two horizontal wavenumbers and using the property of the delta function, the frequency domain free field solution for the moving distributed rectangular load has the form:

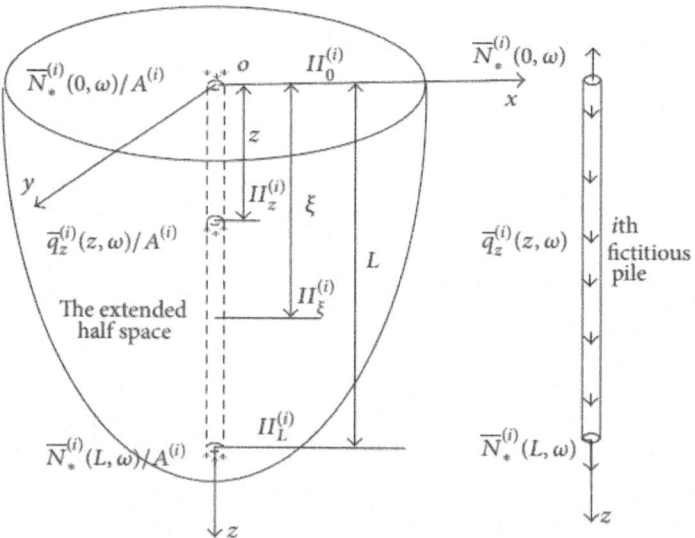

Figure 2: The model for decomposition of the pile-half space system. (a) an extended poro-elastic half-space (b) a fictitious pile.

$$\overline{\Omega}(x, y, z, \omega)$$

$$= \frac{2q_z}{\pi} \frac{\sin\left[b\left(\omega - \omega_0\right)/c\right]}{\omega - \omega_0} e^{i((\omega-\omega_0)/c)(y-y_0)}$$

$$\times \int_{-\infty}^{+\infty} \frac{\sin\left(k_x a\right)}{k_x} \widehat{\Omega}^*\left(k_x, \frac{\omega - \omega_0}{c}, z, \omega\right) e^{ik_x(x+d_s)} dk_x.$$

(8)

For the moving point load, the frequency domain free field solution for all variables is represented by

$$\overline{\Omega}(x, y, z, \omega)$$

$$= -\frac{F_z}{2\pi c} e^{i((\omega-\omega_0)/c)(y-y_0)}$$

$$\times \int_{-\infty}^{+\infty} \widehat{\Omega}^*\left(k_x, \frac{\omega - \omega_0}{c}, z, \omega\right) e^{ik_x(x+d_s)} dk_x.$$

(9)

To establish the integral equations for the pile rows, the frequency domain fundamental solution for a uniform vertical patch load applied in the viscoelastic half space is required. As the problem for a viscoelastic half space subjected to a uniform vertical patch load over a circular area with a radius R (Figure 2) is axisymmetric with respect to the center of the circular area, thus, it is more convenient to consider the problem in the cylindrical coordinate system (r, θ, z).

For a uniform vertical patch load applied in the viscoelastic half space, the surface of the half space is stress-free. Therefore, the boundary conditions for the half-space surface are as follows:

$$\widetilde{\sigma}_{zz}^{[0]}(\xi, 0, \omega) = 0, \qquad \widetilde{\sigma}_{zr}^{[1]}(\xi, 0, \omega) = 0,$$

(10)

where $[m]$ denotes the mth order first kind of Bessel function and ξ denotes the Hankel transform parameter.

It is assumed that the circular patch load is applied at the depth $z=h$. Accordingly, the half space is divided by the plane $z=h$ into an upper and a lower region and, therefore, the continuity conditions at the plane $z=h$ are as follows:

$$\tilde{\bar{u}}_r^{[1]}\left(\xi, h^-, \omega\right) = \tilde{\bar{u}}_r^{[1]}\left(\xi, h^+, \omega\right),$$

$$\tilde{\bar{u}}_z^{[0]}\left(\xi, h^-, \omega\right) = \tilde{\bar{u}}_z^{[0]}\left(\xi, h^+, \omega\right),$$

$$\tilde{\bar{\sigma}}_{zz}^{[0]}\left(\xi, h^+, \omega\right) - \tilde{\bar{\sigma}}_{zz}^{[0]}\left(\xi, h^-, \omega\right) = -\frac{R J_1\left(R\xi\right)}{A\xi},$$

$$\tilde{\bar{\sigma}}_{zr}^{[1]}\left(\xi, h^-, \omega\right) = \tilde{\bar{\sigma}}_{zr}^{[1]}\left(\xi, h^+, \omega\right).$$

(11)

Using the general expressions for the displacement and the stress and (10) and (11), the arbitrary constants involved in the expressions for the potentials, the displacement, and the stress can be determined. The frequency fundamental solution can be obtained by performing inverse Hankel transform on the solutions for the displacement, the stress in the frequency-wavenumber domain [18].

FREDHOLM INTEGRAL EQUATIONS DESCRIBING DYNAMIC INTERACTION BETWEEN PILES AND THE VISCOELASTIC HALF SPACE

As shown in Figure 1, the pile rows embedded in the viscoelastic half space are used to isolate the vibration generated by a moving load. The number of the total pile is $m = \sum_{k=1}^{K} n_k$, where K and n_k denote the number of pile rows and the number of the piles in the kth row. The spacing between two neighboring piles in each pile row is denoted by s. The spacing between two neighboring pile rows is q. Also, it is assumed that each pile has the same diameter d ($d = 2R$) and the same length L ($d/L \ll 1$). A moving load with a constant speed c and an oscillating frequency $\omega 0$ is acting on the surface of the viscoelastic half space and moves along the negative direction of the y-axis (Figure 1).

When the pile-half-space system is subjected to a vertical moving load, generally, the pile will undergo both vertical and horizontal response. However, as the influence of the horizontal interaction between the pile-soil system is relatively smaller, the horizontal interaction between piles and the half space is neglected in this paper. Thus, only the vertical interaction between the piles and the half space is considered in the paper. Following Muki and Sternberg [7] and Pak and Jennings [19], the current problem is decomposed into two subproblems: an extended viscoelastic half space and the multiple fictitious piles. The extended viscoelastic half space is described by the continuum dynamic theory for a viscoelastic medium, while the fictitious piles are described by the 1D bar vibration theory. The decomposition procedure is illustrated in Figure 2 by the ith pile of the pile rows.

It is assumed that the axial force of the ith fictitious pile is $\overline{N}_*^{(i)}(z, \omega)$ and the vertical distributed load along the ith fictitious pile is $\overline{q}_z^{(i)}(z, \omega)$ (Figure 2(b)). The top and the bottom of the ith fictitious pile are subjected to forces $\overline{N}_*^{(i)}(0, \omega)$ and $\overline{N}_*^{(i)}(L, \omega)$, respectively. The viscoelastic half space is subjected to the following loads (Figure 2(a)): $\overline{q}_z^{(i)}(z, \omega)$ which is distributed over the region occupied by the ith pile; $\overline{N}_*^{(i)}(0, \omega)/A^{(i)}$ and $\overline{N}_*^{(i)}(L, \omega)/A^{(i)}$ which are applied to the circular areas $\Pi_0^{(i)}$ and $\Pi_L^{(i)}$, respectively. Note that $^{(i)}$ denotes the cross-section area of the ith pile. For the ith fictitious pile, the displacement $\overline{u}_{zp*}^{(i)}(z, \omega)$, the distributed vertical force $\overline{q}_z^{(i)}(z, \omega)$, and the axial force satisfy the following relations:

$$\overline{q}_z^{(i)}(z, \omega) = -\frac{d\overline{N}_*^{(i)}(z, \omega)}{dz} - \rho_p^{(i)} A^{(i)} \omega^2 \overline{u}_{zp*}^{(i)}(z, \omega),$$

$$i = 1, 2, \ldots, m, \tag{12a}$$

$$\overline{u}_{zp*}^{(i)}(z, \omega) = \overline{u}_{zp*}^{(i)}(0, \omega) + \frac{1}{E_{p*}^{(i)} A^{(i)}} \int_0^z \overline{N}_*^{(i)}(\eta, \omega) \, d\eta,$$

$$i = 1, 2, \ldots, m \tag{12b}$$

in which $\overline{u}_{zp*}^{(i)}(z, \omega)$ is the vertical displacement of the ith pile

The vertical strain of the extended half space along the axis of the ith pile is composed of two parts: the first part is due to the free wave field, while the second part is due to the force applied to the extended half space by the fictitious piles. Thus, the vertical strain of the extended half space along the axis of the ith pile can be written as

$$\overline{\varepsilon}_{zs}^{(i)}(z, \omega) = \overline{\varepsilon}_{zf}^{(i)}(z, \omega)$$

$$+ \sum_{j=1}^m \left[\overline{N}_*^{(j)}(0, \omega) \overline{\varepsilon}_z^{(G)}(r_{ij}, 0, z, \omega) \right.$$

$$- \overline{N}_*^{(j)}(L, \omega) \overline{\varepsilon}_z^{(G)}(r_{ij}, L, z, \omega)$$

$$\left. - \int_0^{L_j} \overline{q}_z^{(j)}(\zeta, \omega) \overline{\varepsilon}_z^{(G)}(r_{ij}, \zeta, z, \omega) \, d\zeta \right],$$

$$i = 1, 2, \ldots, m. \tag{13}$$

In (13), the superscript and subscript i and j denote the ith and the jth pile, respectively, $\bar{\varepsilon}_{zf}^{(i)}(z,\omega)$ is the free field vertical strain at the axis of the ith pile, which is determined by the free field frequency domain solution for the moving load, and $\bar{\varepsilon}_z^{(G)}(r_{ij},\zeta,z,\omega)$ represents the vertical strain at the center of $\Pi_z^{(i)}$ due to a unit patch load applied at $\Pi_\zeta^{(j)}$ (Figure 2(a)) and r_{ij} is the horizontal distance between the axis of the ith and jth pile. It is worth noting that, for the case $i = j$, r_{ij} is vanishing

$$\bar{\varepsilon}_{zs}^{(i)}(z,\omega)$$

$$= \bar{\varepsilon}_{zf}^{(i)}(z,\omega)$$

$$- \overline{N}_*^{(i)}(z,\omega)\left[\bar{\varepsilon}_z^{(G)}(r_{ii},z^+,z,\omega) - \bar{\varepsilon}_z^{(G)}(r_{ii},z^-,z,\omega)\right]$$

$$- \int_0^{L_i} \overline{N}_*^{(i)}(z,\omega)\frac{\partial\bar{\varepsilon}_z^{(G)}(r_{ii},\zeta,z,\omega)}{\partial\zeta}d\zeta$$

$$+ \rho_{p^*}^{(i)}A^{(i)}\omega^2\int_0^{L_i}\bar{u}_{zp^*}^{(i)}(\zeta,\omega)\bar{\varepsilon}_z^{(G)}(r_{ii},\zeta,z,\omega)\,d\zeta$$

$$+ \sum_{j=1}^{m(j\neq i)}\left[-\int_0^{L_j}\overline{N}_*^{(j)}(\zeta,\omega)\frac{\partial\bar{\varepsilon}_z^{(G)}(r_{ij},\zeta,z,\omega)}{\partial\zeta}d\zeta\right.$$

$$\left.+ \rho_{p^*}^{(j)}A^{(j)}\omega^2\int_0^{L_j}\bar{u}_{zp^*}^{(j)}(\zeta,\omega)\bar{\varepsilon}_z^{(G)}(r_{ij},\zeta,z,\omega)\,d\zeta\right],$$

$$i = 1,2,\dots,m, \tag{14}$$

where $\bar{\varepsilon}_z^{(G)}(r_{ii},z^-,z,\omega)$ and $\bar{\varepsilon}_z^{(G)}(r_{ii},z^+,z,\omega)$ denote the vertical strain of the viscoelastic half space at the center of $\Pi(i)\,z$ of the ith pile when the patch load $\Pi_\xi^{(i)}$ approaches $\Pi_z^{(i)}$ from up and down side, respectively

In this study, the compatibility condition between the ith pile and the viscoelastic half space is fulfilled by requiring the vertical strain of the ith fictitious pile and that of the extended half space along the axis of the ith fictitious pile to be equal to

$$\bar{\varepsilon}_{zp^*}^{(i)}(z,\omega) = \bar{\varepsilon}_{zs}^{(i)}(z,\omega), \quad 0 \leq z \leq L, \; i = 1,2,\dots,m, \tag{15}$$

where $\bar{\varepsilon}_{zp^*}^{(i)}(z,\omega)$ represents the vertical strain of the ith fictitious pile

Using (12a) and (12b), (14) and (15), the Fredholm integral equation in the frequency domain describing the vertical interaction between the ith pile and the half space has the form

$$\frac{\overline{N}_*^{(i)}(z,\omega)}{E_{p^*}^{(i)} A^{(i)}}$$

$$+ \overline{N}_*^{(i)}(z,\omega) \left[\overline{\varepsilon}_z^{(G)}(r_{ii}, z^+, z, \omega) - \overline{\varepsilon}_z^{(G)}(r_{ii}, z^-, z, \omega) \right]$$

$$+ \sum_{j=1}^{m} \left[\int_0^{L_j} \overline{N}_*^{(j)}(\zeta,\omega) \frac{\partial \overline{\varepsilon}_z^{(G)}(r_{ij}, \zeta, z, \omega)}{\partial \zeta} d\zeta \right.$$

$$- \int_0^{L_j} \overline{N}_*^{(j)}(\zeta,\omega) \overline{\chi}_{ij}^{(a)}(\zeta, z, \omega) d\zeta$$

$$\left. - \overline{u}_{zp^*}^{(j)}(0,\omega) \overline{\chi}_{ij}^{(b)}(z,\omega) \right]$$

$$= \overline{\varepsilon}_{zf}^{(i)}(z,\omega), \quad i = 1, 2, \ldots, m, \tag{16}$$

where

$$\overline{\chi}_{ij}^{(a)}(\zeta, z, \omega) = \left(\frac{\rho_{p^*}^{(j)} \omega^2}{E_{p^*}^{(j)}} \right) \int_\zeta^{L_j} \overline{\varepsilon}_z^{(G)}(r_{ij}, \eta, z, \omega) d\eta,$$

$$\overline{\chi}_{ij}^{(b)}(z, \omega) = \rho_{p^*}^{(j)} A^{(j)} \omega^2 \int_0^{L_j} \overline{\varepsilon}_z^{(G)}(r_{ij}, \eta, z, \omega) d\eta. \tag{17}$$

Following the similar procedures, the surface vertical displacement $\overline{u}_z(\mathbf{x}_\perp, z = 0, \omega)$ for the viscoelastic half space in the presence of the pile rows can be calculated as follows:

$$\overline{u}_z(\mathbf{x}_\perp, 0, \omega)$$

$$= \overline{u}_{zf}(\mathbf{x}_\perp, 0, \omega)$$

$$+ \sum_{j=1}^{m} \left[- \int_0^{L_j} \overline{N}_*^{(j)}(\zeta, \omega) \frac{\partial \overline{u}^{(G)}(r_{\mathbf{x}_\perp j}, \zeta, 0, \omega)}{\partial \zeta} d\zeta \right.$$

$$\left. + \rho_{p^*}^{(j)} A^{(j)} \omega^2 \int_0^{L_j} \overline{u}_{zp^*}^{(j)}(\zeta, \omega) \overline{u}^{(G)}(r_{\mathbf{x}_\perp j}, \zeta, 0, \omega) d\zeta \right],$$

where $\overline{u}_{zf}^{(S)}(\mathbf{x}_\perp, 0, \omega)$ represents the free field vertical displacement, $\overline{u}^{(G)}(r_{\mathbf{x}_\perp j}, \zeta, 0, \omega)$ denotes the vertical displacement at the surface point $\mathbf{x}_\perp$ ($\mathbf{x}_\perp = xi + yj$) due to a unit patch load applied at $^t\Pi_\xi^{(j)}$, and $r_{\mathbf{x}_\perp j}$ is the horizontal distance between the surface point $\mathbf{x}_\perp$ and the axis of the jth pile

In (11), the vertical displacement of the ith pile top $\overline{u}_{zp^*}^{(i)}(0,\omega)$ is also unknown. The unknown $\overline{u}_{zp^*}^{(i)}(0,\omega)$ can be represented by the axial force of the fictitious piles if the vertical displacement of the ith pile top and the surface vertical displacement of the extended half space at the ith pile top are assumed to be equal; that is, $\overline{u}_{zp^*}^{(i)}(0,\omega) = \overline{u}_z^{(i)}(0,\omega)$.

Note that $\overline{u}_z^{(i)}(0,\omega)$ can be obtained via (18) by setting $x_\perp$ coinciding with the center of the ith pile. Thus, using (12a) and (12b) and (18), the following supplementary equations for $\overline{u}_z^{(i)}(0,\omega)$ are derived:

$$\sum_{j=1}^{m}\left[-\int_0^{L_j} \overline{N}_*^{(j)}(\zeta,\omega)\, \frac{\partial \overline{u}^{(G)}\left(r_{ij},\zeta,0,\omega\right)}{\partial \zeta} \right] d\zeta$$

$$+ \sum_{j=1}^{m} \int_0^{L_j} \overline{N}_*^{(j)}(\zeta,\omega)\, \overline{\chi}_{ij}^{(c)}(\zeta,0,\omega)\, d\zeta$$

$$+ \sum_{j=1}^{m} \overline{u}_{zp^*}^{(j)}(0,\omega)\left[\overline{\chi}_{ij}^{(d)}(0,\omega) - \delta_{ij} \right]$$

$$= -\overline{u}_{zf}^{(i)}(0,\omega), \quad i = 1,2,\ldots,m,$$

(19)

where δ_{ij} is the Kronecker delta and

$$\overline{\chi}_{ij}^{(c)}(\zeta,z,\omega) = \frac{\rho_{pj^*}^{(j)}\omega^2}{E_{pj^*}^{(j)}} \int_\zeta^{L_j} \overline{u}^{(G)}\left(r_{ij},\eta,z,\omega\right) d\eta,$$

$$\overline{\chi}_{ij}^{(d)}(z,\omega) = \rho_{p^*}^{(j)} A^{(j)} \omega^2 \int_0^{L_j} \overline{u}^{(G)}\left(r_{ij},\eta,z,\omega\right) d\eta.$$

(20)

NUMERICAL RESULTS AND DISCUSSIONS

The integral equations in the frequency domain accounting for the vertical interaction between pile rows and the half space can be solved numerically. The methodology for solving integral equation (15) was detailed in [19]. After discretization of (16) and (19), the following linear algebraic equations in the frequency domain are obtained:

$$A(\omega) X(\omega) = b(\omega),$$
(21)

where $A(\omega)$ is the coefficient matrix determined by discrete integral equations which is associated with the fundamental solution, $b(\omega)$ is the right-handed term which is determined by the free field solution, such as $\bar{\varepsilon}_z^{(f)}(x, y, z, \omega)$, and $X(\omega)$ is the discrete unknowns of the integral equations.

In order to recover the solution in the time domain, a series of frequency domain solutions at discrete sample points need to be determined first. Assuming the number of the frequency domain sample points is $2N + 1$, then (21) for the sample points $i = 1, 2, \ldots, N, N + 1$ has the following form:

$$A(\omega)|_{\omega=(i-1)\Delta\omega} X(\omega)|_{\omega=(i-1)\Delta\omega} = b(\omega)|_{\omega=(i-1)\Delta\omega},$$

$$i = 1, 2, \ldots, N, N + 1,$$
(22)

where $\Delta\omega$ is the frequency increment for the sample points in the frequency domain and is given by

$$\Delta\omega = \frac{2\pi}{T}, \qquad T = \frac{2y_0}{c}.$$
(23)

Due to the vibration frequency ω_0 of the moving load in (2a), (2b), and (2c) and (4), the right-handed term $b(\omega)$ for the sample points $i = N+2, \ldots, 2N+1$ should be determined by the following equation:

$$[b(\omega)]_i = b(\omega)|_{\omega=-[(2N+2)-i]\Delta\omega}, \quad i = N + 2, \ldots, 2N + 1.$$
(24)

After numerical solution of the integral equation (16) and (19) for the sample points $i = 1, 2, \ldots, N+1, N+2, \ldots, 2N+1$, all the variables in the frequency domain are obtained. The time domain solution for the variables can be obtained by performing inverse Fourier transform on the corresponding frequency domain solutions, which is implemented by the FFT method in this study [20].

For verification purposes, Section 4.1 demonstrates the comparison between the solution of a special case from the proposed method and published results. In Section 4.2, some numerical examples and corresponding analysis are presented.

Comparison of Our Results with Known Results.

In this section, the method developed in this study is justified by comparing a special case of our model with existing results. As shown in Figure 1, the vibration source is a moving distributed load with a constant speed c in the negative direction of the y-axis. The intensity of the load is 100 kN, and it is uniformly distributed over a rectangle region $2a \times 2b = 0.8$ m $\times$ 0.8 m with a vibration frequency $f = 50$ Hz. A single 8-pile row with circular cross-sections is used as the passive isolation vibration system. All piles are featured by a diameter of $d = 1.0$ m, a length of $L = 5.0$ m, Young's modulus of $E_p = 3.3 \times 10^{10}$ N/m^2, and a density of $pp = 2.4 \times 10^3$ kg/m^3. The net spacing between two neighboring piles is $s = 0.5$ m. The distance between y-axis and the moving load is $ds = 7.5$ m.

According to [14], if the parameters ak $(k = p, s)$ for the viscoelastic half space are assumed to tend to 1, then the viscoelastic half space is reduced to an elastic half space. Moreover, if the speed of the vertical moving distributed load approaches zero, the moving vibration load is reduced to a fixed time-harmonic force. In this paper, the parameter for the viscoelastic is $a_p = a_s = 0.98$, $\alpha_p = \alpha_s = 0.5$, $\eta_p = \eta_s = 0.1$s, $M_{\infty s} = 1.32 \times 10^8$ N/m^2, $M_{\infty p} = 3.96 \times 10^8$ N/m^2, and $\rho = 2.0 \times 10^3$ kg/m^3 The moving load speed is $c = 0.001$ m/s. The corresponding vibration isolation against the fixed timeharmonic force using pile rows in an elastic medium was reported by Kattis et al. [5]. In calculation, the wavelength for the Rayleigh wave of the reduced elastic medium is $\lambda_R = 5.0$ m.

To assess the vibration isolation effect of pile rows, the amplitude reduction ratio A_r at point $x_\perp$, which is the ratio between the amplitude of the surface vertical displacement of the half space in the presence of the pile rows and that of the free field solution, is defined as follows:

$$A_r(x_\perp, t) = \frac{|u_z(x_\perp, z = 0, t)|}{|u_{zf}(x_\perp, z = 0, t)|},$$

(25)

where $|u_z(x_\perp, z = 0, t)|$ is the amplitude of the vertical displacement of the half space in the presence of the pile rows and $|u_{zf}(x_\perp, z = 0, t)|$ is the amplitude of the vertical displacement of the soil given by the free field solution

Woods [21] proposed an average amplitude reduction ratio $\overline{A}_r$ for the evaluation of vibration isolation effect, which is defined as follows:

$$A_{rv} = \frac{1}{A} \int_A A_r dA,$$

(26)

where A is the rectangle with its width and length determined by a reference length and the width of a pile row. In this study, $A = L_r \times \lambda_R$, where L_r is the width of the first pile row and λ_R is the wavelength for the Rayleigh wave of the reduced elastic medium. Figure 3 shows the contour of the amplitude reduction ratio A_r for the single pile row embedded in the elastic half space according to the present method when the moving load with a speed $c = 0.001$ m/s is located at the point $(x, y, z) = (-7.5\text{ m}, 0\text{ m}, 0\text{ m})$. It can be observed that the amplitude reduction ratio Ar right behind the pile rows is much smaller than those at other areas. According to the calculation in this study, the average amplitude reduction ratio A_{rv} for the elastic half space is 0.718, while the result of Kattis et al. [5] for the elastic half space is 0.712.The difference between the present solution and that of Kattis et al. [5] is only 0.842%.

Numerical Results for the Vibration Isolation of Pile Rows

In this section, the influences of the moving load speeds (c), the number of the pile rows (K), Young's modulus (E_p) of the pile, the pile length (L), and the net spacing (s) between two neighboring piles in a pile row as well as the spacing between neighboring pile rows (q) on the vibration isolation effect will be discussed.

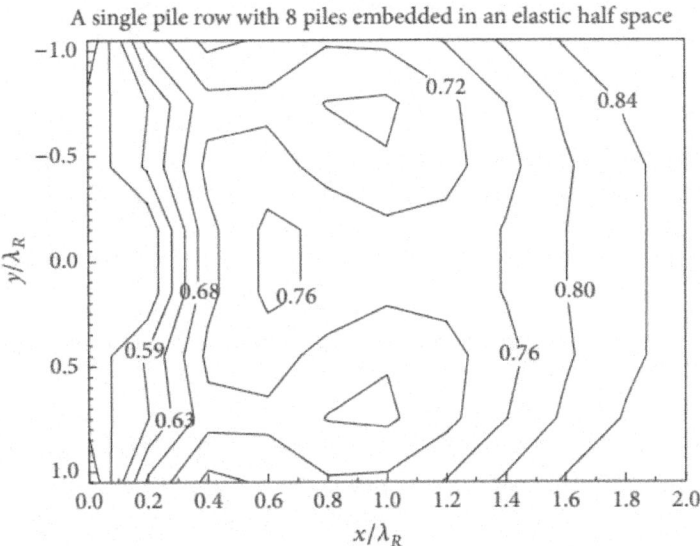

Figure 3: The contour of the amplitude reduction ratio A_r for a single pile row with eight piles as vibration isolation system for the reduced elastic half space.

Typical   values   of   the   parameters   for   the   viscoelastic half   space   are $a_p = a_s = 0.7$, $\alpha_p = \alpha_s = 0.7$, $\eta_p = \eta_s = 0.1\,\mathrm{s}$, $M_{\infty s} = 2.0 \times 10^7\,\mathrm{N/m^2}$, $M_{\infty p} = 6.0 \times 10^7\,\mathrm{N/m^2}$, and $\rho = 2.0 \times 10^3\,\mathrm{kg/m^3}$. The reference length is $\lambda_R = 5.0$ m. The vibration source is a moving distributed load with a constant speed c in the negative direction of the y-axis. The intensity of the load is 100 kN, and it is uniformly distributed over a rectangle $2a \times 2b$ = 0.8 m×0.8 m with a vibration frequency f=0. The load moves along a line parallel to the y-axis at a constant speed $c = 0.5V_{s0}$ in the negative direction of the y-axis, where $V_{s0} = \sqrt{M_{cos}/\rho}$. The distance between the center of the first rows piles and the center of the distributed load is $d_c = 3.0$ m. Each pile has the same diameter $d = 0.8$ m and the same density $\rho_P = 3.9 \times 10^3$ kg/m³. . Each pile has the same length $L = 10.0$ m and the same Young's modulus $E_P = 100E_{s0}$ $(E_{s0} = M_{cos} \times (3M_{cop} - 4M_{cos})/(M_{cop} - M_{cos}))$. Also, the net spacing between two neighboring piles is $s = 0.25$ m and the spacing between the two adjacent pile rows is $q = 0.5$ m.

Note that when the influence of one parameter is examined, all the other parameters will take the typical values as given above.

Effects of the Speed of the Moving Load

In this section, the influences of the speed of the moving load (c) on the vibration isolation effect of the pile rows will be examined. In the calculation, four different moving load speeds $c = 0.2v_{s0}$, $c = 0.5v_{s0}$, $c = 0.9v_{s0}$, and $c = 1.2v_{s0}$ are considered, where $v_{s0} = \sqrt{M_{cos}/\rho}$, while the remainder parameters for the vibration source and the viscoelastic half space take the typical values as given above. Also, two piles rows embedded in the viscoelastic half space are used as the passive vibration isolation system. Thus, the number of pile rows is K=2 and the numbers of piles in the pile rows are $n_1 = 9$ and $n_2 = 8$, respectively.

Figures 4(a)–4(d) illustrate the variation of the amplitude reduction ratio A_r on the surface of the half space at the instant when the moving load passing the point $(x, y, z) = (-3.0$ m, 0 m, 0 m$)$ with speeds $c = 0.2V_{s0}$, $c = 0.5V_{s0}$, $c = 0.9V_{s0}$, and $c = 1.2V_{s0}$, respectively. From Figure 4, one can see that moving load speed has some influence on the effect of the isolation vibration. With the increasing moving load speed, the vibration isolation effectiveness is enhanced pronouncedly when the speed of the moving load is lower than the shear speed of the viscoelastic half space. It is also shown in Figure 4(d) that the amplitude reduction ratio at the instant when the moving load is located at the point $(x, y, z) = (-3.0$ m, 0 m, 0 m$)$ becomes asymmetrical with respect to x axis: for the case of $c = 1.2V_{s0}$, the vibration isolation effect for the domain y>0 is better than that for y<0. According to our model, the average amplitude reduction ratios A_{rv} for the present Cole-Cole viscoelastic half space and the reduced elastic medium corresponding to the instant when the moving load with different velocity passing through the point (x, y, z) = (−3.0 m, 0 m, 0 m) is derived. For the case of the Cole-Cole viscoelastic half-space model, the average amplitude reduction ratios A_{rV} are equal to 0.8491, 0.8346, 0.8024, and 0.4941 for the moving load with velocities $c = 0.2v_{s0}$, $c = 0.5v_{s0}$, $c = 0.9v_{s0}$, and $c = 1.2v_{s0}$, respectively, while for the case of the reduced elastic medium, the corresponding ArV is equal to 0.852, 0.8476, 0.7526, and 0.6213 for $c = 0.2V_{s0}$, $c = 0.5V_{s0}$, $c = 0.9V_{s0}$, and $c = 1.2V_{s0}$, respectively. Thus, we can see that for the same pile row and the same vibration source, the vibration isolation effect of the Cole-Cole viscoelastic half-space model is better than that for the elastic medium, particularly on the case of moving load with high velocity

Effects of the Number of the Pile Rows.

In this example, we consider the following three pile rows embedded in the viscoelastic half space: a single pile row with 9 piles ($K=1$, $n_1 = 9$); two pile rows with 9 and 8 piles ($K=2$, $n_1 = 9$, $n_2 = 8$), respectively; three pile rows with 9, 8, and 9 piles ($K=3$, $n_1 = 9$, $n_2 = 8$, $n_3 = 9$), respectively (Figure 6)

Figures 6(a) and 6(b) plot the variation of the amplitude reduction ratio Ar on the surface of the viscoelastic half space at the instant when the moving load with the velocity $c = 0.5V_{s0}$ passing through the point$(x, y, z) = $ (−3.0 m, 0 m, 0 m) for the two kinds of pile rows ($K=1$ and $K=3$). Note that the results for the case of two pile rows with 9 and 8 piles ($K=2$, $n_1 = 9$, and $n_2 = 8$) have already been presented in Section 4.2.1 (see Figure 4(b)). As expected, it follows from Figure 6 that the increase of the number of pile rows will enhance the vibration isolation effect of the pile rows. The average amplitude reduction ratios at the instant when the moving load with the velocity $c = 0.5V_{s0}$ passing the point $(x, y, z) = $ (−3.0 m, 0 m, 0 m) for the single pile row, the two pile rows and the three pile rows are $A_{rV} = 0.9189$, 0.8346, and 0.7769, respectively.

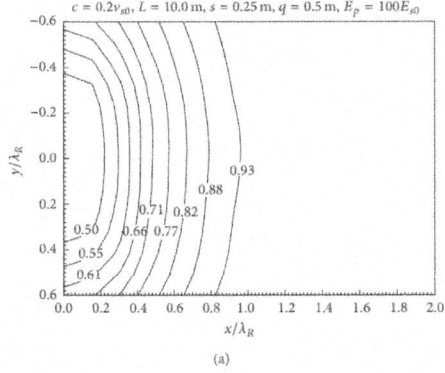

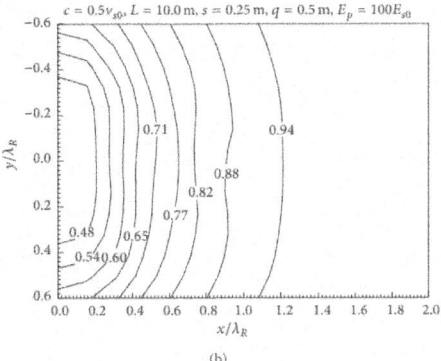

(a)   (b)

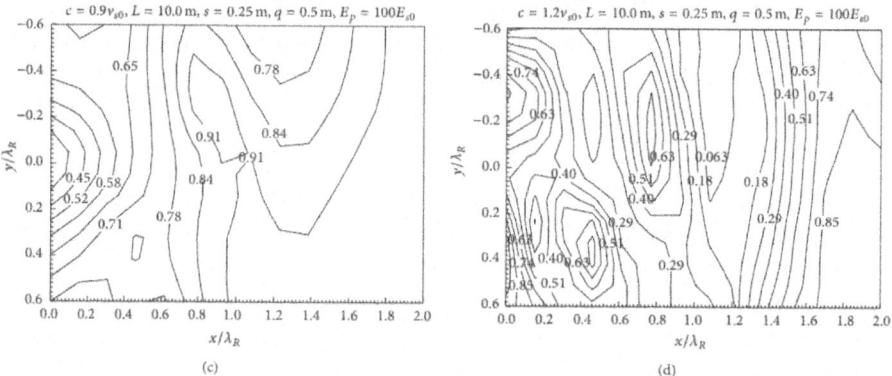

Figure 4: The variation of the amplitude reduction ratio Ar on the surface for the viscoelastic half space at the instant when the moving load passing the point $(x, y, z) = (-3.0$ m, 0 m, 0 m) with four different speeds: (a) $c = 0.2V_{s0}$; (b) $c = 0.5V_{s0}$; (c) $c = 0.9V_{s0}$; (d) $c = 1.2V_{s0}$, where $V_{s0} = v_{s0} = \sqrt{M_{\cos}/\rho}$.

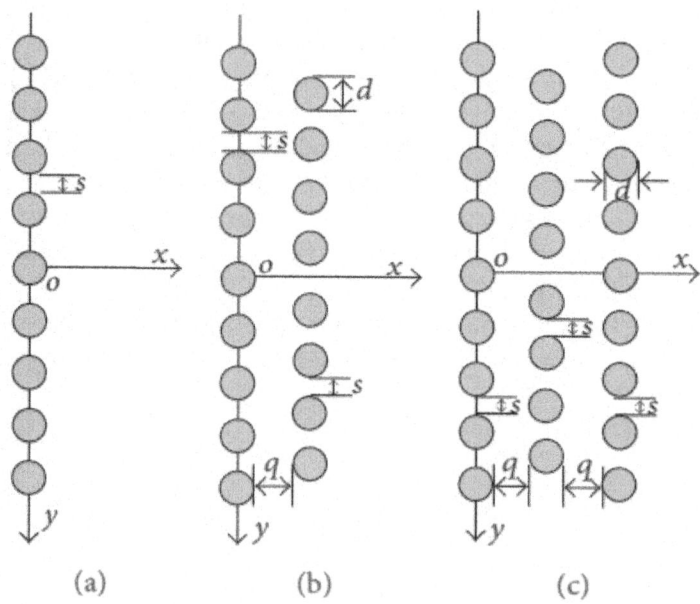

Figure 5: Multiple pile rows embedded in a viscoelastic half space as vibration isolation system against the vibration due to a harmonic moving load: (a) a single pile row; (b) two pile rows; (c) three pile rows.

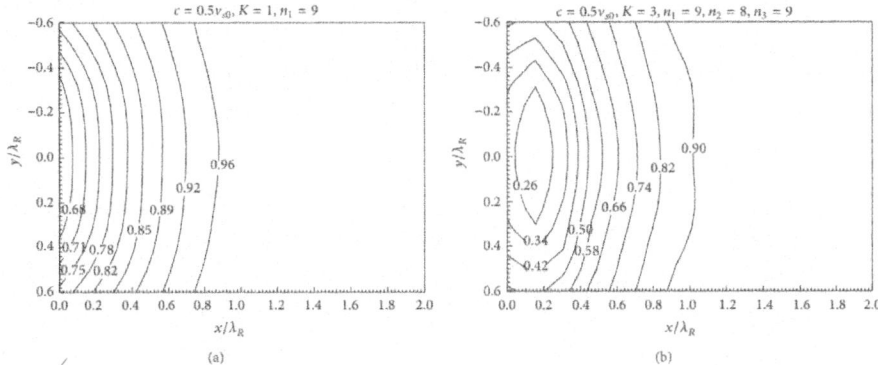

Figure 6: The variation of the amplitude reduction ratio A_r on the surface of the visco-elastic half space at the instant when the moving load passing the point $(x, y, z) = (-3.0$ m, 0 m, 0 m) with the speed $c = 0.5V_{s0}$ for the two kinds of pile rows: (a) a single pile row ($K = 1$, $n_1 = 9$); (b) three piles rows ($K = 3$, $n_1 = 9$, $n2 = 8$, $n_3 = 9$).

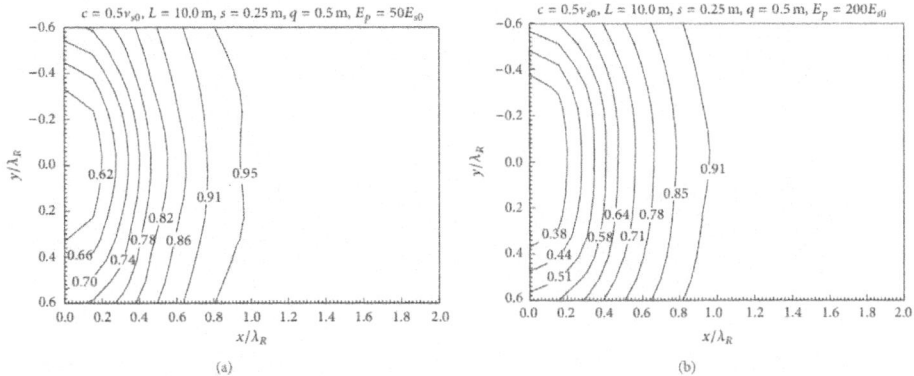

Figure 7: The variation of the amplitude reduction ratio Ar on the surface of the viscoelastic half space at the instant when the moving load passing the point $(x, y, z) = (-3.0$ m, 0 m, 0 m) with speed $c = 0.5v_{s0}$ for two cases of E_p/E_{s0}: (a) $E_p/E_{s0} = 50$; (b) $E_p/E_{s0} = 200$, ,

where $E_{s0} = M_{\infty s} \times (3M_{\infty p} - 4M_{\infty s})/(M_{\infty p} - M_{\infty s})$.

Effects of Young's Modulus (E_p) of the Piles.

Young's modulus (Ep) of the pile is an important parameter for the design of the pile row vibration isolation system. Herein, the influence of Young's modulus of pile on the vibration isolation effect will be investigated. As previously mentioned, in this example, two pile rows ($K = 2$, $n_1 = 9$, $n_2 = 8$) embedded in the viscoelastic half space are used to screen the vibration due to the moving load. Young's modulus of the piles takes three different values to make $E_p/E_{s0} = 50$, 100, and 200, respectively, with $E_{s0} = M_{cos} \times (3M_{cop} - 4M_{cos})/(M_{cop} - M_{cos})$.

Figures 7(a) and 7(b) show the variation of the amplitude reduction ratio Ar on the surface of the viscoelastic half space for $E_p/E_{s0} = 50$ and 200 when the moving load with the velocity $c = 0.5V_{s0}$ passing the point $(x, y, z) = (-3.0$ m, 0 m, 0 m)$. Note that the result for $E_p/E_{s0} = 100$ is presented in Section 4.2.1 (see Figure 4(b)). The average amplitude reduction ratio ArV at the instant when the moving load with the velocity $c = 0.5Vs0$ passing the point $(x, y, z) = (-3.0$ m, 0 m, 0 m)$ for the three modulus ratios is. for $E_p/E_{s0} = 50$, $A_{rV} = 0.8896$; for $E_p/E_{s0} = 100$, $A_{rV} = 0.8346$; for $E_p/E_{s0} = 200$, $A_{rV} = 0.8146$. One can see clearly

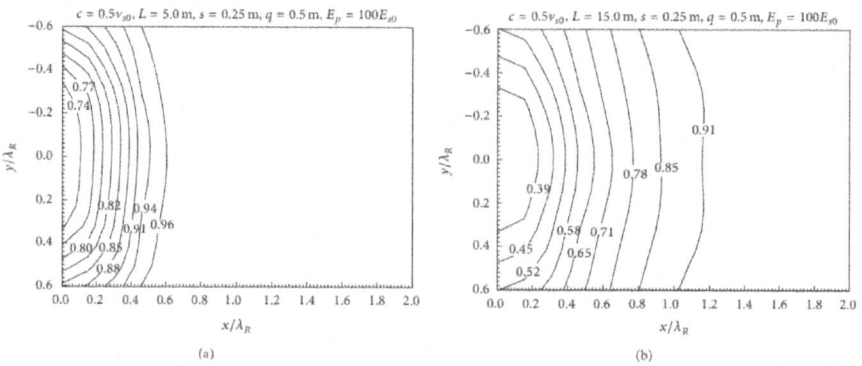

(a)   (b)

Figure 8: The variation of the amplitude reduction ratio A_r on the surface of the viscoelastic half space at the instant when the moving load with speed $c = 0.5V_{s0}$ passing the point $(x, y, z) = (-3.0$ m, 0 m, 0 m)$ for the two cases of pile length (a) $L = 5.0$ m; (b) $L = 15.0$ m.

that the average amplitude reduction ratio A_{rV} decreases with increasing E_p/E_{s0}, which suggest stiffer pile rows will produce a better vibration isolation effect.

Effects of the Length of the Piles

The influence of the length of the piles on the vibration isolation effect of pile rows is examined in this section. As in Section 4.2.1, two pile rows ($K = 2$, $n_1 = 9$, $n_2 = 8$) embedded in the viscoelastic half space are used to isolate the vibration due to the moving load. To check the influence of the pile length, the pile length takes the following values: $L = 5.0$ m, 10.0 m, and 15.0 m, respectively Figures 8(a) and 8(b) show the variation of the amplitude reduction ratio Ar on the surface of the viscoelastic half space for $L = 5.0$ m and 15.0 m when the moving load with the velocity $c = 0.5V_{s0}$ passing the point (x, y, z) $= (-3.0$ m, 0 m, 0 m). Note that the result for $L = 10.0$ m has been presented in Section 4.2.1 (see Figure 4(b)). The average amplitude reduction ratio A_{rv} at the instant when the moving load with the velocity $c = 0.5V_{s0}$ passing the point $(x, y, z) = (-3.0$ m, 0 m, 0 m) for $L = 5.0$ m, 10.0 m and 15.0 m is equal to 0.9463, 0.8346, and 0.7927, respectively. It follows that the length of piles has a significant influence on the average amplitude reduction ratios A_{rv}: pile rows with a larger pile length usually produce a better vibration isolation effect than shorter ones.

Effects of the Pile Net Spacing.

The net spacing (s) between neighboring piles in a pile row is an important parameter for the design of pile rows. In this section, two rows piles ($K = 2$, $n_1 = 9$, $n_2 = 8$) embedded in the viscoelastic half space are used to investigate the influence of the net spacing. To examine the influence of the net spacing s between neighboring piles, the net spacing s takes the values 0.25 m, 0.5 m, and 1.0 m, respectively

Figures 9(a) and 9(b) show the variation of the amplitude reduction ratio Ar on the surface of the viscoelastic half space for $s = 0.5$ m and 1.0 m when the moving load with the velocity $c = 0.5V_{s0}$ is passing the point $(x, y, z) =$ $(-3.0$ m, 0 m, 0 m). The result for the case of $s = 0.25$ m has already been given in Section 4.2.1 (see Figure 4(b)). It clearly indicates the decay of the vibration isolation effect behind row pile with increasing net spacing s. The average amplitude reduction ratio ArV at the instant when the moving load with the velocity $c = 0.5V_{s0}$ passing the point $(x, y, z) = (-3.0$ m, 0 m, 0 m) for $s = 0.25$ m, 0.5 m and 1.0 m is $ArV = 0.8346$, 0.8804, and 0.9103, respectively, which shows that the average amplitude reduction ratio A_{rv} has considerable increase with increasing net spacing s. Thus, it can be concluded that a smaller net pile spacing usually leads to a better vibration isolation effect.

Effects of the Spacing between Adjacent Pile Rows

To examine the effect of the spacing (q) between neighboring pile rows, as in Section 4.2.1, two pile rows ($K=2$, $n_1 = 9$, $n_2 = 8$) embedded in the viscoelastic half space are used as an example. To examine the influence of the spacing between neighboring pile rows, q takes 0.5 m, 1.0 m, and 1.5 m, respectively

Figures 10(a) and 10(b) show the variation of the amplitude reduction ratio A_r on the surface of the viscoelastic half space for $q = 1.0$ m and 1.5 m when the moving load with the velocity $c = 0.5V_{s0}$ is passing the point $(x, y, z) = (-3.0 \text{ m}, 0 \text{ m}, 0 \text{ m})$. The result for the case of $q = 0.5$ m has already been given in Section 4.2.1 (see Figure 4(b)). The average amplitude reduction ratio A_{rv} at the instant when the moving load located at the point $(x, y, z) = (-3.0 \text{ m}, 0 \text{ m}, 0 \text{ m})$ for $q = 0.5$ m, 1.0 m and 1.5 m is equal to $A_{rv} = 0.8346$, 0.8352, and 0.8356, respectively, which indicates that, compared with other parameters, the space between two adjacent pile rows has a relatively smaller effect on the vibration isolation effect of pile rows.

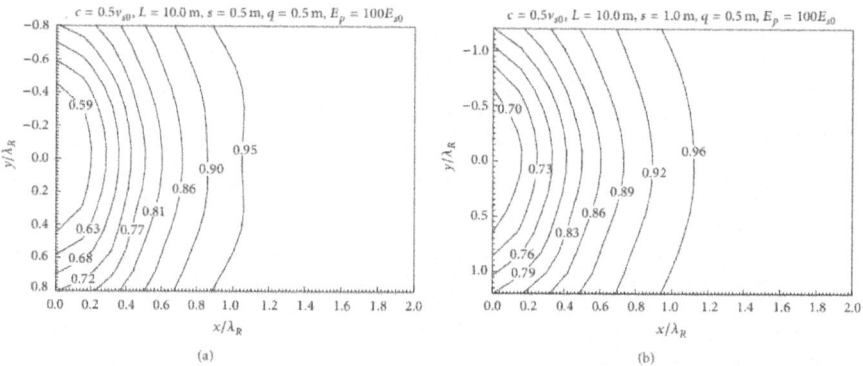

Figure 9: The variation of the amplitude reduction ratio Ar on the surface of the viscoelastic half space at the instant when the moving load passing the point $(x, y, z) = (-3.0 \text{ m}, 0 \text{ m}, 0 \text{ m})$ with speed $c = 0.5V_{s0}$ for the two cases of the net spacing: (a) $s = 0.5$ m; (b) $s = 1.0$ m

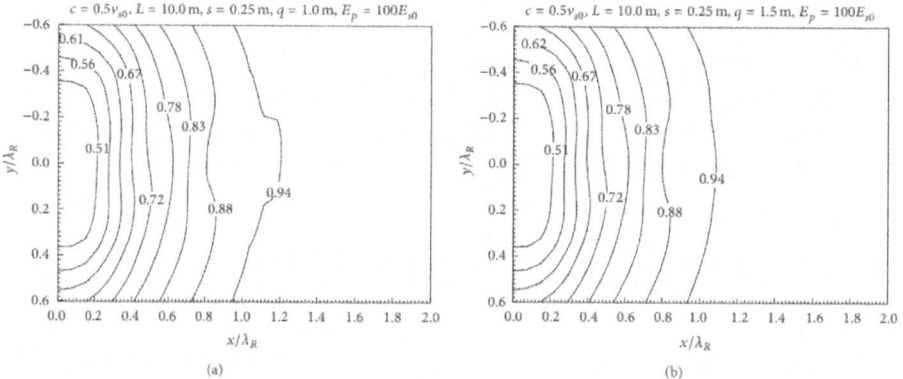

Figure 10: The variation of the amplitude reduction ratio A_r on the surface of the viscoelastic half space at the instant when the moving load passing the point $(x, y, z) = (-3.0 \text{ m}, 0 \text{ m}, 0 \text{ m})$ with speed $c = 0.5V_{s0}$ for the two cases of the net spacing: (a) $q = 1.0$ m; (b) $q = 1.5$ m.

CONCLUSIONS

The numerical simulation of the isolation of vibration due to a harmonic moving load applied on the surface of a viscoelastic half space by pile rows has been carried out in this study (Figure 5). The viscoelastic half space is characterized by the Cole-Cole model. The semianalytical nature of the proposed method circumvents the discretization of the whole calculation domain, thus reducing the CPU time for the current problem substantially. Consequently, the proposed method is important for the design of vibration isolation and can be used to assess the efficiency of pile rows as a barrier system against ground vibrations. To verify the proposed method, result of this study was compared with existing results, which shows the proposed solution is in a good agreement with the existing result.

To study the vibration isolation effect of pile rows, the influences of the moving load speed, Young's modulus of the pile, the pile length, and the spacing between piles as well as the spacing between neighboring pile rows have been investigated. Based on the numerical simulations performed in this study, the following conclusions can be drawn.(i)The same pile rows can achieve a better vibration isolation effect for higher speed loads than for lower speed loads.(ii)Young's modulus of the pile is an important parameter affecting the vertical vibration isolation effects of pile rows. Increasing Young's modulus of the pile rows will enhance the vibration isolation effect.(iii)Pile length is the most important factor for the vertical vibration isolation effect of pile rows. Generally, pile rows with a longer pile length have a better vibration isolation effect than those with a shorter pile length.(iv)The net spacing between neighboring piles is crucial for the vibration isolation effect of pile rows. Generally, to obtain a better vibration isolation effect for high speed loads, the net spacing should take a smaller value.(v)A good vibration isolation effect can be realized by increasing the number of pile rows and reducing the net spacing between neighboring piles.

CONFLICT OF INTERESTS

The authors declare that there is no conflict of interests regarding the publication of this paper.

ACKNOWLEDGMENTS

The project is supported by the National Natural Science Foundation of China with Grant no. 51269021 and Key Project of Natural Science Foundation of Jiangxi Province with Grant no. 20133ACB20006.

REFERENCES

1. K. Emad and G. D. Manolis, "Shallow trenches and propagation of surface waves," Journal of Engineering Mechanics, vol. 111, no. 2, pp. 279–282, 1985.

2. Z. Cao, Y. Cai, A. Boström, and J. Zheng, "Semi-analytical analysis of the isolation to moving-load induced ground vibrations by trenches on a poroelastic half-space," Journal of Sound and Vibration, vol. 331, no. 4, pp. 947–961, 2012.

3. Y. Cai, G. Ding, C. Xu, and J. Wang, "Vertical amplitude reduction of Rayleigh waves by a row of piles in a poroelastic half-space," International Journal for Numerical and Analytical Methods in Geomechanics, vol. 33, no. 16, pp. 1799–1821, 2009.

4. S. E. Kattis, D. Polyzos, and D. E. Beskos, "Modelling of pile wave barriers by effective trenches and their screening effectiveness," Soil Dynamics and Earthquake Engineering, vol. 18, no. 1, pp. 1–10, 1999.

5. S. E. Kattis, D. Polyzos, and D. E. Beskos, "Vibration isolation by a row of piles using a 3-D frequency domain BEM," International Journal for Numerical Methods in Engineering, vol. 46, no. 5, pp. 713–728, 1999.

6. P. Tsai, Z. Feng, and T. Jen, "Three-dimensional analysis of the screening effectiveness of hollow pile barriers for foundation-induced vertical vibration," Computers and Geotechnics, vol. 35, no. 3, pp. 489–499, 2008.

7. R. Muki and E. Sternberg, "Elastostatic load-transfer to a half-space from a partially embedded axially loaded rod," International Journal of Solids and Structures, vol. 6, no. 1, pp. 69–90, 1970.

8. J. Lu, B. Xu, and J. Wang, "A numerical model for the isolation of moving-load induced vibrations by pile rows embedded in layered porous media," International Journal of Solids and Structures, vol. 46, no. 21, pp. 3771–3781, 2009.

9. J. Lu, D. Jeng, J. Wan, and J. Zhang, "A new model for the vibration isolation via pile rows consisting of infinite number of piles," International Journal for Numerical and Analytical Methods in Geomechanics, vol. 37, pp. 2394–2426, 2013.

10. Y. N. Rabotnov, Elements of Hereditary Solid Mechanics, Mir Publication, Moscow, Russia, 1980.

11. B. Nolte, S. Kempfle, and I. Schäfer, "Does a real material behave fractionally? Applications of fractional differential operators to the damped structure borne sound in viscoelastic solids," Journal of

Computational Acoustics, vol. 11, no. 3, pp. 451–489, 2003.

12. K. S. Cole and R. H. Cole, "Dispersion and absorption in dielectrics I. Alternating current characteristics," The Journal of Chemical Physics, vol. 9, no. 4, pp. 341–351, 1941.

13. R. L. Bagley and P. J. Torvik, "On the fractional calculus model of viscoelastic behavior," Journal of Rheology, vol. 30, no. 1, pp. 133–155, 1986.

14. J. Lu and A. Hanyga, "Numerical modelling method for wave propagation in a linear viscoelastic medium with singular memory," Geophysical Journal International, vol. 159, no. 2, pp. 688–702, 2004

15. M. Soula, T. Vinh, and Y. Chevalier, "Transient responses of polymers and elastomers deduced from harmonic responses," Journal of Sound and Vibration, vol. 205, no. 2, pp. 185–203, 1997.

16. T. D. Jones, "Pore fluids and frequency-dependent wave propagation in rocks," Geophysics, vol. 51, no. 10, pp. 1939–1953, 1986.

17. M. Batzle, R. Hofmann, D. Han, and J. Castagna, "Fluids and frequency dependent seismic velocity of rocks," Leading Edge, vol. 20, no. 2, pp. 168–171, 2001.

18. G. F. Fowler and G. B. Sinclair, "The longitudinal harmonic excitation of a circular bar embedded in an elastic half-space," International Journal of Solids and Structures, vol. 14, no. 12, pp. 999–1012, 1978.

19. R. Y. S. Pak and P. C. Jennings, "Elasticdynamic response of pile under transverse excitations," Journal of Engineering Mechanics, vol. 113, no. 7, pp. 1101–1116, 1987.

20. A. V. Oppenheim and R. W. Schafer, Discrete-Time Signal Processing, Prentice-Hall, Englewood Cliffs, NJ, USA, 1999.

21. R. D. Woods, "Screening of surface waves in soils," Journal of Soil Mechanics and Foundation Engineering, vol. 94, no. 4, pp. 951–979, 1968.

Chapter 4

MATHEMATICAL MODEL OF THE THREE-PHASE INDUCTION MACHINE FOR THE STUDY OF STEADY-STATE AND TRANSIENT DUTY UNDER BALANCED AND UNBALANCED STATES

Alecsandru Simion, Leonard Livadaru and Adrian Munteanu

"Gh. Asachi" Technical University of Iaşi, Electrical Engineering Faculty, Romania

INTRODUCTION

A proper study of the induction machine operation, especially when it comes to transients and unbalanced duties, requires effective mathematical models above all. The mathematical model of an electric machine represents all the equations that describe the relationships between electromagnetic torque and the main electrical and mechanical quantities.

The theory of electrical machines, and particularly of induction machine, has mathematical models withdistributed parameters and with concentrated parameters respectively. The first mentioned models start with the cognition of the magnetic field of the machine components. Their most important advantages consist in the high generality degree and accuracy. However, two major disadvantages have to be mentioned. On one hand, the computing time is rather high, which somehow discountenance their use for the real-time control. On the other hand, the distributed parameters models do not take into consideration the influence of the temperature variation or mechanical processing upon the material properties, which can vary up to 25% in comparison to the initial state. Moreover, particular constructive details (for example slots or air-gap dimensions), which essentially affects the parameters evaluation, cannot be always realized from technological point of view.

The mathematical models with concentrated parameters are the most popular and consequently employed both in scientific literature and practice. The equations stand on resistances and inductances, which can be used further for defining magnetic fluxes, electromagnetic torque, and et.al. These models offer results, which are globally acceptable but cannot detect important information concerning local effects (Ahmad, 2010; Chiasson, 2005; Krause et al., 2002; Ong, 1998; Sul, 2011).

The family of mathematical models with concentrated parameters comprises different approaches but two of them are more popular: the phase coordinate model and the orthogonal (dq) model (Ahmad, 2010; Bose, 2006; Chiasson, 2005; De Doncker et al., 2011; Krause et al., 2002; Marino et al., 2010;Ong, 1998; Sul, 2011; Wach, 2011).

The first category works with the real machine. The equations include, among other parameters, the mutual stator-rotor inductances with variable values according to the rotor position. As consequence, the model becomes non-linear and complicates the study of dynamic processes (Bose, 2006; Marino et al., 2010; Wach, 2011).

The orthogonal (dq) model has begun with Park's theory nine decades ago. These models use parameters that are often independent to rotor position. The result is a significant simplification of the calculus, which became more convenient with the defining of the space phasor concept (Boldea & Tutelea, 2010; Marino et al., 2010; Sul, 2011).

Starting with the "classic" theory we deduce in this contribution a mathematical model that exclude the presence of the currents and angular velocity in voltage equations and uses total fluxes alone. Based on this approach, we take into discussion two control strategies of induction motor by principle of constant total flux of the stator and rotor, respectively.

The most consistent part of this work is dedicated to the study of unbalanced duties generated by supply asymmetries. It is presented a comparative analysis, which confronts a balanced duty with two unbalanced duties of different unbalance degrees. The study uses as working tool the Matlab-Simulink environment and provides variation characteristics of the electric, magnetic and mechanical quantities under transient operation.

THE EQUATIONS OF THE THREE-PHASE INDUCTION MACHINE IN PHASE COORDINATES

The structure of the analyzed induction machine contains: 3 identical phase windings placed on the stator in an 120 electric degrees angle of phase difference configuration; 3 identical phase windings placed on the rotor with a similar difference of phase; a constant air-gap (close slots in an ideal approach); an unsaturated (linear) magnetic circuit that allow to each winding to be characterized by a main and a leakage inductance. Each phase winding has Ws turns on stator and WR turns on rotor and a harmonic distribution. All inductances are considered constant. The schematic view of the machine is presented in Fig. 1a.

The voltage equations that describe the 3+3 circuits are:

$$u_{as} = R_s i_{as} + \frac{d\psi_{as}}{dt}, \quad u_{bs} = R_s i_{bs} + \frac{d\psi_{bs}}{dt}, \quad u_{cs} = R_s i_{cs} + \frac{d\psi_{cs}}{dt}$$

(1)

$$u_{AR} = R_R i_{AR} + \frac{d\psi_{AR}}{dt}, \quad u_{BR} = R_R i_{BR} + \frac{d\psi_{BR}}{dt}, \quad u_{CR} = R_R i_{CR} + \frac{d\psi_{CR}}{dt}$$

(2)

In a matrix form, the equations become:

$$\left[u_{abcs} \right] = \left[R_s \right]\left[i_{abcs} \right] + \frac{d\left[\psi_{abcs} \right]}{dt}$$

(3)

$$\left[u_{ABCR} \right] = \left[R_R \right]\left[i_{ABCR} \right] + \frac{d\left[\psi_{ABCR} \right]}{dt}$$

(4)

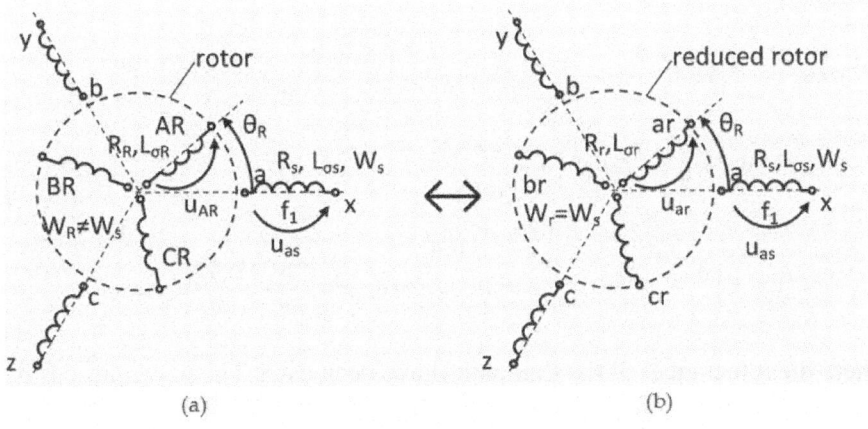

(a)   (b)

Figure 1: Schematic model of three-phase induction machine: a. real; b. reduced rotor

The quantities in brackets represent the matrices of voltages, currents, resistances and total flux linkages for the stator and rotor. Obviously, the total fluxes include both main and mutual components. Further, we define the self-phase inductances, which have a leakage and a main component: $L_{jj}=L_{\sigma s}+L_{hs}$ for stator and $L_{JJ}=L_{\Sigma R}+L_{HR}$ for rotor. The mutual inductances of two phases placed on the same part (stator or rotor) have negative values, which are equal to half of the maximum mutual inductances and with the main self-phase component: $M_{jk}=L_{jk}=L_{hj}=L_{hk}$. The expressions in matrix form are:

$$[L_{ss}] = \begin{bmatrix} L_{\sigma s} + L_{hs} & -(1/2)L_{hs} & -(1/2)L_{hs} \\ -(1/2)L_{hs} & L_{\sigma s} + L_{hs} & -(1/2)L_{hs} \\ -(1/2)L_{hs} & -(1/2)L_{hs} & L_{\sigma s} + L_{hs} \end{bmatrix}$$

(5)

$$[L_{RR}] = \begin{bmatrix} L_{\Sigma R} + L_{HR} & -(1/2)L_{HR} & -(1/2)L_{HR} \\ -(1/2)L_{HR} & L_{\Sigma R} + L_{HR} & -(1/2)L_{HR} \\ -(1/2)L_{HR} & -(1/2)L_{HR} & L_{\Sigma R} + L_{HR} \end{bmatrix}$$

(6)

$$[L_{sR}] = [L_{Rs}]_t = L_{sR} \cdot \begin{bmatrix} \cos\theta_R & \cos(\theta_R + u) & \cos(\theta_R + 2u) \\ \cos(\theta_R + 2u) & \cos\theta_R & \cos(\theta_R + u) \\ \cos(\theta_R + u) & \cos(\theta_R + 2u) & \cos\theta_R \end{bmatrix}$$

(7)

where u denotes the angle of 120^0 (or $2\pi/3$ rad).

The analysis of the induction machine usually reduces the rotor circuit to the stator one. This operation requires the alteration of the rotor quantities with the coefficient $k = W_s/W_R$ by complying with the conservation rules. The new values are:

$$u_{abcr} = k \cdot u_{ABCR}; \quad \psi_{abcr} = k \cdot \psi_{ABCR}; \quad i_{abcr} = (1/k) \cdot i_{ABCR};$$

(8)

$$R_r = k^2 \cdot R_R; \quad L_{hr} = k^2 \cdot L_{HR} = \left(\frac{W_s}{W_R}\right)^2 \cdot \frac{W_R^2}{\Re_h} = \frac{W_s^2}{\Re_h} = L_{hs};$$

(9)

$$L_{\sigma r} = k^2 L_{\Sigma R} = \left(\frac{W_s}{W_R}\right)^2 \frac{W_R^2}{\Re_{\sigma R}} = \frac{W_s^2}{\Re_{\sigma r}} \approx L_{\sigma s}; \quad L_{sr} = kL_{sR} = \left(\frac{W_s}{W_R}\right) \frac{W_s W_R}{\Re_h} = L_{hs}$$

(10)

where the reluctances of the flux paths have been used. The new matrices, with rotor quantities denoted with lowercase letters are:

$$[L_{rr}] = k^2[L_{RR}] = \begin{bmatrix} L_{\sigma r} + L_{hs} & -(1/2)L_{hs} & -(1/2)L_{hs} \\ -(1/2)L_{hs} & L_{\sigma r} + L_{hs} & -(1/2)L_{hs} \\ -(1/2)L_{hs} & -(1/2)L_{hs} & L_{\sigma r} + L_{hs} \end{bmatrix}$$

(11)

$$[L_{sr}] = k[L_{sR}] = [L_{rs}]_t = L_{hs} \cdot \begin{bmatrix} \cos\theta_R & \cos(\theta_R + u) & \cos(\theta_R + 2u) \\ \cos(\theta_R + 2u) & \cos\theta_R & \cos(\theta_R + u) \\ \cos(\theta_R + u) & \cos(\theta_R + 2u) & \cos\theta_R \end{bmatrix}$$

(12)

By virtue of these transformations, the voltage equations become:

$$\begin{cases} \left[u_{abcs}\right] = \left[R_s\right]\left[i_{abcs}\right] + \dfrac{d\left[\psi_{abcs}\right]}{dt} = \left[R_s\right]\left[i_{abcs}\right] + \left[L_{ss}\right]\dfrac{d\left[i_{abcs}\right]}{dt} + \dfrac{d\left\{\left[L_{sr}\right]\left[i_{abcr}\right]\right\}}{dt} \\[4mm] \left[u_{abcr}\right] = \left[R_r\right]\left[i_{abcr}\right] + \dfrac{d\left[\psi_{abcr}\right]}{dt} = \left[R_r\right]\left[i_{abcr}\right] + \left[L_{rr}\right]\dfrac{d\left[i_{abcr}\right]}{dt} + \dfrac{d\left\{\left[L_{sr}\right]_t\left[i_{abcs}\right]\right\}}{dt} \end{cases}$$

(13)

By using the notations:

$$\begin{aligned}
\left(\Sigma L_\Pi\right) &= L_{\sigma r}\left(3L_{hs} + L_{\sigma s}\right) + L_{\sigma s}\left(3L_{hs} + L_{\sigma r}\right) \\
\left(\Pi L_s\right) &= L_{\sigma r}\left(L_{hs} + L_{\sigma s}\right) + L_{\sigma s}\left(3L_{hs} + L_{\sigma r}\right) \\
\left(\Pi L_r\right) &= L_{\sigma s}\left(L_{hs} + L_{\sigma r}\right) + L_{\sigma r}\left(3L_{hs} + L_{\sigma s}\right)
\end{aligned}$$

(14)

and after the separation of the currents derivatives, (8) can be written under operational form as follows:

$$\left(\bar{s} + \frac{R_s(\Pi L_s)}{L_{\sigma s}(\Sigma L_\Pi)}\right)\bar{i}_{as} = -\frac{R_s L_{hs} L_{\sigma r}}{L_{\sigma s}(\Sigma L_\Pi)}\left(\bar{i}_{bs} + \bar{i}_{cs}\right) + \frac{2 R_r L_{hs}}{(\Sigma L_\Pi)}\left[\bar{i}_{ar}\cos\theta_R + \bar{i}_{br}\cos(\theta_R + u) + \bar{i}_{cr}\cos(\theta_R + 2u)\right] +$$
$$+\dot\theta_R\frac{L_{hs}(3L_{hs} + 2L_{\sigma r})}{(\Sigma L_\Pi)}\left[\bar{i}_{ar}\sin\theta_R + \bar{i}_{br}\sin(\theta_R + u) + \bar{i}_{cr}\sin(\theta_R + 2u)\right] +$$
$$+2,6\dot\theta_R\frac{L_{hs}^2}{(\Sigma L_\Pi)}\left(\bar{i}_{bs} - \bar{i}_{cs}\right) + \frac{L_{\sigma r} L_{hs}}{L_{\sigma s}(\Sigma L_\Pi)}\left(\bar{u}_{as} + \bar{u}_{bs} + \bar{u}_{cs}\right) + \frac{3L_{hs} + 2L_{\sigma r}}{(\Sigma L_\Pi)}\bar{u}_{as} -$$
$$-\frac{2L_{hs}}{(\Sigma L_\Pi)}\left[\bar{u}_{ar}\cos\theta_R + \bar{u}_{br}\cos(\theta_R + u) + \bar{u}_{cr}\cos(\theta_R + 2u)\right],$$

$$\left(\bar{s} + \frac{R_s(\Pi L_s)}{L_{\sigma s}(\Sigma L_\Pi)}\right)\bar{i}_{bs} = -\frac{R_s L_{hs} L_{\sigma r}}{L_{\sigma s}(\Sigma L_\Pi)}\left(\bar{i}_{cs} + \bar{i}_{as}\right) + \frac{2 R_r L_{hs}}{(\Sigma L_\Pi)}\left[\bar{i}_{br}\cos\theta_R + \bar{i}_{cr}\cos(\theta_R + u) + \bar{i}_{ar}\cos(\theta_R + 2u)\right] +$$
$$+\dot\theta_R\frac{L_{hs}(3L_{hs} + 2L_{\sigma r})}{(\Sigma L_\Pi)}\left[\bar{i}_{br}\sin\theta_R + \bar{i}_{cr}\sin(\theta_R + u) + \bar{i}_{ar}\sin(\theta_R + 2u)\right] +$$
$$+2,6\dot\theta_R\frac{L_{hs}^2}{(\Sigma L_\Pi)}\left(\bar{i}_{cs} - \bar{i}_{as}\right) + \frac{L_{\sigma r} L_{hs}}{L_{\sigma s}(\Sigma L_\Pi)}\left(\bar{u}_{as} + \bar{u}_{bs} + \bar{u}_{cs}\right) + \frac{3L_{hs} + 2L_{\sigma r}}{(\Sigma L_\Pi)}\bar{u}_{bs} -$$
$$-\frac{2L_{hs}}{(\Sigma L_\Pi)}\left[\bar{u}_{br}\cos\theta_R + \bar{u}_{cr}\cos(\theta_R + u) + \bar{u}_{ar}\cos(\theta_R + 2u)\right],$$

$$\left(\bar{s} + \frac{R_s(\Pi L_s)}{L_{\sigma s}(\Sigma L_\Pi)}\right)\bar{i}_{cs} = -\frac{R_s L_{hs} L_{\sigma r}}{L_{\sigma s}(\Sigma L_\Pi)}\left(\bar{i}_{as} + \bar{i}_{bs}\right) + \frac{2 R_r L_{hs}}{(\Sigma L_\Pi)}\left[\bar{i}_{cr}\cos\theta_R + \bar{i}_{ar}\cos(\theta_R + u) + \bar{i}_{br}\cos(\theta_R + 2u)\right] +$$
$$+\dot\theta_R\frac{L_{hs}(3L_{hs} + 2L_{\sigma r})}{(\Sigma L_\Pi)}\left[\bar{i}_{cr}\sin\theta_R + \bar{i}_{ar}\sin(\theta_R + u) + \bar{i}_{br}\sin(\theta_R + 2u)\right] +$$
$$+2,6\dot\theta_R\frac{L_{hs}^2}{(\Sigma L_\Pi)}\left(\bar{i}_{as} - \bar{i}_{bs}\right) + \frac{L_{\sigma r} L_{hs}}{L_{\sigma s}(\Sigma L_\Pi)}\left(\bar{u}_{as} + \bar{u}_{bs} + \bar{u}_{cs}\right) + \frac{3L_{hs} + 2L_{\sigma r}}{(\Sigma L_\Pi)}\bar{u}_{cs} -$$
$$-\frac{2L_{hs}}{(\Sigma L_\Pi)}\left[\bar{u}_{cr}\cos\theta_R + \bar{u}_{ar}\cos(\theta_R + u) + \bar{u}_{br}\cos(\theta_R + 2u)\right],$$

(15)

$$\left(\bar{s} + \frac{R_r(\Pi L_r)}{L_{\sigma r}(\Sigma L_\Pi)}\right)\bar{i}_{ar} = \frac{2L_{hs}R_s}{(\Sigma L_\Pi)}\left[\bar{i}_{as}\cos\theta_R + \bar{i}_{bs}\cos(\theta_R + 2u) + \bar{i}_{cs}\cos(\theta_R + u)\right] + 2,6\dot\theta_R\frac{L_{hs}^2}{(\Sigma L_\Pi)}\left(-\bar{i}_{br} + \bar{i}_{cr}\right) -$$
$$-\frac{2L_{hs}}{(\Sigma L_\Pi)}\left[\bar{u}_{as}\cos\theta_R + \bar{u}_{bs}\cos(\theta_R + 2u) + \bar{u}_{cs}\cos(\theta_R + u)\right] + \frac{L_{\sigma s} L_{hs}}{L_{\sigma r}(\Sigma L_\Pi)}\left(\bar{u}_{ar} + \bar{u}_{br} + \bar{u}_{cr}\right) +$$
$$+\frac{2L_{\sigma s} + 3L_{hs}}{(\Sigma L_\Pi)}\bar{u}_{ar} + \dot\theta_R L_{hs}\frac{2L_{\sigma s} + 3L_{hs}}{(\Sigma L_\Pi)}\left[\bar{i}_{as}\sin\theta_R + \bar{i}_{bs}\sin(\theta_R + 2u) + \bar{i}_{cs}\sin(\theta_R + u)\right] -$$
$$-\frac{L_{\sigma s} L_{hs} R_r}{L_{\sigma r}(\Sigma L_\Pi)}\left(\bar{i}_{br} + \bar{i}_{cr}\right),$$

$$\left(s+\frac{R_r\left(\Pi L_r\right)}{L_{\sigma r}\left(\Sigma L_{11}\right)}\right)\bar{i}_{br}=\frac{2L_{hs}R_s}{\left(\Sigma L_{11}\right)}\left[\bar{i}_{bs}\cos\theta_R+\bar{i}_{cs}\cos\left(\theta_R+2u\right)+\bar{i}_{as}\cos\left(\theta_R+u\right)\right]+2,6\dot{\theta}_R\frac{L_{hs}^2}{\left(\Sigma L_{11}\right)}\left(-\bar{i}_{cr}+\bar{i}_{ar}\right)-$$

$$-\frac{2L_{hs}}{\left(\Sigma L_{11}\right)}\left[\bar{u}_{bs}\cos\theta_R+\bar{u}_{cs}\cos\left(\theta_R+2u\right)+\bar{u}_{as}\cos\left(\theta_R+u\right)\right]+\frac{L_{\sigma s}L_{hs}}{L_{\sigma r}\left(\Sigma L_{11}\right)}\left(\bar{u}_{ar}+\bar{u}_{br}+\bar{u}_{cr}\right)+$$

$$+\frac{2L_{\sigma s}+3L_{hs}}{\left(\Sigma L_{11}\right)}\bar{u}_{br}+\dot{\theta}_R L_{hs}\frac{2L_{\sigma s}+3L_{hs}}{\left(\Sigma L_{11}\right)}\left[\bar{i}_{bs}\sin\theta_R+\bar{i}_{cs}\sin\left(\theta_R+2u\right)+\bar{i}_{as}\sin\left(\theta_R+u\right)\right]-$$

$$-\frac{L_{\sigma s}L_{hs}R_r}{L_{\sigma r}\left(\Sigma L_{11}\right)}\left(\bar{i}_{cr}+\bar{i}_{ar}\right),$$

(16)

$$\left(s+\frac{R_r\left(\Pi L_r\right)}{L_{\sigma r}\left(\Sigma L_{11}\right)}\right)\bar{i}_{cr}=\frac{2L_{hs}R_s}{\left(\Sigma L_{11}\right)}\left[\bar{i}_{cs}\cos\theta_R+\bar{i}_{as}\cos\left(\theta_R+2u\right)+\bar{i}_{bs}\cos\left(\theta_R+u\right)\right]+2,6\dot{\theta}_R\frac{L_{hs}^2}{\left(\Sigma L_{11}\right)}\left(-\bar{i}_{ar}+\bar{i}_{br}\right)-$$

$$-\frac{2L_{hs}}{\left(\Sigma L_{11}\right)}\left[\bar{u}_{cs}\cos\theta_R+\bar{u}_{as}\cos\left(\theta_R+2u\right)+\bar{u}_{bs}\cos\left(\theta_R+u\right)\right]+\frac{L_{\sigma s}L_{hs}}{L_{\sigma r}\left(\Sigma L_{11}\right)}\left(\bar{u}_{ar}+\bar{u}_{br}+\bar{u}_{cr}\right)+$$

$$+\frac{2L_{\sigma s}+3L_{hs}}{\left(\Sigma L_{11}\right)}\bar{u}_{cr}+\dot{\theta}_R L_{hs}\frac{2L_{\sigma s}+3L_{hs}}{\left(\Sigma L_{11}\right)}\left[\bar{i}_{cs}\sin\theta_R+\bar{i}_{as}\sin\left(\theta_R+2u\right)+\bar{i}_{bs}\sin\left(\theta_R+u\right)\right]-$$

$$-\frac{L_{\sigma s}L_{hs}R_r}{L_{\sigma r}\left(\Sigma L_{11}\right)}\left(\bar{i}_{ar}+\bar{i}_{br}\right),$$

(17)

Besides (10), the equations concerning mechanical quantities must be added. To this end, the electromagnetic torque has to be calculated. To this effect, we start from the coenergy expression, W'_m, of the 6 circuits (3 are placed on stator and the other 3 on rotor) and we take into consideration that the leakage fluxes, which are independent of rotation angle of the rotor, do not generate electromagnetic torque, that is:

$$W'_m=\frac{1}{2}\left[i_{abcs}\right]_t\left(\left[L_{ss}\right]-L_{\sigma s}\left[1\right]\right)\left[i_{abcs}\right]+\frac{1}{2}\left[i_{abcr}\right]_t\left(\left[L_{rr}\right]-L_{\sigma r}\left[1\right]\right)\left[i_{abcr}\right]+\left[i_{abcs}\right]_t\left[L_{sr}\left(\theta_R\right)\right]\left[i_{abcr}\right]$$

(18)

The magnetic energy of the stator and the rotor does not depend on the rotation angle and consequently, for the electromagnetic torque calculus nothing but the last term of (11) is used. One obtains:

$$T_e=\frac{1}{2}p\left[i_{abcs}\right]_t\frac{d\left[L_{sr}\left(\theta_R\right)\right]}{d\theta_R}\left[i_{abcr}\right]=$$

$$=\frac{1}{2}pL_{hs}\sin\theta_R\left[i_{as}\left(-2i_{ar}+i_{br}+i_{cr}\right)+i_{bs}\left(+i_{ar}-2i_{br}+i_{cr}\right)+i_{cs}\left(+i_{ar}+i_{br}-2i_{cr}\right)\right]+$$

$$+\frac{\sqrt{3}}{2}pL_{hs}\cos\theta_R\left[i_{as}\left(i_{cr}-i_{br}\right)+i_{bs}\left(i_{ar}-i_{cr}\right)+i_{cs}\left(i_{br}-i_{ar}\right)\right]$$

(19)

The equation of torque equilibrium can now be written under operational form as:

$$\bar{\omega}_R\left(\frac{Js+k_z}{p}\right)=\frac{1}{2}pL_{hs}\left\{\sin\theta_R\cdot\left[i_{as}\left(-2i_{ar}+i_{br}+i_{cr}\right)+i_{bs}\left(i_{ar}-2i_{br}+i_{cr}\right)+\right.\right.$$

$$\left.+i_{cs}\left(i_{ar}+i_{br}-2i_{cr}\right)\right]+\sqrt{3}\cos\theta_R\cdot\left[i_{as}\left(i_{cr}-i_{br}\right)+i_{bs}\left(i_{ar}-i_{cr}\right)+i_{cs}\left(i_{br}-i_{ar}\right)\right]\left.\right\}-T_{st}$$

(20)

$$s\overline{\theta}_R = \omega_R = \dot{\theta}_R$$

$$(21)$$

where ω_R represents the rotational pulsatance (or rotational pulsation).

The simulation of the induction machine operation in Matlab-Simulink environment on the basis of the above equations system is rather complicated. Moreover, since all equations depend on the angular speed than the precision of the results could be questionable mainly for the study of rapid transients. Consequently, the use of other variables is understandable. Further, we shall use the total fluxes of the windings (3 motionless windings on stator and other rotating 3 windings on rotor).

It is well known that the total fluxes have a self-component and a mutual one. Taking into consideration the rules of reducing the rotor circuit to the stator one, the matrix of inductances can be written as follows:

$$[L_{abcabc}] = L_{hs} \cdot \begin{bmatrix} 1+l_{\sigma s} & -(1/2) & -(1/2) & \cos\theta_R & \cos(\theta_R+u) & \cos(\theta_R+2u) \\ -(1/2) & 1+l_{\sigma s} & -(1/2) & \cos(\theta_R+2u) & \cos\theta_R & \cos(\theta_R+u) \\ -(1/2) & -(1/2) & 1+l_{\sigma s} & \cos(\theta_R+u) & \cos(\theta_R+2u) & \cos\theta_R \\ \cos\theta_R & \cos(\theta_R+2u) & \cos(\theta_R+u) & 1+l_{\sigma r} & -(1/2) & -(1/2) \\ \cos(\theta_R+u) & L_{hs}\cos\theta_R & \cos(\theta_R+2u) & -(1/2) & 1+l_{\sigma r} & -(1/2) \\ \cos(\theta_R+2u) & \cos(\theta_R+u) & \cos\theta_R & -(1/2) & -(1/2) & 1+l_{\sigma r} \end{bmatrix}$$

$$(22)$$

Now, the equation system (8) can be written shortly as:

$$[u_{abcabc}] = [R_{s,r}][i_{abcabc}] + \frac{d[\psi_{abcabc}]}{dt}, \quad where: \quad [\psi_{abcabc}] = [L_{abcabc}][i_{abcabc}]$$

$$(23)$$

By using the multiplication with the reciprocal matrix:

$$[L_{abcabc}]^{-1}[\psi_{abcabc}] = [L_{abcabc}]^{-1}[L_{abcabc}][i_{abcabc}], \quad or \quad [i_{abcabc}] = [L_{abcabc}]^{-1}[\psi_{abcabc}]$$

$$(24)$$

than (15) becomes:

$$[u_{abcabc}] = [R_{s,r}][L_{abcabc}]^{-1}[\psi_{abcabc}] + \frac{d[\psi_{abcabc}]}{dt}$$

$$(25)$$

This is an expression that connects the voltages to the total fluxes with no currents involvement. Now, practically the reciprocal matrix must be found. To this effect, we suppose that the reciprocal matrix has a similar form with the direct matrix. If we use the condition: $[L_{abcabc}]^{-1}[L_{abcabc}] = [1]$, than through term by term identification is obtained:

$$\left[L_{abcabc}\right]^{-1} = \frac{1}{(\Pi LD)} \cdot$$

$$\begin{bmatrix} \Pi L_{s\sigma} & L_{hs}L_{\sigma r}^2 & L_{hs}L_{\sigma r}^2 & \Gamma\cos\theta_R & \Gamma\cos(\theta_R+u) & \Gamma\cos(\theta_R+2u) \\ L_{hs}L_{\sigma r}^2 & \Pi L_{s\sigma} & L_{hs}L_{\sigma r}^2 & \Gamma\cos(\theta_R+2u) & \Gamma\cos\theta_R & \Gamma\cos(\theta_R+u) \\ L_{hs}L_{\sigma r}^2 & L_{hs}L_{\sigma r}^2 & \Pi L_{s\sigma} & \Gamma\cos(\theta_R+u) & \Gamma\cos(\theta_R+2u) & \Gamma\cos\theta_R \\ \Gamma\cos\theta_R & \Gamma\cos(\theta_R+2u) & \Gamma\cos(\theta_R+u) & \Pi L_{r\sigma} & L_{hs}L_{\sigma s}^2 & L_{hs}L_{\sigma s}^2 \\ \Gamma\cos(\theta_R+u) & \Gamma\cos\theta_R & \Gamma\cos(\theta_R+2u) & L_{hs}L_{\sigma s}^2 & \Pi L_{r\sigma} & L_{hs}L_{\sigma s}^2 \\ \Gamma\cos(\theta_R+2u) & \Gamma\cos(\theta_R+u) & \Gamma\cos\theta_R & L_{hs}L_{\sigma s}^2 & L_{hs}L_{\sigma s}^2 & \Pi L_{r\sigma} \end{bmatrix}$$

(26)

where the following notations have been used:

$$(\Pi LD) = \left(3L_{hs}L_{\sigma s} + 3L_{hs}L_{\sigma r} + 2L_{\sigma r}L_{\sigma s}\right)L_{\sigma r}L_{\sigma s}; \qquad \Gamma = -2L_{hs}L_{\sigma s}L_{\sigma r};$$
$$\Pi L_{s\sigma} = \left(L_{hs}L_{\sigma r} + 3L_{hs}L_{\sigma s} + 2L_{\sigma r}L_{\sigma s}\right)L_{\sigma r}; \quad \Pi L_{r\sigma} = \left(L_{hs}L_{\sigma s} + 3L_{hs}L_{\sigma r} + 2L_{\sigma r}L_{\sigma s}\right)L_{\sigma s}$$

(27)

Further, the matrix product is calculated: $\left[R_{s,r}\right]\left[L_{abcabc}\right]^{-1}\left[\psi_{abcabc}\right]$, which is used in (17). After a convenient grouping, the system becomes:

$$\frac{d\psi_{as}}{dt} + \frac{\Pi L_{s\sigma}R_s}{(\Pi LD)}\psi_{as} = u_{as} - \frac{L_{hs}L_{\sigma r}^2 R_s}{(\Pi LD)}\left(\psi_{bs} + \psi_{cs}\right) + \frac{L_{hs}L_{\sigma s}L_{\sigma r}R_s}{(\Pi LD)} \times$$
$$\times\left[\left(2\psi_{ar} - \psi_{br} - \psi_{cr}\right)\cos\theta_R + \sqrt{3}\left(\psi_{cr} - \psi_{br}\right)\sin\theta_R\right]$$

(28)

$$\frac{d\psi_{bs}}{dt} + \frac{\Pi L_{s\sigma}R_s}{(\Pi LD)}\psi_{bs} = u_{bs} - \frac{L_{hs}L_{\sigma r}^2 R_s}{(\Pi LD)}\left(\psi_{cs} + \psi_{as}\right) + \frac{L_{hs}L_{\sigma s}L_{\sigma r}R_s}{(\Pi LD)} \times$$
$$\times\left[\left(-\psi_{ar} + 2\psi_{br} - \psi_{cr}\right)\cos\theta_R + \sqrt{3}\left(\psi_{ar} - \psi_{cr}\right)\sin\theta_R\right]$$

(29)

$$\frac{d\psi_{cs}}{dt} + \frac{\Pi L_{s\sigma}R_s}{(\Pi LD)}\psi_{cs} = u_{cs} - \frac{L_{hs}L_{\sigma r}^2 R_s}{(\Pi LD)}\left(\psi_{as} + \psi_{bs}\right) + \frac{L_{hs}L_{\sigma s}L_{\sigma r}R_s}{(\Pi LD)} \times$$
$$\times\left[\left(-\psi_{ar} - \psi_{br} + 2\psi_{cr}\right)\cos\theta_R + \sqrt{3}\left(\psi_{br} - \psi_{ar}\right)\sin\theta_R\right]$$

(30)

$$\frac{d\psi_{ar}}{dt} + \frac{\Pi L_{r\sigma}R_r}{(\Pi LD)}\psi_{ar} = u_{ar} - \frac{L_{hs}L_{\sigma s}^2 R_r}{(\Pi LD)}\left(\psi_{br} + \psi_{cr}\right) + \frac{L_{hs}L_{\sigma s}L_{\sigma r}R_r}{(\Pi LD)} \times$$
$$\times\left[\left(2\psi_{as} - \psi_{bs} - \psi_{cs}\right)\cos\theta_R + \sqrt{3}\left(\psi_{bs} - \psi_{cs}\right)\sin\theta_R\right]$$

(31)

$$\frac{d\psi_{br}}{dt} + \frac{\Pi L_{r\sigma}R_r}{(\Pi LD)}\psi_{br} = u_{br} - \frac{L_{hs}L_{\sigma s}^2 R_r}{(\Pi LD)}\left(\psi_{cr} + \psi_{ar}\right) + \frac{L_{hs}L_{\sigma s}L_{\sigma r}R_r}{(\Pi LD)} \times$$
$$\times\left[\left(-\psi_{as} + 2\psi_{bs} - \psi_{cs}\right)\cos\theta_R + \sqrt{3}\left(\psi_{cs} - \psi_{as}\right)\sin\theta_R\right]$$

(32)

$$\frac{d\psi_{cr}}{dt} + \frac{\Pi L_{r\sigma}R_r}{(\Pi LD)}\psi_{cr} = u_{cr} - \frac{L_{hs}L_{\sigma s}^2 R_r}{(\Pi LD)}\left(\psi_{ar} + \psi_{br}\right) + \frac{L_{hs}L_{\sigma s}L_{\sigma r}R_r}{(\Pi LD)} \times$$
$$\times\left[\left(-\psi_{as} - \psi_{bs} + 2\psi_{cs}\right)\cos\theta_R + \sqrt{3}\left(\psi_{as} - \psi_{bs}\right)\sin\theta_R\right]$$

(33)

For the calculation of the electromagnetic torque we can use the principle of energy conservation or the expression of stored magnetic energy. The expression of the electromagnetic torque corresponding to a multipolar machine (p is the number of pole pairs) can be written in a matrix form as follows:

$$T_e = -\frac{p}{2} \cdot \left\{\left[\psi_{abcabc}\right]_t \cdot \frac{d\left[L_{abcabc}\right]^{-1}}{d\theta_R} \cdot \left[\psi_{abcabc}\right]\right\}$$

(34)

To demonstrate the validity of (21), one uses the expression of the matrix $\left[L_{abcabc}\right]^{-1}$, (18), in order to calculate its derivative:

$$\frac{d}{d\theta_R}\left[L_{abcabc}\right]^{-1} = \Lambda_3 \cdot$$

$$\begin{bmatrix}
0 & 0 & 0 & \sin\theta_R & \sin(\theta_R + u) & \sin(\theta_R + 2u) \\
0 & 0 & 0 & \sin(\theta_R + 2u) & \sin\theta_R & \sin(\theta_R + u) \\
0 & 0 & 0 & \sin(\theta_R + u) & \sin(\theta_R + 2u) & \sin\theta_R \\
\sin\theta_R & \sin(\theta_R + 2u) & \sin(\theta_R + u) & 0 & 0 & 0 \\
\sin(\theta_R + u) & \sin\theta_R & \sin(\theta_R + 2u) & 0 & 0 & 0 \\
\sin(\theta_R + 2u) & \sin(\theta_R + u) & \sin\theta_R & 0 & 0 & 0
\end{bmatrix}$$

(35)

where the following notation has been used:

$$\Lambda_3 = \frac{1}{(3/2)(L_{\sigma s} + L_{\sigma r}) + L_{\sigma r}L_{\sigma s}/L_{hs}}$$

(36)

This expression defines the permeance of a three-phase machine for the mathematical model in total fluxes.

Observation: One can use the general expression of the electromagnetic torque where the direct and reciprocal matrices of the inductances (which link the currents with the fluxes) should be replaced, that is:

$$T_e = \frac{1}{2}p\left[i_{abcabc}\right]_t \frac{d\left[L_{abcabc}\right]}{d\theta_R}\left[i_{abcabc}\right] = \frac{1}{2}p\cdot\left[\psi_{abcabc}\right]_t\left[L\right]_t^{-1}\cdot\frac{d\left[L_{abcabc}\right]}{d\theta_R}\cdot\left[L\right]^{-1}\left[\psi_{abcabc}\right]$$

$$T_e = -\frac{1}{2}p\left[\psi_{abcabc}\right]_t\frac{d\left[L_{abcabc}\right]^{-1}}{d\theta_R}\left[\psi_{abcabc}\right]$$

$$(37)$$

A more convenient expression that depends on $\sin\theta_R$ and $\cos\theta_R$, leads to the electromagnetic torque equation in fluxes alone:

$$T_e = -(1/2)p\Lambda_3\left\{\left[\psi_{as}\left(2\psi_{ar}-\psi_{br}-\psi_{cr}\right)+\psi_{bs}\left(2\psi_{br}-\psi_{cr}-\psi_{ar}\right)+\psi_{cs}\left(2\psi_{cr}-\psi_{ar}-\psi_{br}\right)\right]\sin\theta_R + \right.$$
$$\left. +\sqrt{3}\left[\psi_{as}\left(\psi_{br}-\psi_{cr}\right)+\psi_{bs}\left(\psi_{cr}-\psi_{ar}\right)+\psi_{cs}\left(\psi_{ar}-\psi_{br}\right)\right]\cos\theta_R\right\}$$

$$(38)$$

Ultimately, by getting together the equations of the 6 electric circuits and the movement equations we obtain an 8 equation system, which can be written under operational form:

$$\overline{\psi}_{as}\left(s+\frac{\Pi L_{s\sigma}R_s}{(\Pi LD)}\right) = \overline{u}_{as} - \frac{L_{hs}L_{\sigma r}^2R_s}{(\Pi LD)}\left(\overline{\psi}_{bs}+\overline{\psi}_{cs}\right)+\frac{L_{hs}L_{\sigma s}L_{\sigma r}R_s}{(\Pi LD)}\times$$
$$\times\left[\left(2\overline{\psi}_{ar}-\overline{\psi}_{br}-\overline{\psi}_{cr}\right)\cos\theta_R+\sqrt{3}\left(\overline{\psi}_{cr}-\overline{\psi}_{br}\right)\sin\theta_R\right]$$

$$(39)$$

$$\overline{\psi}_{bs}\left(s+\frac{\Pi L_{s\sigma}R_s}{(\Pi LD)}\right) = \overline{u}_{bs} - \frac{L_{hs}L_{\sigma r}^2R_s}{(\Pi LD)}\left(\overline{\psi}_{cs}+\overline{\psi}_{as}\right)+\frac{L_{hs}L_{\sigma s}L_{\sigma r}R_s}{(\Pi LD)}\times$$
$$\times\left[\left(-\overline{\psi}_{ar}+2\overline{\psi}_{br}-\overline{\psi}_{cr}\right)\cos\theta_R+\sqrt{3}\left(\overline{\psi}_{ar}-\overline{\psi}_{cr}\right)\sin\theta_R\right]$$

$$(40)$$

$$\overline{\psi}_{cs}\left(s+\frac{\Pi L_{s\sigma}R_s}{(\Pi LD)}\right) = \overline{u}_{cs} - \frac{L_{hs}L_{\sigma r}^2R_s}{(\Pi LD)}\left(\overline{\psi}_{as}+\overline{\psi}_{bs}\right)+\frac{L_{hs}L_{\sigma s}L_{\sigma r}R_s}{(\Pi LD)}\times$$
$$\times\left[\left(-\overline{\psi}_{ar}-\overline{\psi}_{br}+2\overline{\psi}_{cr}\right)\cos\theta_R+\sqrt{3}\left(\overline{\psi}_{br}-\overline{\psi}_{ar}\right)\sin\theta_R\right]$$

$$(41)$$

$$\overline{\psi}_{ar}\left(s+\frac{\Pi L_{r\sigma}R_r}{(\Pi LD)}\right) = \overline{u}_{ar} - \frac{L_{hs}L_{\sigma s}^2R_r}{(\Pi LD)}\left(\overline{\psi}_{br}+\overline{\psi}_{cr}\right)+\frac{L_{hs}L_{\sigma s}L_{\sigma r}R_r}{(\Pi LD)}\times$$
$$\times\left[\left(2\overline{\psi}_{as}-\overline{\psi}_{bs}-\overline{\psi}_{cs}\right)\cos\theta_R+\sqrt{3}\left(\overline{\psi}_{bs}-\overline{\psi}_{cs}\right)\sin\theta_R\right]$$

$$(42)$$

$$\bar{\psi}_{br}\left(s+\frac{\Pi L_{r\sigma}R_r}{(\Pi LD)}\right)=\bar{u}_{br}-\frac{L_{hs}L_{\sigma s}^2 R_r}{(\Pi LD)}\left(\bar{\psi}_{cr}+\bar{\psi}_{ar}\right)+\frac{L_{hs}L_{\sigma s}L_{\sigma r}R_r}{(\Pi LD)}\times$$
$$\times\left[\left(-\bar{\psi}_{as}+2\bar{\psi}_{bs}-\bar{\psi}_{cs}\right)\cos\theta_R+\sqrt{3}\left(\bar{\psi}_{cs}-\bar{\psi}_{as}\right)\sin\theta_R\right]$$

(43)

$$\bar{\psi}_{cr}\left(s+\frac{\Pi L_{r\sigma}R_r}{(\Pi LD)}\right)=\bar{u}_{cr}-\frac{L_{hs}L_{\sigma s}^2 R_r}{(\Pi LD)}\left(\bar{\psi}_{ar}+\bar{\psi}_{br}\right)+\frac{L_{hs}L_{\sigma s}L_{\sigma r}R_r}{(\Pi LD)}\times$$
$$\times\left[\left(-\bar{\psi}_{as}-\bar{\psi}_{bs}+2\bar{\psi}_{cs}\right)\cos\theta_R+\sqrt{3}\left(\bar{\psi}_{as}-\bar{\psi}_{bs}\right)\sin\theta_R\right]$$

(44)

$$\dot{\theta}_R\left(s+k_z/J\right)=\left(p/J\right)\langle-(1/2)p\Lambda_3\{\sin\theta_R\left[\bar{\psi}_{as}\left(2\bar{\psi}_{ar}-\bar{\psi}_{br}-\bar{\psi}_{cr}\right)+\right.$$
$$+\bar{\psi}_{bs}\left(2\bar{\psi}_{br}-\bar{\psi}_{cr}-\bar{\psi}_{ar}\right)+\bar{\psi}_{cs}\left(2\bar{\psi}_{cr}-\bar{\psi}_{ar}-\bar{\psi}_{br}\right)\right]+\sqrt{3}\cos\theta_R$$
$$\cdot\left[\bar{\psi}_{as}\left(\bar{\psi}_{br}-\bar{\psi}_{cr}\right)+\bar{\psi}_{bs}\left(\bar{\psi}_{cr}-\bar{\psi}_{ar}\right)+\bar{\psi}_{cs}\left(\bar{\psi}_{ar}-\bar{\psi}_{br}\right)\right]\left.\right\}-T_{st}\rangle$$

(45)

$$\frac{d\theta_R}{dt}=\dot{\theta}_R=\omega_R$$

(46)

This equation system, (26-1)-(26-8) allows the study of any operation duty of the three-phase induction machine: steady state or transients under balanced or unbalanced condition, with simple or double feeding.

MATHEMATICAL MODELS USED FOR THE STUDY OF STEADY-STATE UNDER BALANCED AND UNBALANCED CONDITIONS

Generally, the symmetrical three-phase squirrel cage induction machine has the stator windings connected to a supply system, which provides variable voltages according to certain laws but have the same pulsation. Practically, this is the case with 4 wires connection, 3 phases and the neutral. The sum of the phase currents gives the current along neutral and the homopolar component can be immediately defined. The analysis of such a machine can use the symmetric components theory. This is the case of the machine with two unbalances as concerns the supply. The study can be done either using the equation system (26-1...8) or on the basis of symmetric components theory with three distinct mathematical models for each component (positive sequence, negative sequence and homopolar).

The vast majority of electric drives uses however the 3 wires connection (no neutral). Consequently, there is no homopolar current component, the homopolar fluxes are zero as well and the sum of the 3 phase total fluxes is null. This is an asymmetric condition with single unbalance, which can be studied by using the direct and inverse sequence components when the transformation from 3 to 2 axes is mandatory. This approach practically replaces the three-phase machine with unbalanced supply with two symmetric three-phase machines. One of them produces the positive torque and the other provides the negative torque. The resultant torque comes out through superposition of the effects.

The ABC–Aβ0 Model in Total Fluxes

The operation of the machine with 2 unbalances can be analyzed by considering certain expressions for the instantaneous values of the stator and rotor quantities (voltages, total fluxes and currents eventually, which can be transformed from (a, b, c) to (α, β, 0) reference frames in accordance with the following procedure:

$$
\begin{bmatrix} \psi_{\alpha s} \\ \psi_{\beta s} \\ \psi_{0s} \end{bmatrix} = \sqrt{\frac{2}{3}} \cdot \begin{bmatrix} 1 & -1/2 & -1/2 \\ 0 & \sqrt{3}/2 & -\sqrt{3}/2 \\ \sqrt{2}/2 & \sqrt{2}/2 & \sqrt{2}/2 \end{bmatrix} \cdot \begin{bmatrix} \psi_{as} \\ \psi_{bs} \\ \psi_{cs} \end{bmatrix}
$$

$$(47)$$

We define the following notations:

$$
\frac{\Pi L_{s\sigma} R_s}{(\Pi LD)} = \frac{\left(L_{hs}L_{\sigma r} + 3L_{hs}L_{\sigma s} + 2L_{\sigma r}L_{\sigma s}\right)}{\left(3L_{hs}L_{\sigma r} + 3L_{hs}L_{\sigma s} + 2L_{\sigma r}L_{\sigma s}\right)}\left(\frac{R_s}{L_{\sigma s}}\right) \cong \frac{2}{3}\left(\frac{R_s}{L_{\sigma s}}\right) = v_{st};
$$

$$
\frac{L_{hs}L_{\sigma r}^2 R_s}{(\Pi LD)} = \frac{L_{hs}L_{\sigma r}R_s}{\left(3L_{hs}L_{\sigma s} + 3L_{hs}L_{\sigma r} + 2L_{\sigma r}L_{\sigma s}\right)L_{\sigma s}} \cong \frac{1}{6}\left(\frac{R_s}{L_{\sigma s}}\right) = v_{sr};
$$

$$
v_{st} - v_{sr} = \frac{3 + 2\left(L_{\sigma r}/L_{hs}\right)}{3\left(L_{\sigma r}/L_{\sigma s}\right) + 3 + 2\left(L_{\sigma r}/L_{hs}\right)}\left(\frac{R_s}{L_{\sigma s}}\right) \cong \frac{1}{2}\left(\frac{R_s}{L_{\sigma s}}\right) = v_s;
$$

$$(48)$$

$$
\frac{\Pi L_{r\sigma} R_r}{(\Pi LD)} = \frac{\left(L_{hs}L_{\sigma s} + 3L_{hs}L_{\sigma r} + 2L_{\sigma r}L_{\sigma s}\right)}{\left(3L_{hs}L_{\sigma r} + 3L_{hs}L_{\sigma s} + 2L_{\sigma r}L_{\sigma s}\right)}\left(\frac{R_r}{L_{\sigma r}}\right) \cong \frac{2}{3}\left(\frac{R_r}{L_{\sigma r}}\right) = v_{rt};
$$

$$
\frac{L_{hs}L_{\sigma s}^2 R_r}{(\Pi LD)} = \frac{L_{hs}L_{\sigma s}R_r}{\left(3L_{hs}L_{\sigma s} + 3L_{hs}L_{\sigma r} + 2L_{\sigma r}L_{\sigma s}\right)L_{\sigma r}} \cong \frac{1}{6}\left(\frac{R_r}{L_{\sigma r}}\right) = v_{rs};
$$

$$
v_{rt} - v_{rs} = \frac{3 + 2\left(L_{\sigma s}/L_{hs}\right)}{3\left(L_{\sigma s}/L_{\sigma r}\right) + 3 + 2\left(L_{\sigma s}/L_{hs}\right)}\left(\frac{R_r}{L_{\sigma r}}\right) \cong \frac{1}{2}\left(\frac{R_r}{L_{\sigma r}}\right) = v_r;
$$

$$(49)$$

$$
\frac{3L_{hs}L_{\sigma s}L_{\sigma r}R_s}{(\Pi LD)} = \frac{3L_{\sigma s}}{\left(3L_{\sigma s} + 3L_{\sigma r} + 2L_{\sigma r}L_{\sigma s}/L_{hs}\right)}\left(\frac{R_s}{L_{\sigma s}}\right) \cong \frac{1}{2}\left(\frac{R_s}{L_{\sigma s}}\right) = v_{\sigma s};
$$

$$(50)$$

$$\frac{3L_{hs}L_{\sigma s}L_{\sigma r}R_r}{(\Pi LD)} = \frac{3L_{\sigma r}}{\left(3L_{\sigma s} + 3L_{\sigma r} + 2L_{\sigma r}L_{\sigma s}/L_{hs}\right)}\left(\frac{R_r}{L_{\sigma r}}\right) \cong \frac{1}{2}\left(\frac{R_r}{L_{\sigma r}}\right) = v_{\sigma r};$$

$$(51)$$

By using these notations in (17) and after convenient groupings we obtain:

$$\frac{d\psi_{as}}{dt} + v_{st}\psi_{as} = u_{as} - v_{sr}\left(\psi_{bs} + \psi_{cs}\right) + \frac{1}{3}v_{\sigma s} \times$$
$$\times\left[\left(2\psi_{ar} - \psi_{br} - \psi_{cr}\right)\cos\theta_R + \sqrt{3}\left(\psi_{cr} - \psi_{br}\right)\sin\theta_R\right]$$

$$(52)$$

$$\frac{d\psi_{bs}}{dt} + v_{st}\psi_{bs} = u_{bs} - v_{sr}\left(\psi_{cs} + \psi_{as}\right) + \frac{1}{3}v_{\sigma s} \times$$
$$\times\left[\left(-\psi_{ar} + 2\psi_{br} - \psi_{cr}\right)\cos\theta_R + \sqrt{3}\left(\psi_{ar} - \psi_{cr}\right)\sin\theta_R\right]$$

$$(53)$$

$$\frac{d\psi_{cs}}{dt} + v_{st}\psi_{cs} = u_{cs} - v_{sr}\left(\psi_{as} + \psi_{bs}\right) + \frac{1}{3}v_{\sigma s} \times$$
$$\times\left[\left(-\psi_{ar} - \psi_{br} + 2\psi_{cr}\right)\cos\theta_R + \sqrt{3}\left(\psi_{br} - \psi_{ar}\right)\sin\theta_R\right]$$

$$(54)$$

$$\frac{d\psi_{ar}}{dt} + v_{rt}\psi_{ar} = u_{ar} - v_{rs}\left(\psi_{br} + \psi_{cr}\right) + \frac{1}{3}v_{\sigma r} \times$$
$$\times\left[\left(2\psi_{as} - \psi_{bs} - \psi_{cs}\right)\cos\theta_R + \sqrt{3}\left(\psi_{bs} - \psi_{cs}\right)\sin\theta_R\right]$$

$$(55)$$

$$\frac{d\psi_{br}}{dt} + v_{rt}\psi_{br} = u_{br} - v_{rs}\left(\psi_{cr} + \psi_{ar}\right) + \frac{1}{3}v_{\sigma r} \times$$
$$\times\left[\left(-\psi_{as} + 2\psi_{bs} - \psi_{cs}\right)\cos\theta_R + \sqrt{3}\left(\psi_{cs} - \psi_{as}\right)\sin\theta_R\right]$$

$$(56)$$

$$\frac{d\psi_{cr}}{dt} + v_{rt}\psi_{cr} = u_{cr} - v_{rs}\left(\psi_{ar} + \psi_{br}\right) + \frac{1}{3}v_{\sigma r} \times$$
$$\times\left[\left(-\psi_{bs} + 2\psi_{cs} - \psi_{as}\right)\cos\theta_R + \sqrt{3}\left(\psi_{as} - \psi_{bs}\right)\sin\theta_R\right]$$

$$(57)$$

Typical for the cage machine or even for the wound rotor after the starting rheostat is short-circuited is the fact that the rotor voltages become zero. The equations of the six circuits get different as a result of certain convenient math operations. (29-2) and (29-3) are multiplied by (-1/2) and afterwards added to (29-1); (29-3) is subtracted from (29-2); (29-1), (29-2) and (29-3) are added together. We obtain three equations that describe the stator. Similarly, (29-4), (29-5) and (29-6) are used for the rotor equations. The new equation system is:

$$
\begin{cases}
\dfrac{d\psi_{\alpha s}}{dt} + v_s \psi_{\alpha s} = u_{\alpha s} + v_{\sigma s}\left(\psi_{\alpha r}\cos\theta_R - \psi_{\beta r}\sin\theta_R\right) \\[2ex]
\dfrac{d\psi_{\beta s}}{dt} + v_s \psi_{\beta s} = u_{\beta s} + v_{\sigma s}\left(\psi_{\alpha r}\sin\theta_R + \psi_{\beta r}\cos\theta_R\right) \\[2ex]
\dfrac{d\psi_{0s}}{dt} + \left(v_{st} + 2v_{sr}\right)\psi_{0s} = u_{0s}
\end{cases}
$$

Further, the movement equation has to be attached. It is necessary to establish the detailed expression of the electromagnetic torque in fluxes alone starting with (25) and using convenient transformations:

$$
T_e = -(3/2)p\Lambda_3\left[\left(\psi_{\alpha s}\psi_{\alpha r} + \psi_{\beta s}\psi_{\beta r}\right)\sin\theta_R + \left(\psi_{\alpha s}\psi_{\beta r} - \psi_{\beta s}\psi_{\alpha r}\right)\cos\theta_R\right] \tag{58}
$$

Ultimately, the 8 equation system under operational form is:

$$
\overline{\psi}_{\alpha s}\left(\overline{s}+v_s\right) = \overline{u}_{\alpha s} + v_{\sigma s}\left(\overline{\psi}_{\alpha r}\cos\theta_R - \overline{\psi}_{\beta r}\sin\theta_R\right) \tag{59}
$$

$$
\overline{\psi}_{\beta s}\left(\overline{s}+v_s\right) = \overline{u}_{\beta s} + v_{\sigma s}\left(\overline{\psi}_{\alpha r}\sin\theta_R + \overline{\psi}_{\beta r}\cos\theta_R\right) \tag{60}
$$

$$
\overline{\psi}_{0s}\left(\overline{s}+v_{st}+2v_{sr}\right) = \overline{u}_{0s} \tag{61}
$$

$$
\overline{\psi}_{\alpha r}\left(\overline{s}+v_r\right) = \overline{u}_{\alpha r} + v_{\sigma r}\left(\overline{\psi}_{\alpha s}\cos\theta_R + \overline{\psi}_{\beta s}\sin\theta_R\right) \tag{62}
$$

$$
\overline{\psi}_{\beta r}\left(\overline{s}+v_r\right) = \overline{u}_{\beta r} + v_{\sigma r}\left(-\overline{\psi}_{\alpha s}\sin\theta_R + \overline{\psi}_{\beta s}\cos\theta_R\right) \tag{63}
$$

$$
\overline{\psi}_{0r}\left(\overline{s}+v_{rt}+2v_{rs}\right) = \overline{u}_{0r} \tag{64}
$$

$$
\dot{\theta}_R\left(\overline{s}+k_z/J\right) = (p/J)\cdot\left\{-(3/2)p\Lambda_3\left[\left(\overline{\psi}_{\alpha s}\overline{\psi}_{\alpha r} + \overline{\psi}_{\beta s}\overline{\psi}_{\beta r}\right)\sin\theta_R + \left(\overline{\psi}_{\alpha s}\overline{\psi}_{\beta r} - \overline{\psi}_{\beta s}\overline{\psi}_{\alpha r}\right)\cos\theta_R\right] - T_{st}\right\} \tag{65}
$$

$$
\frac{d\theta_R}{dt} = \dot{\theta}_R = \omega_R \tag{66}
$$

These equations allow the study of three-phase induction machine for any duty. It has to be mentioned that the electromagnetic torque expression has no homopolar components of the total fluxes.

The Abc-Dq Model In Total Fluxes

For the study of the single unbalance condition is necessary to consider expressions of the instantaneous values of the stator and rotor quantities (voltages, total fluxes and eventually currents in a,b,c reference frame) whose sum is null. The real quantities can be transformed to (d,q) reference frame (Simion et al., 2011). By using the notations (28-1), (28-2), (28-3) and (28-4) then after convenient grouping we obtain (Simion, 2010):

$$\frac{d\psi_{\alpha s}}{dt} + v_s\psi_{\alpha s} = u_{\alpha s} + v_{\sigma s}\left(\psi_{\alpha r}\cos\theta_R - \psi_{\beta r}\sin\theta_R\right)$$

$$\frac{d\psi_{\beta s}}{dt} + v_s\psi_{\beta s} = u_{\beta s} + v_{\sigma s}\left(\psi_{\alpha r}\sin\theta_R + \psi_{\beta r}\cos\theta_R\right)$$

Further, the movement equation (31) must be attached. The operational form of the equation system (4 electric circuits and 2 movement equations) is:

$$\overline{\psi}_{\alpha s}\left(s+v_s\right) = \overline{u}_{\alpha s} + v_{\sigma s}\left(\overline{\psi}_{\alpha r}\cos\theta_R - \overline{\psi}_{\beta r}\sin\theta_R\right)$$

(67)

$$\overline{\psi}_{\beta s}\left(s+v_s\right) = \overline{u}_{\beta s} + v_{\sigma s}\left(\overline{\psi}_{\alpha r}\sin\theta_R + \overline{\psi}_{\beta r}\cos\theta_R\right)$$

(68)

$$\overline{\psi}_{\alpha r}\left(s+v_r\right) = v_{\sigma r}\left(\overline{\psi}_{\alpha s}\cos\theta_R + \overline{\psi}_{\beta s}\sin\theta_R\right)$$

(69)

$$\overline{\psi}_{\beta r}\left(s+v_r\right) = v_{\sigma r}\left(-\overline{\psi}_{\alpha s}\sin\theta_R + \overline{\psi}_{\beta s}\cos\theta_R\right)$$

(70)

$$\dot{\theta}_R\left(s+k_z/J\right) = (p/J)\cdot\left\{-(3/2)p\Lambda_3\left[\left(\overline{\psi}_{\alpha s}\overline{\psi}_{\alpha r}+\overline{\psi}_{\beta s}\overline{\psi}_{\beta r}\right)\sin\theta_R +\left(\overline{\psi}_{\alpha s}\overline{\psi}_{\beta r}-\overline{\psi}_{\beta s}\overline{\psi}_{\alpha r}\right)\cos\theta_R\right]-T_{st}\right\}$$

(71)

$$\frac{d\theta_R}{dt} = \dot{\theta}_R = \omega_R$$

(72)

The equation sets (33-1...4) and (34-1...6) prove that a three-phase induction machine connected to the supply system by 3 wires can be studied similarly to a two-phase machine (two-phase mathematical model). Its parameters can be deduced by linear transformations of the original parameters including the supply voltages (Fig. 2a).

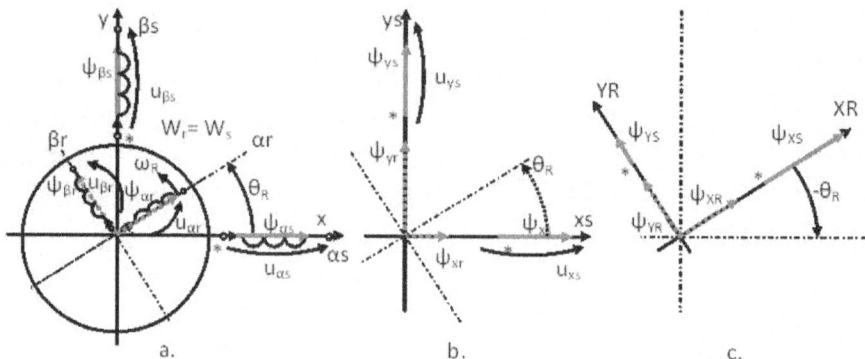

Figure 2: Induction machine schematic view: a.Two-phase model; b. Simplified view of the total fluxes in stator reference frame; c. Idem, but in rotor reference frame

The windings of two-phase model are denoted with (αs, βs) and (αr, βr) in order to trace a correspondence with the real two-phase machine, whose subscripts are (as, bs) and (ar, br) respectively. We shall use the subscripts xs and ys for the quantities that corresponds to the three-phase machine but transformed in its two-phase model. This is a rightful assumption since (αs, βs) axes are collinear with (x, y) axes, which are commonly used in analytic geometry. Further, new notations (35) for the flux linkages of the right member of the equations (33-1...4) will be defined by following the next rules:

- Projection sums corresponding to rotor flux linkages from (αr, βr) axes along the two stator axes (denoted with x and y that is ψ_{xr}, ψ_{yr}) when they refer to the flux linkages from the right member of the first two equations, Fig. 2b.

- Projection sums corresponding to stator flux linkages from (αs, βs) axes along the two rotor axes (denoted with X and Y that is ψXS, ψYS) when they refer to the flux linkages from the last two equations, Fig. 2c.

$$\begin{cases} \psi_{xr} = \psi_{\alpha r} \cos\theta_R - \psi_{\beta r} \sin\theta_R, & \psi_{yr} = \psi_{\alpha r} \sin\theta_R + \psi_{\beta r} \cos\theta_R \\ \psi_{XS} = \psi_{\alpha s} \cos\theta_R + \psi_{\beta s} \sin\theta_R, & \psi_{YS} = -\psi_{\alpha s} \sin\theta_R + \psi_{\beta s} \cos\theta_R \end{cases}$$

$$(73)$$

Some aspects have to be pointed out. When the machine operates under motoring duty, the pulsation of the stator flux linkages from (αs, βs) axes is equal to ω_s. Since the rotational pulsation is ω_R then the pulsation of the rotor quantities from (αr, βr) axes is equal to $\omega r = s\omega_s = \omega_s - \omega_R$. The pulsation of the rotor quantities projected along the stator axes with the subscripts x and y is equal to ω_s. The pulsation of the stator quantities projected along the rotor axes with the subscripts XS and YS is equal to ω_r. The equations (33-1...4) become:

$$\overline{\psi}_{\alpha s}\left(\overline{s}+v_s\right)=\overline{u}_{\alpha s}+V_{\sigma s}\overline{\psi}_{xr}$$
(74)

$$\overline{\psi}_{\beta s}\left(\overline{s}+v_s\right)=\overline{u}_{\beta s}+V_{\sigma s}\overline{\psi}_{yr}$$
(75)

$$\overline{\psi}_{\alpha r}\left(\overline{s}+v_r\right)=V_{\sigma r}\overline{\psi}_{XS}$$
(76)

$$\overline{\psi}_{\beta r}\left(\overline{s}+v_r\right)=V_{\sigma r}\overline{\psi}_{YS}$$
(77)

The first two equations join the quantities with the pulsation ωs and the other two, the quantities with the pulsation $\omega r = s\omega s$. The expression of the magnetic torque, in total fluxes and rotor position anglebecomes:

$$T_e=-\left(3/2\right)p\Lambda_3\left(\overline{\psi}_{\alpha s}\overline{\psi}_{yr}-\overline{\psi}_{\beta s}\overline{\psi}_{xr}\right)$$
(78)

or a second equivalent expression:

$$T_e=(3/2)p\Lambda_3\left(\overline{\psi}_{\alpha r}\overline{\psi}_{YS}-\overline{\psi}_{\beta r}\overline{\psi}_{XS}\right)$$
(79)

which shows the "total symmetry" of the two-phase model of the three-phase machine regarding both stator and rotor. The equations of the four circuits together with the movement equation (37) under operational form give:

$$\overline{\psi}_{\alpha s}\left(\overline{s}+v_s\right)=\overline{u}_{\alpha s}+V_{\sigma s}\overline{\psi}_{xr}$$
(80)

$$\overline{\psi}_{\beta s}\left(\overline{s}+v_s\right)=\overline{u}_{\beta s}+V_{\sigma s}\overline{\psi}_{yr}$$
(81)

$$\overline{\psi}_{\alpha r}\left(\overline{s}+v_r\right)=V_{\sigma r}\overline{\psi}_{XS}$$
(82)

$$\overline{\psi}_{\beta r}\left(\overline{s}+v_r\right)=V_{\sigma r}\overline{\psi}_{YS}$$
(83)

$$\dot{\theta}_R\left(s + k_z / J\right) = (p/J)\cdot\left\{\ (3/2)p\Lambda_3\left(\overline{\psi}_{\beta s}\overline{\psi}_{xr} - \overline{\psi}_{\alpha s}\overline{\psi}_{yr}\right) - T_{st}\right\} \tag{84}$$

$$\frac{d\theta_R}{dt} = \dot{\theta}_R = \omega_R \tag{85}$$

This last equation system allows the study of transients under single unbalance condition. It is similar with the frequently used equations (Park) but contains as variables only total fluxes and the rotation angle. There are no currents or angular speed in the voltage equations.

EXPRESSIONS OF ELECTROMAGNETIC TORQUE

For the steady state analysis of the symmetric three-phase induction machine, one can define the simplified space phasor of the stator flux, which is collinear to the total flux of the (αs) axis and has a $\sqrt{3}$ times higher modulus. In a similar way can be obtained the space phasors of the stator voltages and rotor fluxes and the system equation (39-1...6) that describe the steady state becomes:

$$U_{sR3} = \left(v_s + j\omega_s\right)\underline{\Psi}_{sR3} - v_{\sigma s}\underline{\Psi}_{rR3} = \left(\omega_s - jv_s\right)\Psi_{sR3}e^{j\alpha_s} + jv_{\sigma s}\Psi_{rR3}e^{j\alpha_r}$$

$$0 = v_{\sigma r}\underline{\Psi}_{sR3} - \left(v_r + js\omega_s\right)\underline{\Psi}_{rR3} = -jv_{\sigma r}\Psi_{sR3}e^{j\alpha_s} + \left(jv_r - s\omega_s\right)\Psi_{rR3}e^{j\alpha_r} \tag{86}$$

$$T_e = \left(3/2\right)p\Lambda_3\Psi_{sR3}\Psi_{rR3}\sin\left(\alpha_s - \alpha_r\right) \tag{87}$$

When the speed regulation of the cage induction machine is employed by means of voltage and/or frequency variation then the simultaneous control of the two total flux space vectors is difficult. As consequence, new strategies more convenient can be chosen. To this effect, we shall deduce expressions of the electromagnetic torque that include only one of the total flux space vectors either from stator or rotor.

Variation of the Torque with the Stator Total Flux Space Vector

One of the methods used for the control of induction machine consists in the operation with constant stator total flux space vector. From (40), the rotor total flux space vector is:

$$\underline{\Psi}_{rR3} = \frac{V_{\sigma r}}{V_r + js\omega_s}\underline{\Psi}_{sR3} = \frac{V_{\sigma r}\,\underline{\Psi}_{sR3}}{\sqrt{\omega_s^2 s^2 + v_r^2}}\left(\frac{V_r}{\sqrt{\omega_s^2 s^2 + v_r^2}} - j\frac{s\omega_s}{\sqrt{\omega_s^2 s^2 + v_r^2}}\right) =$$

$$= \frac{V_{\sigma r}\,\underline{\Psi}_{sR3}}{\sqrt{\omega_s^2 s^2 + v_r^2}}e^{-j\theta};(\theta = \alpha_s - \alpha_r);\sin\theta = \frac{s\omega_s}{\sqrt{\omega_s^2 s^2 + v_r^2}};\cos\theta = \frac{V_r}{\sqrt{\omega_s^2 s^2 + v_r^2}}$$

(88)

where θ is the angle between stator and rotor total flux space vectors. This angle has the meaning of an internal angle of the machine.

The expression of the magnetic torque that depends with the stator total flux space vector becomes:

$$T_e = -\left(\frac{3}{2}\right)p\Lambda_3\,\text{Re}\left(j\underline{\Psi}_{sR3}\underline{\Psi}^*_{rR3}\right) = -\left(\frac{3}{2}\right)p\Lambda_3\,\text{Re}\left\{j\underline{\Psi}_{sR3}\frac{V_{\sigma r}}{\sqrt{\omega_s^2 s^2 + v_r^2}}\cdot\underline{\Psi}^*_{sR3}(\cos\theta + j\sin\theta)\right\} =$$

$$= \frac{3}{2}\frac{V_{\sigma r}}{V_r}p\Lambda_3\Psi^2_{sR3}\frac{s\omega_s V_r}{\omega_s^2 s^2 + v_r^2} = \frac{3}{4}\frac{V_{\sigma r}}{V_r}p\Lambda_3\Psi^2_{sR3}\sin 2\theta.$$

(89)

Assuming the ideal hypothesis of maintaining constant the stator flux, for example equal to the no-load value, then the pull-out torque, T_{emax}, corresponds to $\sin 2\theta = 1$ that is:

$$T_e = -\left(\frac{3}{2}\right)p\Lambda_3\,\text{Re}\left(j\underline{\Psi}_{sR3}\underline{\Psi}^*_{rR3}\right) = -\left(\frac{3}{2}\right)p\Lambda_3\,\text{Re}\left\{j\underline{\Psi}_{sR3}\frac{V_{\sigma r}}{\sqrt{\omega_s^2 s^2 + v_r^2}}\cdot\underline{\Psi}^*_{sR3}(\cos\theta + j\sin\theta)\right\} =$$

$$= \frac{3}{2}\frac{V_{\sigma r}}{V_r}p\Lambda_3\Psi^2_{sR3}\frac{s\omega_s V_r}{\omega_s^2 s^2 + v_r^2} = \frac{3}{4}\frac{V_{\sigma r}}{V_r}p\Lambda_3\Psi^2_{sR3}\sin 2\theta.$$

(90)

Now an observation can be formulated. Let us suppose an ideal static converter that operates with a U/f=constant=k1 strategy. For low supply frequencies, the pull-out torque decreases in value since the denominator increases with the pulsatance decrease, ωs (Fig. 3). Within certain limits at low frequencies, an increase of the supply voltage is necessary in order to maintain the pull-out torque value. In other words, U/f = k2, and k2>k1.

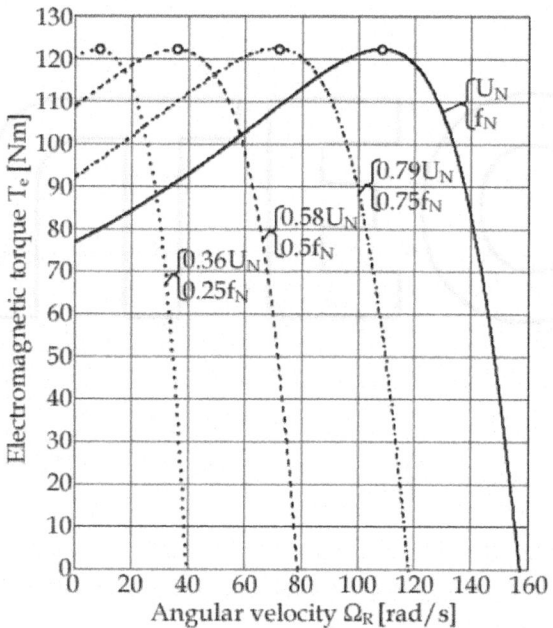

Figure 3: Mechanical characteristics, Me=f(ΩR) at ΨsR3=const.

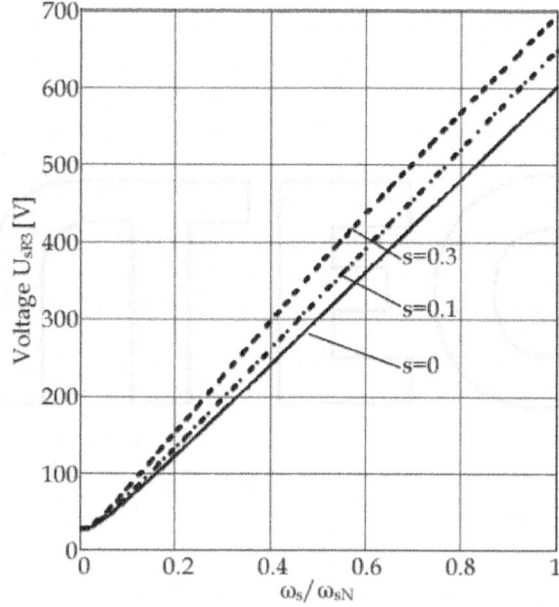

Figure 4: Resultant stator voltage vs. pulsatance UsR3=f(ωs) at ΨsR3=const. (1,91Wb)

A proper control of the induction machine requires a strategy based on U/f = variable. More precisely, for low frequency values it is necessary to increase the supply voltage with respect to the values that result from U/f = const. strategy. At a pinch, when the frequency becomes zero, the supply voltage must have a value capable to compensate the voltage drops upon the equivalent resistance of the windings. Lately, the modern static converters can be parameterized on the basis of the catalog parameters of the induction machine or on the basis of some laboratory tests results.

From (40) we can deduce:

$$\underline{\Psi}_{sR3} = -j\frac{U_{sR3}}{\omega_s}\frac{(s\omega_s - jv_r)}{(s\omega_s - v_{tt}) - j(v_r + sv_s)} \leftrightarrow U^2_{sR3} = \frac{\Psi^2_{sR3}\omega^2_s(As^2 + 2Bs + C)}{\omega^2_s s^2 + v^2_r} \tag{91}$$

and further:

$$\frac{U^2_{sR3}}{\omega^2_s} = \Psi^2_{sR3}\left[1 + \frac{v^2_s s^2 + 2v_{\sigma s}v_{\sigma r}s + v^2_{tt}}{\omega^2_s s^2 + v^2_r}\right] \leftrightarrow \Psi_{sR3} = \frac{U_{sR3}}{\omega_s}\sqrt{\frac{F(s)}{F(s) + sG(s)}}$$

$$where: \quad F(s) = \omega^2_s s^2 + v^2_r; \quad G(s) \approx v^2_s s + 2v_{\sigma s}v_{\sigma r} \tag{92}$$

if the term vtt was neglected. By inspecting the square root term, which is variable with the slip (and load as well), we can point out the following observations.

- Constant maintaining of the stator flux for low pulsations (that is low angular velocity values including start-up) can be obtained with a significant increase of the supply voltage. The "additional" increasing of the voltage depends proportionally on the load value. Analytically, this fact is caused by the predominance of the term G against F, (45). From the viewpoint of physical phenomena, a higher voltage in case of severe start-up or low frequency operation is necessary for the compensation of the leakage fluxes after which the stator flux must keep its prescribed value.

- Constant maintaining of the stator flux for high pulsations (that is angular speeds close or even over the rated value) requires an insignificant rise of the supply voltage. The U/f ratio is close to its rated value (rated values of U and f) especially for low load torque values. However, a certain increase of the voltage is required proportionally with the load degree. Analytically, this fact is now caused by the predominance of the term F against G, (45).

- In conclusion, the resultant stator flux remain constant for U/f =constant=k1 strategy if the load torque is small. For high loads (especially if the operation is close to the pull-out point), the maintaining of the stator flux requires an increase of the U/f ratio, which means a significant rise of the voltage and current.

If the machine parameters are established, then a variation rule of the supply voltage can be settled in order to have a constant stator flux (equal, for example, to its no-load value) both for frequency and load variation.

Fig. 4 presents (for a machine with predetermined parameters: supply voltage with the amplitude of 490 V (U_{as}=346.5V); R_s=R_r=2; L_{hs}=0,09; $L_{\sigma s}$= $L_{\sigma r}$=0,01; J=0,05; p=2; k_z=0,02; ω_1=314,1 (SI units)) the variation of the resultant stator voltage with the pulsatance (in per unit description) for three constant slip values. The variation is a straight line for reduced loads and has a certain inflection for low frequency values (a few Hz). For under-load operation, a significant increase of the voltage with the frequency is necessary. This fact is more visible at high slip values, close to pull-out value (in our example the pull-out slip is of 0,33).

The variation rule based on UsR=f(ωs) strategy (applied to the upper curve from Fig. 4) provide an operation of the motor within a large range of angular speeds (from start-up to rated point) under a developed torque, whose value is close to the pull-out one. Obviously, the input current is rather high (4-5 I1N) and has to be reduced. Practically, the operation points must be placed within the upper and the lower curves, Fig. 4. It is also easy to notice that the operation with higher frequency values than the rated one does not generally require an increase of the supply voltage but the developed torque is lower and lower. In this case, the output power keeps the rated value.

Variation of the Torque with the Rotor Total Flux Space Vector

Usually, the electric drives that demand high value starting torque use constant rotor total flux space vector strategy. The stator total flux space vector can be written from (41) as:

$$\underline{\Psi}_{sR3} = \frac{v_r + js\omega_g}{v_{\sigma r}}\underline{\Psi}_{rR3} \Leftrightarrow \Psi_{sR3} = \Psi_{rR3}\frac{\sqrt{\omega_s^2 s^2 + v_r^2}}{v_{\sigma r}};\Psi_{sR3} = \frac{\sqrt{\omega_s^2 s^2 + v_r^2}}{v_{\sigma r}}\underline{\Psi}_{rR3}e^{j\theta};$$

$$(\theta = \alpha_s - \alpha_r);\sin\theta = \frac{s\omega_s}{\sqrt{\omega_s^2 s^2 + v_r^2}};\cos\theta = \frac{v_r}{\sqrt{\omega_s^2 s^2 + v_r^2}}$$

(93)

and the expression of the electromagnetic torque on the basis of rotor flux alone becomes:

$$T_e = -\left(\frac{3}{2}\right)p\Lambda_3 \operatorname{Re}\left(j\underline{\Psi}_{sR3}\underline{\Psi}^*_{rR3}\right) = \frac{3}{2}\frac{p\Lambda_3}{v_{\sigma r}}\Psi^2_{rR3}s\omega_s$$

(94)

Assuming the ideal hypothesis of maintaining constant the rotor flux, for example equal to the no-load value, then the electromagnetic torque expression is:

$$T_e = \frac{3}{2}\frac{p\Lambda_3}{V_{\sigma r}}\Psi^2_{rR30}s\omega_s \approx \frac{3}{2}\frac{p\Lambda_3}{V_{\sigma r}}\left(\frac{V_{\sigma r}U_{sR3}}{V_r\omega_s}\right)^2 s\omega_s = \frac{3}{2}\frac{p\Lambda_3 V_{\sigma r}}{V_r^2}\left(\frac{U^2_{sR3}}{\omega_s^2}\right)(\omega_s - p\Omega_R)$$

(95)

where the voltage and pulsation is supposed to have rated values. Taking into discussion a machine with predetermined parameters (supply voltage with the amplitude of 490 V (U_{as}=346.5V); R_s=R_r=2; L_{hs}=0,09; $L_{\sigma s}$= $L_{\sigma r}$=0,01; J=0,05; p=2; k_z=0,02; ω_1=314,1 (SI units)) then the expression of the mechanical characteristic is:

$$T_e = \frac{3}{2}\frac{2\cdot 32,14\cdot 96,43}{103,57^2}\left(\frac{U^2_{sR3N}}{\omega_{sN}^2}\right)(\omega_s - 2\Omega_R) = 3,17(\omega_s - 2\Omega_R)$$

(96)

which is a straight line, A1 in Fig. 5. The two intersection points with the axes correspond to synchronism (T_e=0, Ω_R=ω_s/2=157) and start-up (T_e=995 Nm, Ω_R=0) respectively.

The pull-out torque is extremely high and acts at start-up. This behavior is caused by the hypothesis of maintaing constant the rotor flux at a value that corresponds to no-load operation (when the rotor reaction is null) no matter the load is. The compensation of the magnetic reaction of the rotor under load is hypothetical possible through an unreasonable increase of the supply voltage. Practically, the pull-out torque is much lower.

Another unreasonable possibility is the maintaining of the rotor flux to a value that corresponds to start-up (s = 1) and the supply voltage has its rated value. In this case the expression of the mechanical characteristic is (50) and the intersection points with the axes (line A2, Fig. 5) correspond to synchronism (T_e=0, Ω_R=ω_s/2=157) and start-up (T_e=78 Nm, Ω_R=0) respectively.

$$T_e = \frac{3}{2}\frac{2\cdot 32,14\cdot}{96,43}\Psi^2_{rRk}(\omega_s - 2\Omega_R) = 0,25(\omega_s - 2\Omega_R)$$

(97)

The supply of the stator winding with constant voltage and rated pulsation determines a variation of the resultant rotor flux within the short-circuit value (Ψ_{rRk}=0,5Wb) and the synchronism value (Ψ_{rR0}=1,78Wb). The operation points lie between the two lines, A1 and A2, on a position that depends on the load torque. When the supply pulsation is two times smaller (and the voltage itself is two times smaller as well) and the resultant rotor flux is maintained constant to the value Ψ_{rR0}=1,78Wb, then the mechanical characteristic is described by the straight line B1, which is parallel to the line A1. Similarly, for $\Psi_{rRk}=0,5\,Wb$, the mechanical characteristic become the line B2, which is parallel to A2.

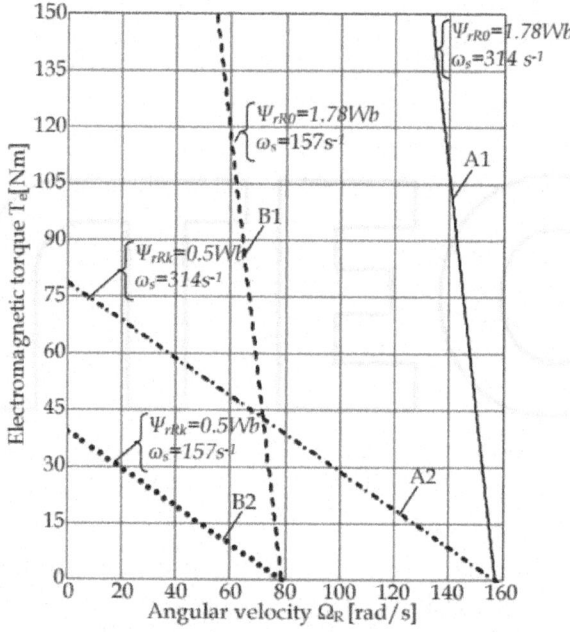

Figure 5: Mechanical characteristics Te=f(ΩR), ΨrR=const.

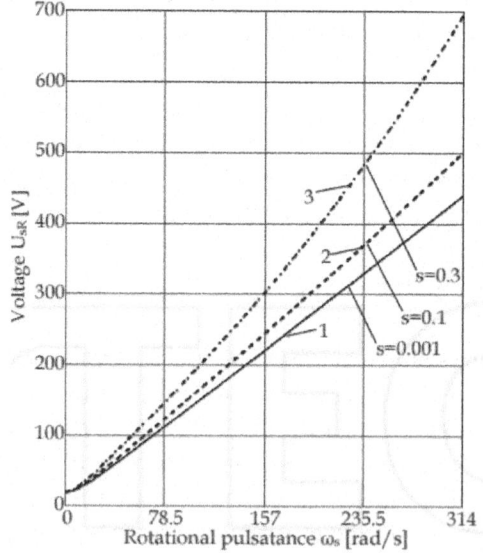

Figure 6: Resultant stator voltage vs. pulsatance, UsR=f(ωs) at ΨrR=const. (1.3Wb)

When the applied voltage and pulsation are two times smaller regarding the rated values then the operation points lie between B1 and B2 since the rotor flux varies within $\Psi_{rRk}=0,5$Wb (short-circuit) and $\Psi_{rR0}=1,78$Wb (synchronism).

The control based on constant rotor flux strategy ensures parallel mechanical characteristics. This is an important advantage since the induction machine behaves like shunt D.C. motor. A second aspect is also favorable in the behavior under this strategy. The mechanical characteristic has no sector of unstable operation as the usual induction machine has.

The modification of the flux value (generally with decrease) leads to a different slope of the characteristics, which means a significant decrease of the torque for a certain angular speed.

The question is "what variation rule of UsR/ωs must be used in order to have constant rotor flux"? The expression of the modulus of the resultant rotor flux can be written as:

$$\frac{U_{sR3}^2}{\omega_s^2} = \Psi_{rR3}^2 \frac{As^2 + 2Bs + C}{v_{\sigma r}^2} \leftrightarrow \frac{U_{sR3}}{\omega_s} = \frac{\Psi_{rR3}}{v_{\sigma r}}\sqrt{As^2 + 2Bs + C}$$

(98)

$$with: \quad A = \omega_s^2 + v_s^2; B = v_{\sigma s}v_{\sigma r}; C = v_r^2 + v_{tt}^2; v_{tt}^2 = (v_s v_r - v_{\sigma s}v_{\sigma r})^2 / \omega_s^2.$$

(99)

Fig. 6 presents the variation of the stator voltage with pulsatance at constant resultant rotor flux (1,3 Wb), which are called the control characteristics of the static converter connected to the induction machine. The presented characteristics correspond to three constant slip values, s=0,001 (no-load)-curve 1, s=0,1 (rated duty)-curve 2 and s=0,3 (close to pull-out point)-curve 3. It can be seen that the operation with high slip values (high loads) require an increased stator voltage for a certain pulsation. As a matter of fact, the ratio UsR3/ωs must be increased with the load when the pulsatance (pulsation) and the angular speed rise as well. Such a strategy is indicated for fans, pumps or load machines with speed-dependent torque.

When the pulsation of the stator voltage is low (small angular velocities) then the torque that has to be overcame is small too, but it will rise with the speed and the frequency along a parabolic variation. Since the upper limit of the torque is given by the limited power of the machine (thermal considerations) then this strategy requires additional precautions as concern the safety devices that protect both the static converter and the supply source itself.

The analysis of the square root term from (51) generates similar remarks as in the above discussed control strategy. Finally is important to say that a control characteristic must be prescribed for the static converter. This characteristic should be simplified and generally reduced to a straight line placed between the curves 1 and 2 from Fig. 6.

STUDY OF THE UNBALANCED DUTIES

The unbalanced duties (generated by supply asymmetries) are generally analyzed by using the theory of symmetric components, according to which any asymmetric three-phase system with single unbalance(the sum of the applied instantaneous voltages is always zero) can be equated with two symmetric systems of opposite sequences: positive (+) (or direct) and negative (-) (or inverse) respectively. There are two possible ways for the analysis of this problem.

- When the amplitudes of the phase voltages are different and/or the angles of phase difference are not equal to $2\pi/3$ then the unbalanced three-phase system can be replaced with an equivalentunbalanced two-phase system, which further is taken apart in two systems, one of direct sequencewith higher two-phase amplitude voltages and the other of inverse sequence with lower two-phase amplitude voltages. Usually, this equivalence process is obtained by using an orthogonal transformation. Not only voltages but also the total fluxes and eventually the currents must be established for the two resulted systems. The quantities of the unbalanced two-phase system can be written as follows:

$$\begin{bmatrix} \underline{U}_{as} \\ \underline{U}_{\beta s} \\ \underline{U}_{0s} \end{bmatrix} = \sqrt{\frac{2}{3}} \begin{bmatrix} 1 & -1/2 & -1/2 \\ 0 & \sqrt{3}/2 & -\sqrt{3}/2 \\ 1/\sqrt{2} & 1/\sqrt{2} & 1/\sqrt{2} \end{bmatrix} \begin{bmatrix} \underline{U}_{as} \\ \underline{U}_{bs} \\ \underline{U}_{cs} \end{bmatrix} \leftrightarrow \begin{cases} \underline{U}_{as} = \sqrt{\frac{3}{2}}\underline{U}_{as}; \underline{U}_{\beta s} = \sqrt{\frac{3}{2}}\frac{\underline{U}_{bs} - \underline{U}_{cs}}{\sqrt{3}} \\ \underline{U}_{0s} = 0; \quad \underline{U}_{as} + \underline{U}_{bs} + \underline{U}_{cs} = 0 \end{cases} \quad (100)$$

Further, the unbalanced quantities are transformed to balanced quantities and we obtain:

$$\begin{bmatrix} \underline{U}_{as(+)} \\ \underline{U}_{as(-)} \end{bmatrix} = \frac{1}{2}\begin{bmatrix} 1 & j \\ 1 & -j \end{bmatrix}\begin{bmatrix} \underline{U}_{as} \\ \underline{U}_{\beta s} \end{bmatrix}, \text{ or: } \begin{cases} \underline{U}_{as(+)} = \left(\underline{U}_{as}e^{j\pi/6} + j\underline{U}_{bs}\right)/\sqrt{2}; \\ \underline{U}_{as(-)} = \left(\underline{U}_{as}e^{-j\pi/6} - j\underline{U}_{bs}\right)/\sqrt{2} \end{cases} \quad (101)$$

The quantities of the three-phase system with single unbalance can be written as follows:

$$u_{as} = U\sqrt{2}\cos\omega t \Leftrightarrow \underline{U}_{as} = Ue^{j0}; \underline{U}_{bs} = kUe^{-j\beta}; \underline{U}_{cs} = -U(1 + ke^{-j\beta}) \quad (102)$$

and further:

$$\underline{U}_{as(+)} = U(e^{j\pi/6} + ke^{j(\pi/2-\beta)})/\sqrt{2}; \underline{U}_{as(-)} = U(e^{-j\pi/6} - ke^{j(\pi/2-\beta)})/\sqrt{2} \quad (103)$$

Modulus of these components can be determined at once with:

$$U_{as(+)} = U\sqrt{1+k^2 + 2k\sin(\beta+\pi/6)}/\sqrt{2};\ U_{as(-)} = U\sqrt{1+k^2 - 2k\sin(\beta-\pi/6)}/\sqrt{2}$$

(104)

For the transformation of the unbalanced two-phase quantities in balanced two-phase components (53) must be used:

$$\begin{cases} \underline{U}_{\alpha s} = \underline{U}_{as(+)} + \underline{U}_{as(-)} \\ \underline{U}_{\beta s} = -j\underline{U}_{as(+)} + j\underline{U}_{as(-)} \end{cases}$$

(105)

The matrix equation of the two-phase model is written in a convenient way hereinafter:

$$\begin{bmatrix} \underline{U}_{\alpha s} \\ \underline{U}_{\beta s} \\ 0 \\ 0 \end{bmatrix} = \begin{bmatrix} v_s + j\omega_s & 0 & 0 & -v_{\sigma s} \\ 0 & v_s + j\omega_s & -v_{\sigma s} & 0 \\ 0 & 0 & v_{\sigma r} & -(v_r + j\omega_s) & \omega_R \\ 0 & v_{\sigma r} & 0 & -\omega_R & -(v_r + j\omega_s) \end{bmatrix} \times \begin{bmatrix} \underline{\Psi}_{\alpha s} \\ \underline{\Psi}_{\beta s} \\ \underline{\Psi}_{yr} \\ \underline{\Psi}_{xr} \end{bmatrix}$$

(106)

Using elementary math (multiplications with constants, addition and subtraction of different equations) we can obtain the equations of the two-phase direct (M2D) and inverse (M2I) models:

(M2D) $$\begin{bmatrix} \underline{U}_{as(+)} \\ 0 \end{bmatrix} = \begin{bmatrix} v_s + j\omega_s & -v_{\sigma s} \\ v_{\sigma r} & -(v_r + js_d\omega_s) \end{bmatrix} \times \begin{bmatrix} \underline{\Psi}_{as(+)} \\ \underline{\Psi}_{xr(+)} \end{bmatrix}$$

(107)

(M2I) $$\begin{bmatrix} \underline{U}_{as(-)} \\ 0 \end{bmatrix} = \begin{bmatrix} v_s + j\omega_s & -v_{\sigma s} \\ v_{\sigma r} & -(v_r + js_i\omega_s) \end{bmatrix} \times \begin{bmatrix} \underline{\Psi}_{as(-)} \\ \underline{\Psi}_{xr(-)} \end{bmatrix}$$

(108)

We have defined the slip values for the direct (+) and respectively inverse (-) machines: $s_d = s = \dfrac{\omega_s - \omega_R}{\omega_s};\ s_i = \dfrac{\omega_s + \omega_R}{\omega_s}$ with the interrelation expression: $s_i = 2 - s$

The two machine-models create self-contained torques, which act simultaneously upon rotor. The resultant torque emerges from superposition effects procedure (Simion et al., 2009; Simion & Livadaru, 2010). The equation set (59), for M2D, gives two equations:

$$\underline{U}_{as(+)} = (v_s + j\omega_s)\underline{\Psi}_{as(+)} - v_{\sigma s}\underline{\Psi}_{xr(+)};\quad 0 = v_{\sigma r}\underline{\Psi}_{as(+)} - (v_r + js\omega_s)\underline{\Psi}_{xr(+)}$$

(109)

which give further

$$\underline{\Psi}_{as(+)} = \frac{(v_r + js\omega_s)\underline{U}_{as(+)}}{\underline{\Delta}_{(+)}}; \underline{\Psi}_{xr(+)} = \frac{v_{\sigma r}\underline{U}_{as(+)}}{\underline{\Delta}_{(+)}}; \underline{\Delta}_{(+)} = (v_s + j\omega_s)(v_r + js\omega_s) - v_{\sigma s}v_{\sigma r}$$

(110)

Similarly, for M2I we obtain:

$$\underline{U}_{as(-)} = (v_s + j\omega_s)\underline{\Psi}_{as(-)} - v_{\sigma s}\underline{\Psi}_{xr(-)}; \quad 0 = v_{\sigma r}\underline{\Psi}_{as(-)} - \left[v_r + j(2-s)\omega_s\right]\underline{\Psi}_{xr(-)}$$

(111)

$$\underline{\Psi}_{as(-)} = \frac{\left[v_r + j(2-s)\omega_s\right]\underline{U}_{as(-)}}{\underline{\Delta}_{(-)}}; \underline{\Psi}_{xr(-)} = \frac{v_{\sigma r}\underline{U}_{as(-)}}{\underline{\Delta}_{(-)}}; \underline{\Delta}_{(-)} = (v_s + j\omega_s)\left[v_r + j(2-s)\omega_s\right] - v_{\sigma s}v_{\sigma r}$$

(112)

To determine the electromagnetic torque developed under unbalanced supply condition we use the symmetric components and the superposition effect. The mean electromagnetic torque M2D results from (25) but transformed in simplified complex quantities:

$$T_{e(+)} = -\frac{3p}{2}\Lambda_3 \cdot 2\mathrm{Re}\left(j\underline{\Psi}_{as(+)} \cdot \underline{\Psi}_{xr(+)}^{\prime*}\right) = \frac{3pv_{\sigma r}\Lambda_3}{2\omega_s} \cdot \frac{2U_{as(+)}^2 s}{As^2 + 2Bs + C}$$

(113)

Similarly, the expression of the mean electromagnetic torque M2D is:

$$T_{e(-)} = -\frac{3p}{2}\Lambda_3 \cdot 2\mathrm{Re}\left(j\underline{\Psi}_{as(-)} \cdot \underline{\Psi}_{xr(-)}^*\right) = \frac{3pv_{\sigma r}\Lambda_3}{2\omega_s} \cdot \frac{2U_{as(-)}^2(2-s)}{A(2-s)^2 + 2B(2-s) + C}$$

(114)

The mean resultant torque, as a difference of the torques produced by M2D and M2I, can be written by using (65) and (66):

$$T_{erez} = \frac{3pv_{\sigma r}\Lambda_3}{2\omega_s}\left[\frac{s \cdot 2U_{as(+)}^2}{As^2 + 2Bs + C} - \frac{(2-s) \cdot 2U_{as(-)}^2}{A(2-s)^2 + 2B(2-s) + C}\right]$$

(115)

where we have defined the notations: $\omega_s^2 + v_s^2 = A; \quad v_{\sigma s}v_{\sigma r} = B; \quad v_r^2 + v_u^2 = C;$ and

$$\sqrt{2}U_{as(+)} = U\sqrt{1 + k^2 + 2k\sin(\beta + \pi/6)}; \sqrt{2}U_{as(-)} = U\sqrt{1 + k^2 - 2k\sin(\beta - \pi/6)}$$

(116)

Finally, the expression of the mean resultant torque with the slip is:

$$T_{erez} = \frac{3pv_{\sigma r}\Lambda_3 U^2}{2\omega_s}\left[\frac{1 + k^2 + 2k\sin(\beta + \pi/6)}{As^2 + 2Bs + C}s - \frac{1 + k^2 - 2k\sin(\beta - \pi/6)}{A(2-s)^2 + 2B(2-s) + C}(2-s)\right]$$

(117)

The influence of the supply unbalances upon Te=f(s) characteristic are presented in Fig. 7. To this effect, let us take again into discussion the machine with the following parameters: supply voltages with the amplitude of 490 V (U_{as}=346.5V) and 2π/3 rad. shifted in phase; R_s=R_r=2; L_{hs}=0,09; $L_{\sigma s}$= $L_{\sigma r}$=0,01; J=0,05; p=2; k_z=0,02; ω_1=314,1 (SI units). The characteristic corresponding to the three-phase symmetric machine is the curve A (the motoring pull-out torque is equal to 124 Nm and obviously Uas(-) = 0). If the voltage on phase b keeps the same amplitude as the voltage in phase a, for example, but the angle of phase difference changes with π/24=7,5 degrees (from 2π/3=16π/24 to 17π/24rad.) then the new characteristic is the B curve. The pull-out torque value decreases with approx. 12% but the pull-out slip keeps its value. Other two unbalance degrees are presented in Fig. 7 as well.

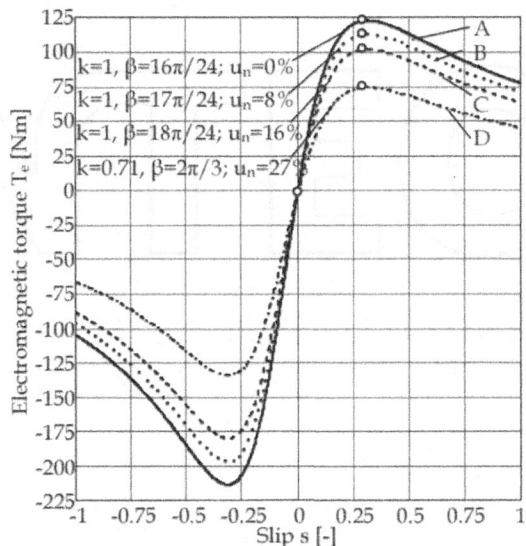

Figure 7: T_e=f(s) characteristic for different unbalance degrees

Usually, the unbalance degree of the supply voltage is defined as the ratio of inverse and direct components:

$$u_n = \frac{U_{as(-)}}{U_{as(+)}} = \frac{\sqrt{1+k^2 - 2k\sin(\beta - \pi/6)}}{\sqrt{1+k^2 + 2k\sin(\beta + \pi/6)}} \cdot 100[\%]$$

(118)

The curves A, B, C, and D from Fig. 7 correspond to the following values of the unbalance degree: un= 0; 8%; 16% and 27%. The highest unbalance degree (27% - curve D) causes a decrease of the pull-out torque by 40%.

- The second approach takes into consideration the following reasoning. When the amplitudes of the three-phase supply system and/or the angles of the phase difference are not equal to $2\pi/3$ then theunbalanced system can be replaced by two balanced three-phase systems that act in opposition. One of them is the direct sequence system and has higher voltages and the other is the inverse sequence system and has lower voltages. A transformation of the unbalanced voltages and total fluxes into two symmetric systems is again necessary. In other words, there is an unbalanced voltage system (Uas, Ubs, Ucs), which is replaced by the direct and inverse symmetric systems. The mean resultant torque is the difference between the torques developed by the two symmetric machine-models. Taking into consideration their slip values (sd = s and si = 2-s) we can deduce the torque expression:

$$T_{erez} = -(3/2)p\Lambda_3 \cdot [3\operatorname{Re}\left(j\underline{\Psi}_{as1}\underline{\Psi}_{ar1}^{'*}\right) - 3\operatorname{Re}\left(j\underline{\Psi}_{as2}\underline{\Psi}_{ar2}^{'*}\right)]$$

(119)

$$T_{erez} = \frac{3pv_{\sigma r}\Lambda_3}{2\omega_s}\left[\frac{3sU_{as1}^2}{As^2 + 2Bs + C} - \frac{3(2-s)U_{as2}^2}{A(2-s)^2 + 2B(2-s) + C}\right]$$

(120)

and this is the same with (69) as we expected.

SIMULATION STUDY UPON SOME TRANSIENT DUTIES OF THE THREE-PHASE INDUCTION MACHINE

Symmetric Supply System

The mathematical model described by the equation system (26-1…8) allows a complete simulation study of the operation of the three-phase induction machine, which include start-up, any sudden change of the load and braking to stop eventually. To this end, the machine parameters (resistances, main and leakage phase inductances, moments of inertia corresponding to the rotor and the load, coefficients that characterize the variable speed and torque, etc.) have to be calculated or experimentally deduced. At the same time, the values of the load torque and the expressions of the instantaneous voltages applied to each stator phase winding are known, as well. The rotor winding is considered short-circuited. Using the above mentioned equation system, the structural diagram in the Matlab-Simulink environment can be carried out. Additionally, for a complete evaluation, virtual oscillographs for the visualization of the main physical parameters such as voltage, current, magnetic flux, torque, speed, rotation angle and current or specific characteristics (mechanical characteristic, angular characteristic or flux hodographs) fill out the structural diagram.

The study of the symmetric three-phase condition in the Matlab-Simulink environment takes into consideration the following parameter values: three identical supply voltages with the amplitude of 490 V (U_{as}=346.5V) and $2\pi/3$ rad. shifted in phase; u_{ar}=u_{br}=u_{cr}=0 since the rotor winding is short-circuited; R_s=R_r=2; L_{hs}=0,09; $L_{\sigma s}$= $L_{\sigma r}$=0,01; J=0,05; p=2; k_z=0,02; ω_1=314,1 (SI units). The equation system becomes:

$$\left(s+135,71\right)\overline{\psi}_{as} = \overline{u}_{as} - 32,14\left(\overline{\psi}_{bs}+\overline{\psi}_{cs}\right)+32,14\left(2\overline{\psi}_{ar}-\overline{\psi}_{br}-\overline{\psi}_{cr}\right)\cos\theta_R + 55,67(\overline{\psi}_{cr}-\overline{\psi}_{br})\sin\theta_R$$

(121)

$$\left(s+135,71\right)\overline{\psi}_{bs} = \overline{u}_{bs} - 32,14\left(\overline{\psi}_{cs}+\overline{\psi}_{as}\right)+32,14\left(2\overline{\psi}_{br}-\overline{\psi}_{cr}-\overline{\psi}_{ar}\right)\cos\theta_R + 55,67(\overline{\psi}_{ar}-\overline{\psi}_{cr})\sin\theta_R$$

(122)

$$\left(s+135,71\right)\overline{\psi}_{cs} = \overline{u}_{cs} - 32,14\left(\overline{\psi}_{as}+\overline{\psi}_{bs}\right)+32,14\left(2\overline{\psi}_{cr}-\overline{\psi}_{ar}-\overline{\psi}_{br}\right)\cos\theta_R + 55,67(\overline{\psi}_{br}-\overline{\psi}_{ar})\sin\theta_R$$

(123)

$$\left(s+135,71\right)\overline{\psi}_{ar} = 0 - 32,14\left(\overline{\psi}_{br}+\overline{\psi}_{cr}\right)+32,14\left(2\overline{\psi}_{as}-\overline{\psi}_{bs}-\overline{\psi}_{cs}\right)\cos\theta_R + 55,67(\overline{\psi}_{bs}-\overline{\psi}_{cs})\sin\theta_R$$

(124)

$$\left(s+135,71\right)\overline{\psi}_{br} = 0 - 32,14\left(\overline{\psi}_{cr}+\overline{\psi}_{ar}\right)+32,14\left(2\overline{\psi}_{bs}-\overline{\psi}_{cs}-\overline{\psi}_{as}\right)\cos\theta_R + 55,67(\overline{\psi}_{cs}-\overline{\psi}_{as})\sin\theta_R$$

(125)

$$\left(s+135,71\right)\overline{\psi}_{cr} = 0 - 32,14\left(\overline{\psi}_{ar}+\overline{\psi}_{br}\right)+32,14\left(2\overline{\psi}_{cs}-\overline{\psi}_{as}-\overline{\psi}_{bs}\right)\cos\theta_R + 55,67(\overline{\psi}_{as}-\overline{\psi}_{bs})\sin\theta_R$$

(126)

$$\theta_R\left(s+0,4\right)=\left(40\right)\langle-(32,14)\{\sin\theta_R\left[\overline{\psi}_{as}\left(2\overline{\psi}_{ar}-\overline{\psi}_{br}-\overline{\psi}_{cr}\right)+\overline{\psi}_{bs}\left(2\overline{\psi}_{br}-\overline{\psi}_{cr}-\overline{\psi}_{ar}\right)+\right.$$
$$\left.+\overline{\psi}_{cs}\left(2\overline{\psi}_{cr}-\overline{\psi}_{ar}-\overline{\psi}_{br}\right)\right]+\sqrt{3}\cos\theta_R\cdot\left[\overline{\psi}_{as}\left(\overline{\psi}_{br}-\overline{\psi}_{cr}\right)+\overline{\psi}_{bs}\left(\overline{\psi}_{cr}-\overline{\psi}_{ar}\right)+\overline{\psi}_{cs}\left(\overline{\psi}_{ar}-\overline{\psi}_{br}\right)\right]\}-T_{st}\rangle$$

(127)

$$\theta_R = \omega_R\frac{1}{s}$$

(128)

$$\overline{u}_{as} \leftrightarrow \frac{490}{\sqrt{2}}e^{j(314,1t)}; \overline{u}_{bs} \leftrightarrow \frac{490}{\sqrt{2}}e^{j(314,1t-2,094)}; \overline{u}_{cs} \leftrightarrow \frac{490}{\sqrt{2}}e^{j(314,1t-4,188)};$$
$$U_{as\max} = U_{bs\max} = U_{cs\max} = 490$$

(129)

It has to be mentioned again that the above equation system allows the analysis of the three-phase induction machine under any condition, that is transients, steady state, symmetric or unbalanced, with one or both windings (from stator and rotor) connected to a supply system. Generally, a supplementary requirement upon the stator supply voltages is not mandatory. The case of short-circuited rotor winding, when the rotor supply voltages are zero, include the wound rotor machine under rated operation since the starting rheostat is short-circuited as well. The presented simulation takes into discussion a varying duty, which consists in a no-load start-up (the load torque derives of frictions and ventilation and is proportional to the angular speed and have a steady state rated value of approx. 3 Nm) followed after 0,25 seconds by a sudden loading with a constant torque of 50 Nm. The simulation results are presented in Fig. 8, 10, 12, 14 and 15 and denoted by the symbol RS-50. A second simulation iterates the presented varying duty but with a load torque of 120 Nm, symbol RS-120, Fig. 9, 11 and 13. Finally, a third simulation takes into consideration a load torque of 125 Nm, which is a value over the pull-out torque. Consequently, the falling out and the stop of the motor in t≈0,8 seconds mark the varying duty (symbol RS-125, Fig. 16, 17, 18 and 19).

The RS-50 simulation shows an upward variation of the angular speed to the no-load value (in t ≈ 0,1 seconds), which has a weak overshoot at the end, Fig. 8. The 50 Nm torque enforcement determines a decrease of the speed corresponding to a slip value of s ≈ 6,5%. In the case of the RS-120 simulation, the start-up is obviously similar but the loading torque determines a much more significant decrease of the angular speed and the slip value gets to s ≈ 25%, Fig. 9.

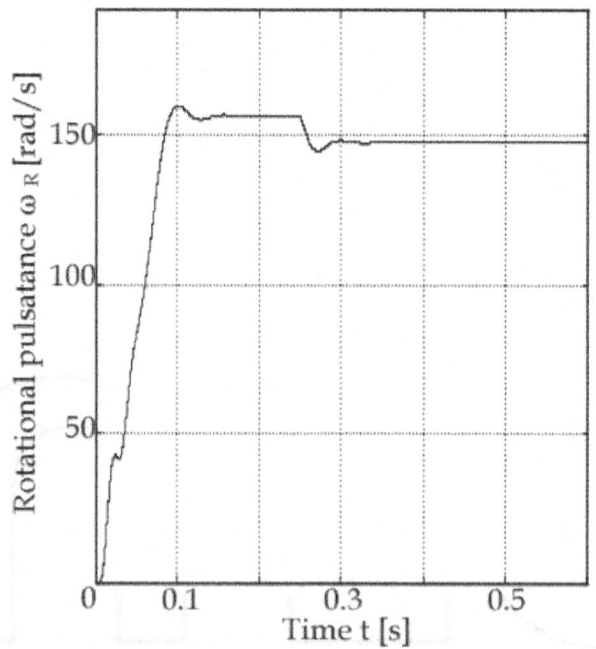

Figure 8: Time variation of rotational pulsatance – RS-50

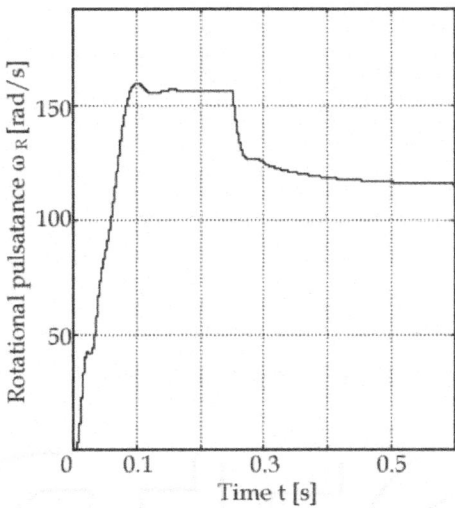

Figure 9: Time variation of rotational pulsatance – RS-120

In the first moments of the start-up, the electromagnetic torque oscillates around 100 Nm and after the load torque enforcement, it gets to approx. 53 Nm for RS-50, Fig. 10 and to approx. 122 Nm for RS-120, Fig. 11. The operation of the motor remains stable for the both duties.

The behavior of the machine is very interesting described by the hodograph of the resultant rotor flux (the locus of the head of the resultant rotor flux phasor), Fig. 12 and 13. With the connecting moment, the rotor fluxes start from 0 (O points on the hodograph) and track a corkscrew to the maximum value that corresponds to synchronism (ideal no-load operation), S points on the hodographs.

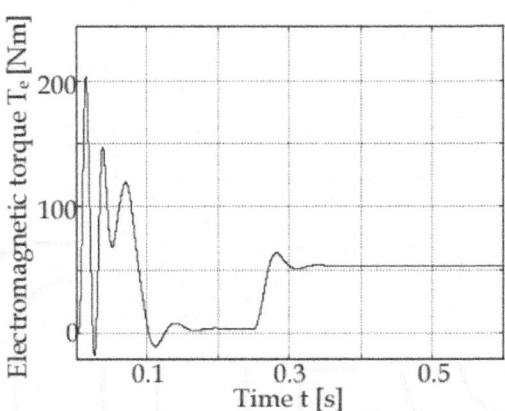

Figure 10: Time variation of electromagnetic torque – RS-50

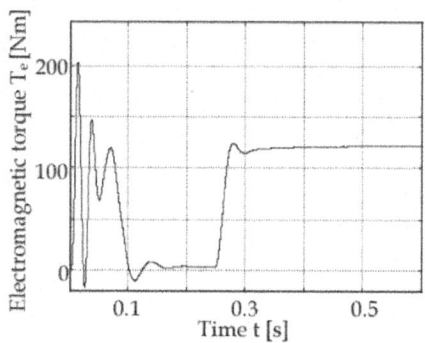

Figure 11: Time variation of electromagnetic torque – RS-120

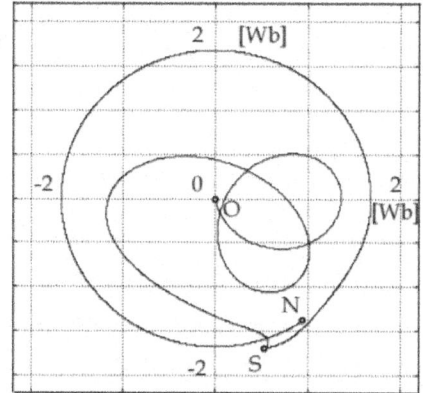

Figure 12: Hodograph of resultant rotor flux – RS-50

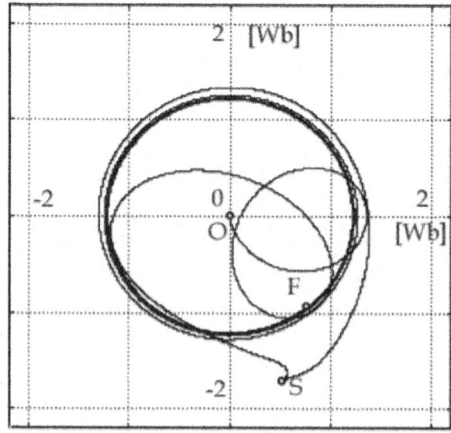

Figure 13: Hodograph of resultant rotor flux – RS-120

The enforcement of the load torque determines a decrease of the resultant rotor flux, which is proportional to the load degree, and is due to the rotor reaction. The locus of the head of the phasor becomes a circle whose radius is proportional to the amplitude of the resultant rotor flux. The speed on this circle is given by the rotor frequency that is by the slip value. It is interesting to notice that the load torque of 50 Nm causes a unique rotation of the rotor flux whose amplitude becomes equal to the segment ON (Fig. 12) whereas the 120 Nm torque causes approx. 4 rotations of the rotor flux and the amplitude OF is significantly smaller (Fig. 13).

If the expressions (1) and (2) are also used in the structural diagram then both stator and rotor phase currents can be plotted. The stator current corresponding to as phase has the frequency f_1=50 Hz and gets a start-up amplitude of approx. 70 A. This value decreases to approx. 6 A (no-load current) and after the torque enforcement (50 Nm) it rises to a stable value of approx. 14 A, Fig. 14. The rotor current on phase ar, which has a frequency value of $f_2 = s\, f_1$, gets a similar (approx. 70 A) start-up variation but in opposition to the stator current, ias. Then, its value decrease and the frequency go close to zero. The loading of the machine has as result an increase of the rotor current up to 13 A and a frequency value of $f_2 \approx 3 Hz$, Fig. 15. The fact that the current variations are sinusoidal and keep a constant frequency is an argument for a stable operation under symmetric supply conditions.

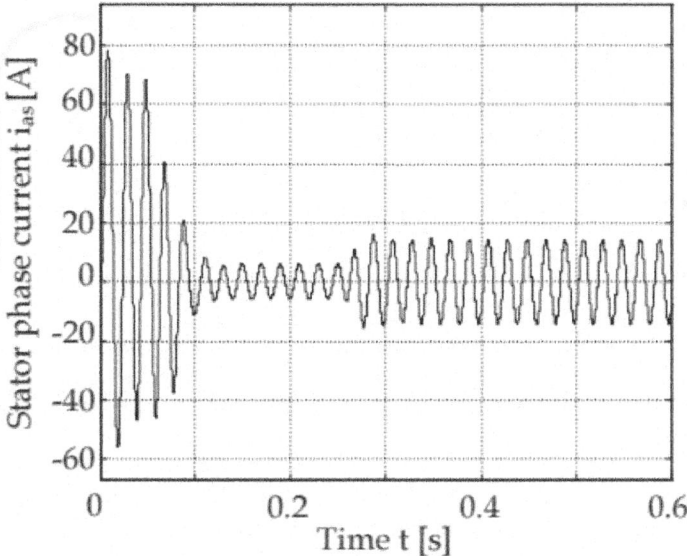

Figure 14: Time variation of stator phase current – RS-50

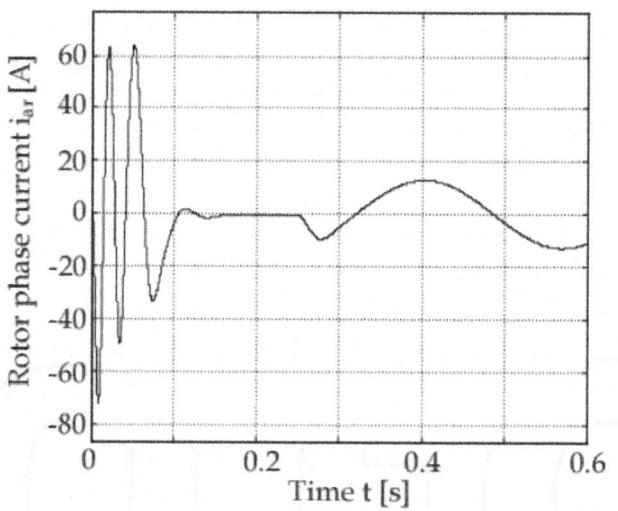

Figure 15: Time variation of rotor phase current – RS-50

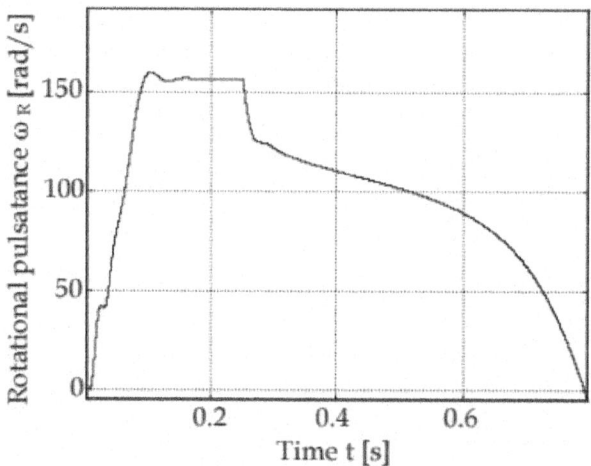

Figure 16: Time variation of rotational pulsatance – RS-125 (start-up to locked-rotor)

The third simulation, RS-125, has a similar start-up but the enforcement of the load torque determines a fast deceleration of the rotor. The pull-out slip (s≈33%) happens in t≈0,5 seconds after which the machine falls out. The angular speed reaches the zero value in t≈0,8 seconds, Fig. 16, and the electromagnetic torque get a value of approx. 78 Nm. This value can be considered the locked-rotor (starting) torque of the machine, Fig. 17.

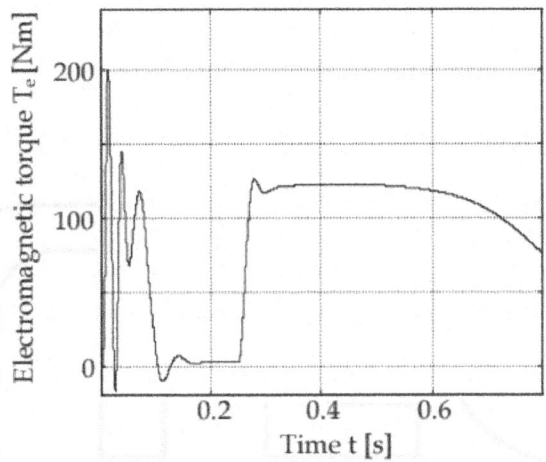

Figure 17: Time variation of electromagnetic torque – RS-125

The described critical duty that involves no-load start-up and operation, overloading, falling out and stop is plotted in terms of resultant rotor flux and angular speed versus electromagnetic torque. The hodograph (Fig. 18) put in view a cuasi corkscrew section, corresponding to the start-up, characterized by its maximum value represented by the segment OS. The falling out tracks the corkscrew SP with a decrease of the amplitude, which is proportional to the deceleration of the rotor.

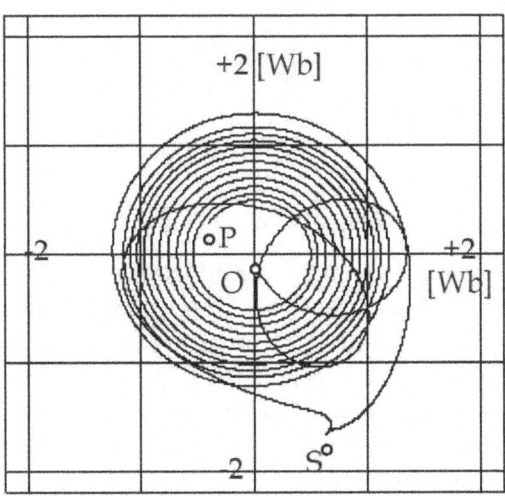

Figure 18: Hodograph of resultant rotor flux – RS-125 (start-up to locked-rotor)

The point P corresponds to the locked-rotor position (s=1). Fig. 19 presents the dynamic mechanical characteristic, which shows the variation of the electromagnetic torque under variable operation condition. During the no-load start-up, the operation point tracks successively the points O, M, L and S, that is from locked-rotor to synchronism with an oscillation of the electromagnetic torque inside certain limits ($\approx$+200Nm to $\approx$-25Nm). The enforcement of the overload torque leads the operation point along the downwardcurve SK characterized by an oscillation section followed by the unstable falling out section, KP. The PKS curve, together with the marked points (Fig. 19) can be considered the natural mechanical characteristic under motoring duty.

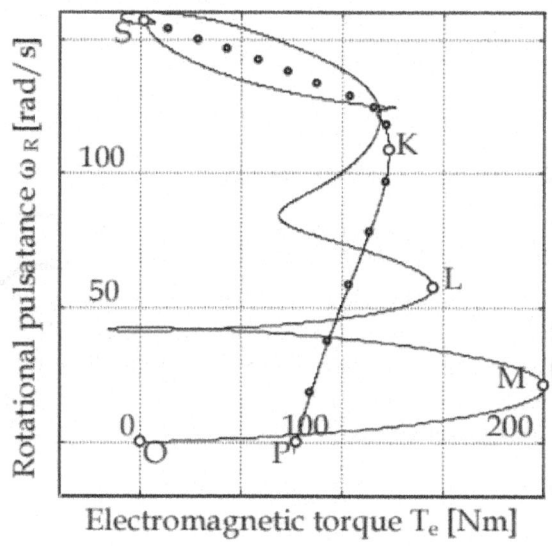

Electromagnetic torque T_e [Nm]

Figure 19: Rotational pulsatance vs. torque – RS-125 (start-up to locked-rotor)

Asymmetric Supply System

A simulation study of the three-phase induction machine under unbalanced supply condition and varying duty (start-up, sudden torque enforcement and braking to stop eventually) is possible by using the same mathematical model described by the equation system (26-1...8). The values of the resistant torques and the expressions of the instantaneous phase voltages have to be stated. Since the rotor winding is short-circuited, the supply rotor voltages are $u_{ar}=u_{br}=u_{cr}=0$. On this basis, the structural diagram has been put into effect in the Matlab-Simulink environment. As regards the unbalanced three-phase supply system, it has to be mentioned that the phase voltages are no more equal in amplitude and the angles of phase difference may have other values than $2\pi/3$ rad. In any event, the sum of the instantaneous values of the applied voltages must be zero, that is $u_{as}+u_{bs}+u_{cs}=0$. As an argument for this seemingly constraint stands the fact that the vast majority of the three-phase induction machines are connected to the industrial system via three supply leads (no neutral).

The simulation presented here takes into discussion an induction machine with the same parameters as above that is: $R_s=R_r=2$; $L_{hs}=0,09$; $L_{\sigma s}=L_{\sigma r}=0,01$; J=0,05; p=2; $k_z=0,02$; $\omega_1=314,1$ (SI units). Consequently, the equations (73-1) - (73-8) keep unchanged. The expressions (73-9) have to be modified in accordance with the asymmetry degree.

Two varying duties under unbalanced condition have been simulated. The first (denoted RNS-1) is characterized by an asymmetry degree, $u_n = 16,5\%$ and the following supply voltages:

$$\bar{u}_{as} \leftrightarrow \frac{490}{\sqrt{2}}e^{j(314,1t)}; \bar{u}_{bs} \leftrightarrow \frac{375}{\sqrt{2}}e^{j(314,1t-1,96)}; \bar{u}_{cs} \leftrightarrow \frac{490}{\sqrt{2}}e^{j(314,1t-3,927)}; u_n = 16,5\%$$

$$(130)$$

The simulation results are presented in Fig. 20, 22, 24, 25 and 28. The second study simulation (denotedRNS-2) has an asymmetry degree of $u_n = 27\%$ given by the following stator voltages:

$$\bar{u}_{as} \leftrightarrow \frac{490}{\sqrt{2}}e^{j(314,1t)}; \bar{u}_{bs} \leftrightarrow \frac{346,43}{\sqrt{2}}e^{j(314,1t-2,357)}; \bar{u}_{cs} \leftrightarrow \frac{346,43}{\sqrt{2}}e^{j(314,1t-3,295)}; u_n = 27\%$$

$$(131)$$

The simulation results are presented in Fig. 21, 23, 26, 27 and 29. The varying duties are similar to those discussed above and consist in a no-load start-up (the load torque derives of frictions and ventilation and is proportional to the angular speed and have a steady state rated value of approx. 3 Nm) followed after 0,25 seconds by a sudden loading with a constant torque of 50 Nm.

In comparison to symmetric supply, the unbalanced voltage system causes a longer start-up time with approx. 20% for RNS-1 (Fig. 20) and with 50% for RNS-2 (Fig. 21). Moreover, the higher asymmetry degree of RNS-2 leads to the cancelation of the overshoot at the end of the start-up process. At the same time, significant speed oscillations are noticeable during the operation (no matter the load degree), which are higher with the increase of the asymmetry degree. These oscillations have a constant frequency, which is twice of the supply voltage frequency. They represent the main cause that determines the specific noise of the machines with unbalanced supply system.

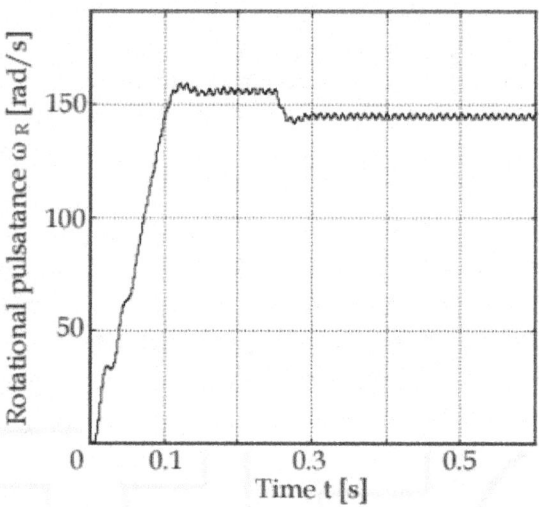

Figure 20: Time variation of rotational pulsatance – RNS-1 (start-up + sudden load)

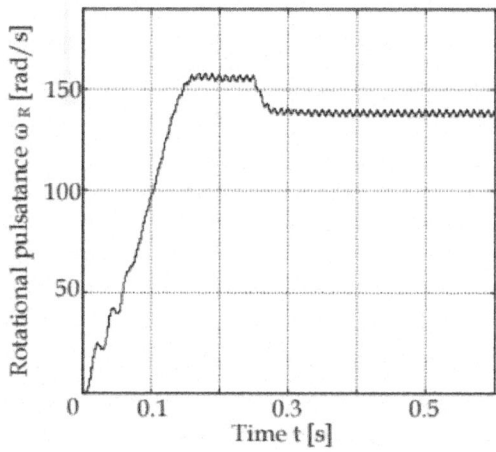

Figure 21: Time variation of rotational pulsatance – RNS-2 (start-up + sudden load)

The inspection of the electromagnetic torque variation (Fig. 22 and 23) shows the presence of a variable oscillating torque, whose frequency is twice the supply voltage frequency (in our case 100 Hz) and overlaps the average torque. This oscillating component is demonstrated by the analytic expression of the instantaneous torque, which is written using nothing but total flux linkages (25). The symmetric components theory, for example, is not capable to provide information about these oscillating torques. At the most, this theory evaluates the average torque, probably with inherent errors. Coming back to the torque variations, one can see that the amplitude oscillations increase with the asymmetry degree, but their frequency keeps unchanged.

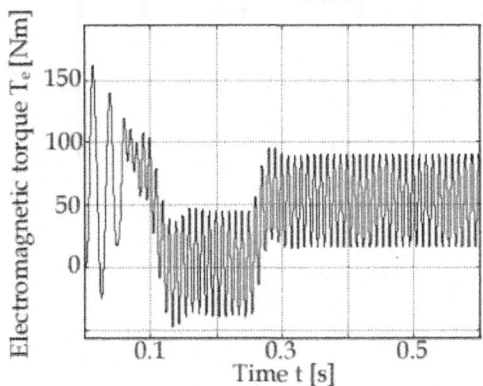

Figure 22: Time variation of electromagnetic torque – RNS-1

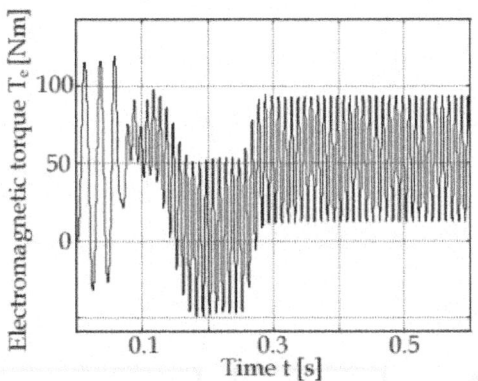

Figure 23: Time variation of electromagnetic torque – RNS-2

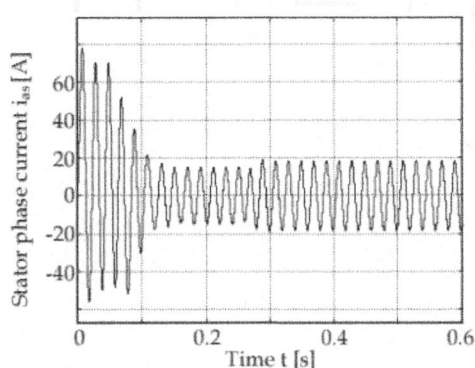

Figure 24: Time variation of stator phase current – RNS-1

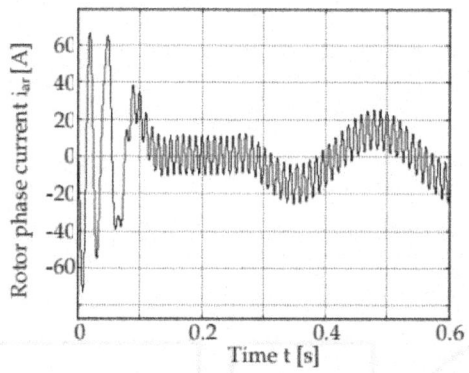

Figure 25: Time variation of rotor phase current – RNS-1

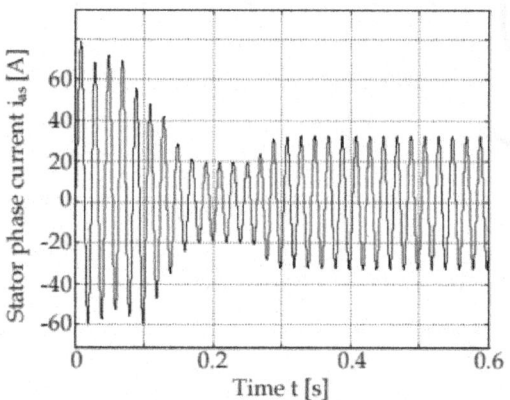

Figure 26: Time variation of stator phase current – RNS-2

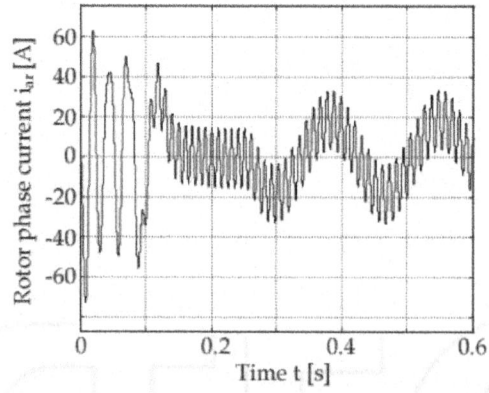

Figure 27: Time variation of rotor phase current – RNS-2

The stator currents variation, Fig. 24 and 26, have a sinusoidal shape and an unmodified frequency of 50 Hz. Their amplitude increases however with the asymmetry degree (approx. 18 A for RNS-1 and approx. 32 A for RNS-2). As a consequence of this fact, both power factor and efficiency decrease. The rotor currents (Fig. 25 and 27) include besides the main component of $f_2 = s\, f_1$ frequency a second oscillating component of high frequency, $f'_2 = (2-s)\, f1$, which is responsible for parasitic torques and vibrations of the rotor. The amplitude of these oscillating currents increases with the asymmetry degree.

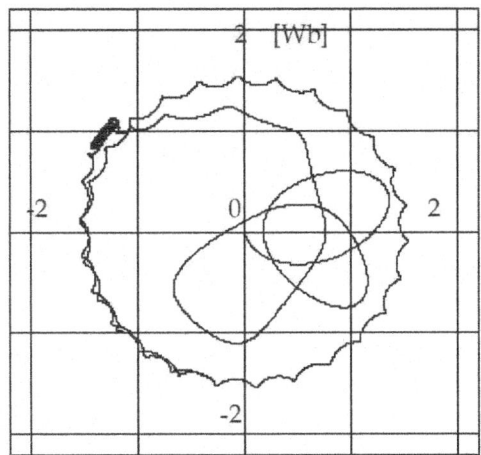

Figure 28: Hodograph of resultant rotor flux – RNS-1

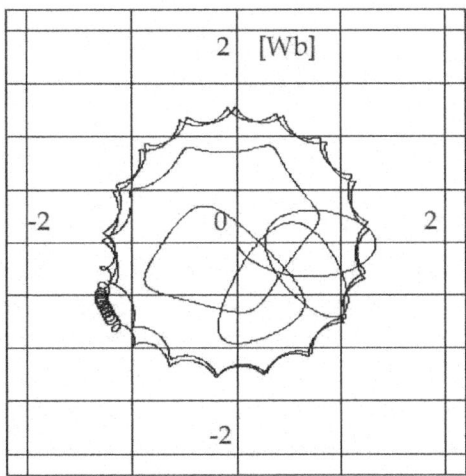

Figure 29: Hodograph of resultant rotor flux – RNS-2

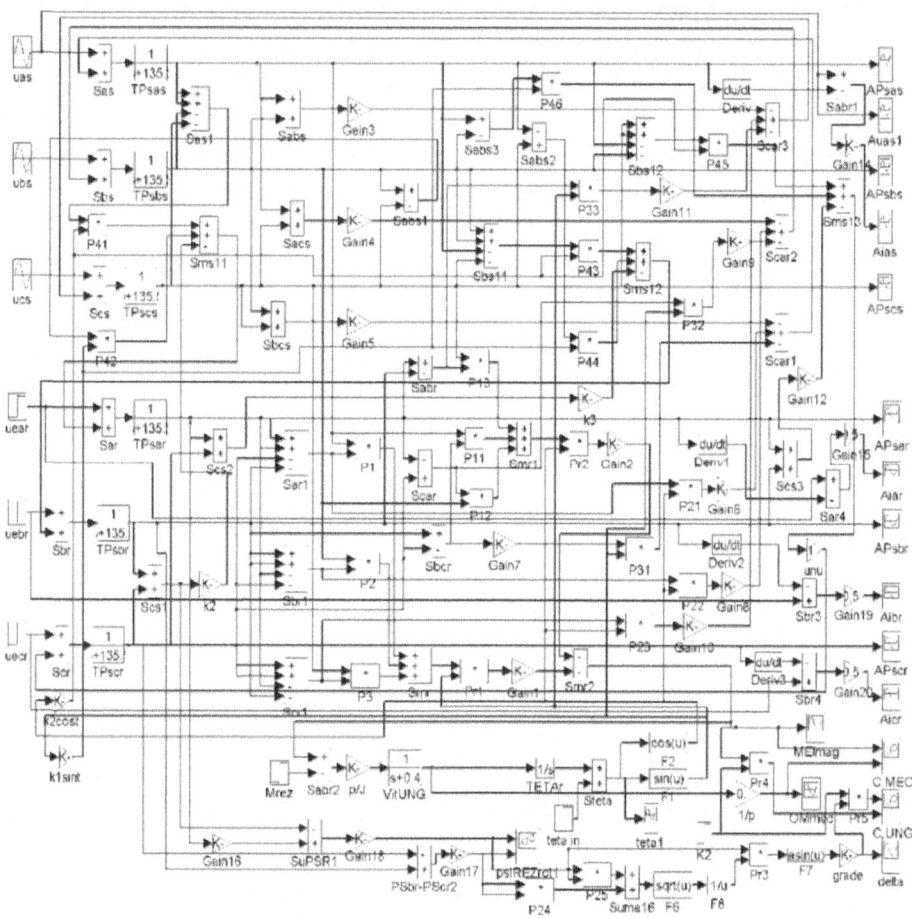

Figure 30: Structural diagram of the three-phase induction machine

The hodographs of the resultant rotor flux show a very interesting behavior of the unbalanced machines, Fig and 29. In comparison to the symmetric supply cases where the hodograph is a circle under steady state, the asymmetric system distort the curve into a „gear wheel" with a lot of teeth placed on a mean diameter whose magnitude depends inverse proportionally with the asymmetry degree. Generally, these curves do not overlap and prove that during the operation the interaction between stator and rotor fluxes is not constant in time since the rotor speed is not constant. Consequently, the rotor vibrations are usually propagated to the mechanical components and working machine.

In order to point out the superiority of the proposed mathematical model, Fig. 30 shows the structural diagram used in Simulink environment. The diagram is capable to simulate any steady-state and transient duty under balanced or unbalanced state of the induction machine including doubly-fed operation as generator or motor by simple modification of the input data. To prove this statement, a simulation of an unbalanced doubly-fed operation has been performed. The operation cycle involves: I. A no-load start-up (the wound rotor winding is short-circuited); II. Application of a supplementary output torque of (-70) Nm (at the moment t=0.4 sec.) which leads the induction machine to the generating duty (over synchronous speed); III. Supply of two series connected rotor phases with d.c. current (U_{ar}=+40V, U_{br}= −40V, U_{cr}=0V), at the moment time t=0.6 sec., which change the operation of the induction generator into a synchronized induction generator (SIG).

Fig. 31 and 32 show the dynamic mechanical characteristic, Te=f(Ω_R) and the hodograph of the resultant rotor flux respectively. The start-up corresponds to A-S1 curve, the over synchronous acceleration is modeled by S1-S curve and the operation under SIG duty corresponds to S-S2 curve. A few observations regarding Fig. 32 are necessary as well. The rotor flux hodograph is rotating in a counterclockwise direction corresponding to motoring duty, in a clockwise direction for generating duty and stands still at synchronism. The "in time" modification and the position of the hodograph corresponding to SIG duty depend on the moment of d.c. supply and the load angle of the machine.

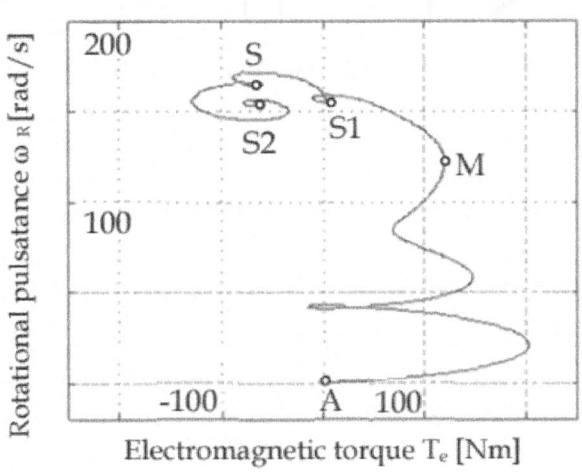

Figure 31: Dynamic mechanical characteristic

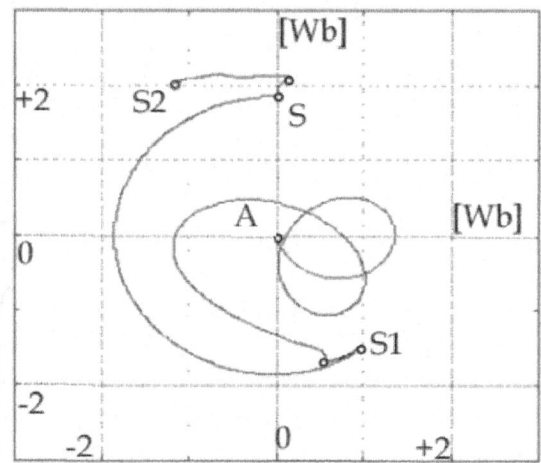

Figure 32: Hodograph of resultant rotor flux

CONCLUSION

The mathematical model presented in this contribution is characterized by the total lack of the winding currents and angular speed in the voltage equations. Since these parameters are differential quantities of other electric parameters, they usually bring supplementary calculus errors mainly for the dynamic duty analysis. Their removal assures a high accuracy of the results. If their variation is however necessary to be known then simple subsequent calculations can be performed.

The use of the mathematical model in total fluxes is appropriate for the study of the electric machines with permanent magnets where the definitive parameter is the magnetic flux and not the electric current.

The coefficients defined by (28.1-4), which depend on resistances and inductances, take into consideration the saturation. Consequently, the study of the induction machine covers more than the linear behavior of the magnetization phenomenon.

The most important advantage of the proposed mathematical model is its generality degree. Any operation duty, such as steady-state or transients, balanced or unbalanced, can be analyzed. In particular, the double feeding duty and the synchronized induction machine operation (feeding with D.C. current of a rotor phase while the other two are short-circuited) can be simulated as well.

The results obtained by simulation are based on the transformation of the equations in structural diagrams under Matlab-Simulink environment. They present the variation of electrical quantities (voltages and currents corresponding to stator and rotor windings), of mechanical quantities (expressed through rotational pulsatance) and of magnetic parameters (electromagnetic torque, resultant rotor and stator fluxes). They put in view the behavior of the induction machine for different transient duties. In particular, they prove that any unbalance of the supply system generates important variations of the electromagnetic torque and rotor speed. This fact causes vibrations and noise

REFERENCES

1.   M. Ahmad, 2010High Performance AC DrivesModeling Analysis and Control, Springer, 978-3-64213-149-3London, UK

2.   I. Boldea, L. Tutelea, 2010Electric Machines.Steady State, Transients and Design with MATLAB, CRC Press, 978-1-42005-572-6Boca Raton, USA

3.   B. Bose, 2006Power Electronics and Motor Drives, Elsevier,978-0-12088-405-6San Diego, USA

4.   J. Chiasson, 2005Modeling and High-Performance Control of Electrical Machines, IEEE Press, Wiley Interscience, 047168449Hoboken, USA

5.   R. De Doncker, D. Pulle, A. Veltman, 2011Advanced Electrical Drives. Analysis, Modeling, Control,Springer, 978-9-40070-179-3Dordrecht, Germany

6.   P. Krause, O. Wasynczuk, S. Sudhoff, 2002Analysis of Electric Machinery and Drive Systemssec. ed.), IEEE Press, 047114326Piscataway, USA

7.   R. Marino, P. Tomei, C. Verrelli, 2010Induction Motor Control DesignSpringer, 978-1-84996-283-4London, UK

8.   C. Ong-M, 1998Dynamic Simulation of Electric Machinery using Matlab/Simulink,Prentice Hall, 0-13723-785-5Jersey, USA

9.   Al. Simion, L. Livadaru, D. Lucache, 2009Computer-Aided Simulation on the Reversing Operation of the Two-Phase Induction MachineInternational Journal of Mathematical Models and Methods in Applied SciencesIss. 1, 337471998-0140

10.   Al. Simion, 2010Study of the Induction Machine Unsymmetrical Condition Using In Total Fluxes EquationsAdvances in Electrical and Computer EngineeringIss. 1 (February 2010), 34411582-7445

11.   Al. Simion, L. Livadaru, 2010On the Unsymmetrical Regime of Induction Machine. Bul. Inst. Polit. Iaşi, Tomul LVI(LX), Fasc.4, 79911223-8139

12. Al. Simion, L. Livadaru, A. Munteanu, 2011New Approach on the Study of the Steady-State and Transient Regimes of Three-Phase Induction Machine. Buletinul AGIR, Nr.4/2011, 16ISSN-L 1224-7928

13. S. Sul-K, 2011Control of Electric Machine Drive SystemsIEEE Press, Wiley Interscience, 978-0-47087-655-8Hoboken, USA

14. P. Wach, 2011Dynamics and Control of Electric Drives,Springer, 978-3-64220-221-6Berlin, Germany

Chapter 5

A SCALE INVARIANT DISTRIBUTION OF THE PRIME NUMBERS

Wayne S. Kendal[1], and Bent Jørgensen[2],

[1]Division of Radiation Oncology, University of Ottawa, 501 Smyth Road, Ottawa, ON K1H 8L6, Canada

[2]Department of Mathematics and Computer Science, University of Southern Denmark, Campusvej 55, Odense M DK-5230, Denmark

ABSTRACT

The irregular distribution of prime numbers amongst the integers has found multiple uses, from engineering applications of cryptography to quantum theory. The degree to which this distribution can be predicted thus has become a subject of current interest. Here, we present a computational analysis of the deviations between the actual positions of the prime numbers and their predicted positions from Riemann's counting formula, focused on the variance function of these deviations from sequential enumerative bins. We show empirically that these deviations can be described by a class of probabilistic models known as the Tweedie exponential dispersion models that are characterized by a power law relationship between the variance and the mean, known by biologists as Taylor's power law and by engineers as fluctuation scaling. This power law behavior of the prime number deviations is remarkable in that the same behavior has been found within the distribution of genes and single nucleotide polymorphisms (SNPs) within the human genome, the distribution of animals and plants within their habitats, as well as within many other biological and physical processes. We explain the common features of this behavior through a statistical convergence effect related to the central limit theorem that also generates $1/f$ noise.

INTRODUCTION

Prediction of the positions of the prime numbers within the sequence of integers has been a goal that has long interested mathematicians, engineers, and physicists [1]. The prime counting function $\pi(x)$ gives the actual number of prime numbers up to the positive real value x and takes the form of a step function that increases by the value of 1 with each new prime number. In 1808, Legendre empirically showed that $\pi(x)$ could be approximated by the formula $x/[\log(x) - B]$; B being a numerical constant [2]. A further improvement in the estimation of $\pi(x)$ was provided by Gauss, who used the logarithmic integral, defined on the positive real numbers $x \neq 1$ for this purpose [3]:

$$\mathrm{li}(x) = \int_0^x dt / \ln(t)$$

(1)

These approximations lead to the prime number theorem:

$$\pi(x) \sim x / \ln(x)$$

(2)

with initial proofs delivered by Hadamard [4] and de la Vallée Poussin [5] in 1896. A more detailed formula for the general trend in the position of prime numbers has been attributed to Bernhard Riemann [6]:

$$R(x) = \sum_{n=1}^{\infty} \frac{\mu(n)}{n} \mathrm{li}(x^{1/n})$$

(3)

expressed in terms of the Möbius function on the integer values n:

$$\mu(n) = \begin{cases} 1 & \text{if } n = 1 \\ 0 & \text{if } n \text{ has one or more repeated prime factors} \\ (-1)^k & \text{if } n \text{ is a product of } k \text{ distinct primes} \end{cases}$$

(4)

The present report focuses on the local behavior of the deviations between $\pi(x)$ and $R(x)$ that have been conjectured to obey the equation [7]:

$$D(x) = R(x) - \pi(x) \sim \sum_{\rho} R(x^{\rho})$$

(5)

The summation here is specified over the nontrivial (complex) zeros ρ of the Riemann zeta function, providing a link to Riemann's hypothesis that the nontrivial zeros of the Riemann zeta function should have as their real component the value of ½ [6]. An equivalent form of this (yet unproven) hypothesis can be stated using the summatory Mertens function [8]: for every positive ε [9].

$$M(n) = \sum_{i=1}^{n} \mu(i)$$

(6)

Namely that:

$$M(n) = O(\, n^{1/2+\varepsilon})$$

(7)

The deviations $D(x)$ empirically reveal chaotic features, as well as long range correlations [10], that would seem to indicate an underlying structure. Indeed, the positional irregularities of prime numbers have the characteristics of $1/f$ noise [11], and can be related to the eigenvalues of random matrices from the Gaussian unitary ensemble that are employed in quantum chaos [12]. It remains unclear, though, why the distribution of prime numbers should exhibit such features. Here these deviations are empirically shown to correspond to a probabilistic model characterized by a power law relationship between the variance and the mean. This model belongs to the class of scale invariant, or Tweedie, exponential dispersion models [13] which appear as weak limits for other models under a certain central limit-like effect [14]. This convergence effect, we will argue, leads to a wide manifestation of these models within complex natural and physical systems, analogous to what is seen with the Gaussian distribution within more restricted systems governed by the central limit theorem itself. Moreover, this Tweedie convergence effect can explain the genesis of $1/f$ noise in such sequences that has implications regarding self-organized criticality [15].

EXPERIMENTAL SECTION

The Riemann prime counting function Equation (3) was estimated here using the Gram series [16]. The absolute values of the prime number deviations $|D(x)|$ were estimated for sequential equal-sized enumerative bins that span the integers. The values of $|D(x)|$ from within each bin were summed and the means and variances of these sums estimated over the region of interest. This process was repeated for successively larger bin sizes so that a log-log plot could be constructed of the sampled variances *versus* the sampled means with different sized bins. A straight line relationship in such a plot would imply a power law relationship between the variance and the mean, variously referred to in the literature as Taylor's ecological power law as well as fluctuation scaling.

Empirical cumulative distribution functions (CDFs) were also constructed from the primary sequences of $|D(x)|$ and then compared to the theoretical CDF from the Tweedie compound Poisson-gamma distribution [13], a distribution which inherently expresses a variance to mean power law with its power law exponent constrained to range between 1 and 2.

RESULTS AND DISCUSSION

The Variance Function of the Prime Number Deviations

We empirically reviewed the absolute values of the deviations $|D(x)|$ for the first 50,000 integers (Figure 1), focusing on their variance function [13]: The mean and variance of the sequence $|D(x)|$ were estimated and, to extend the range of values for the mean, the sequence $|D(x)|$ was divided into non-overlapping and adjacent counting bins, 10 integers long. The values within each bin were summed and the mean and variance of these sums were estimated over all the bins. This process was repeated for successively larger and larger bin sizes. Figure 2a demonstrates the log-log plot of the empirical variances *versus* their respective means. A linear relationship was evident on the log-log plot, indicative of a power law relationship $\sigma^2 = a \cdot \overline{m}^p$ between the variance σ^2 and the mean $\overline{m}$ with exponent $p = 1.83$.

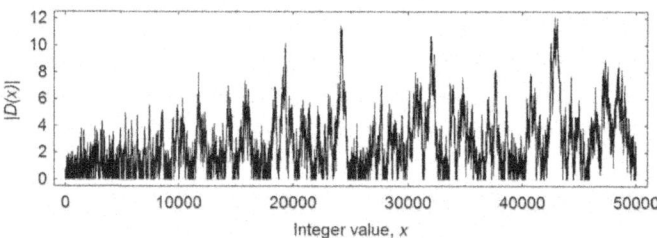

Figure 1: The absolute value $|D(x)|$ of the discrepancy between the prime counting function $\pi(x)$ and Riemann's formula $R(x)$ for the first 50,000 integers.

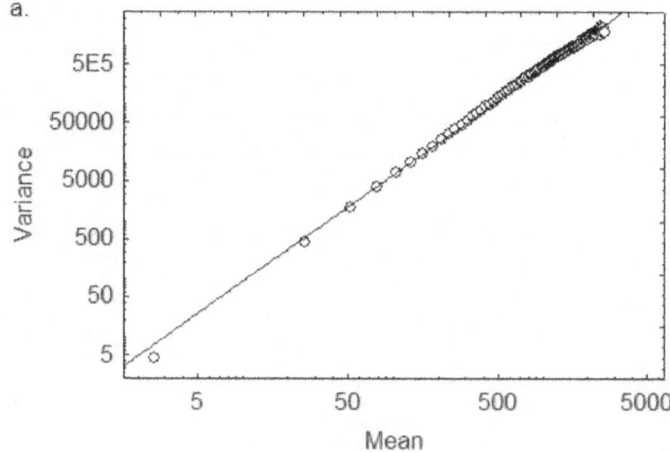

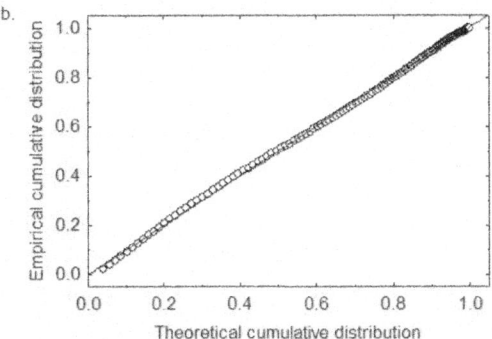

Figure 2: (a) The variance function for the deviations of the prime numbers. A power law relationship was obtained with exponent $p = 1.83$ and constant $a = 1.37$; (b) Probability-probability plot. A frequency histogram was constructed for the values $|D(x)|$. The empirical CDF obeyed a Tweedie compound Poisson-gamma distribution with $\theta = -0.837$, $\lambda = 0.934$, $\alpha = -0.586$ and $p = 1.63$.

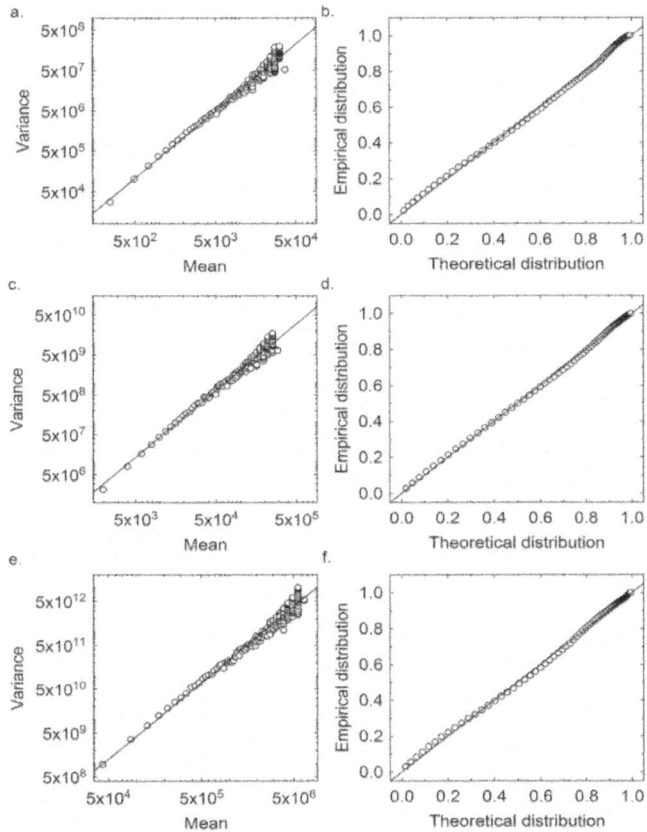

Figure 3: (a) Variance function of $|D(x)|$ for $x = 10^6 \ldots 10^7$ in steps of 10^3. Linear re-

gression gave $p = 1.66$ and $a = 3.53$; **(b)** CDF with $x = 10^6...10^7$ in steps of 10^3. Here $\theta = -0.096$, $\lambda = 0.62$, $\alpha = -0.38$ and $p = 1.73$; **(c)** Variance function of $|D(x)|$ for $x = 10^8...10^9$ in steps of 10^5. Here $p = 1.67$ and $a = 9.28$; **(d)** CDF with $x = 10^8...10^9$ in steps of 10^5. Here $\theta = -9.94 \times 10^{-3}$, $\lambda = 0.34$, $\alpha = -0.30$ and $p = 1.77$; **(e)** Variance function of $|D(x)|$ for $x = 10^{11}...10^{12}$ in steps of 10^8. Here $p = 1.74$ and $a = 8.57$; **(f)** CDF with $x = 10^{11}...10^{12}$ in steps of 10^8. Here $\theta = -5.37 \times 10^{-4}$, $\lambda = 0.067$, $\alpha = -0.41$ and $p = 1.71$.

We sought to determine the behavior of the power law for the deviations $|D(x)|$ beyond the first 50,000 integers in order to make sure that this behavior was not simply a loco-regional artifact. Figure 3a provides this function determined from the integers $x = 10^6...10^7$, in steps of 10^3. The variance to mean power law fitted this sequence well with $p = 1.66$. In Figure 3c,e this process was repeated for the integer sequences $x = 10^8...10^9$ in steps of 10^5 and $x = 10^{11}...10^{12}$ in steps of 10^8, to reveal that variance to mean power law was evident for over 12 orders of magnitude.

Prime Number Deviations as a Self-Similar Process

In our analysis we examined the sequence of prime number deviations $Y = (Y_i = |D(i)|: i = 1, 2,..., N)$ using expanding enumerative bins. Let the mean and variance of this initial sequence be represented by the constants $\hat{\mu} = E[Y]$ and $\hat{\sigma}^2 = var[Y]$, respectively. One can construct a sequence of equal-sized and non-overlapping enumerative bins of integer size m to yield a new set of sequences $Y^{(m)}$ with the reproductive property:

$$Y_i^{(m)} = \frac{1}{m}(Y_{im-m+1} + ... + Y_{im}), \quad i > 1 \tag{8}$$

The values of m were chosen here so that N/m was an integer. We then have $E[Y(m)]=E[Y]=\mu\hat{}$, which is a constant provided that the initial sequence remains unaltered.

Alternatively, the enumerative bins can be used to construct a set of additive sequences $Z^{(m)}$ so that:

$$Z_i^{(m)} = (Y_{im-m+1} + ... + Y_{im}) \tag{9}$$

as was done with the prime number deviations $|D(x)|$. These additive and reproductive sequences are related to each other by the equation $Z(m)_i=mY(m)_i$. We have for their means and variances, $E[Z(m)]=mE[Y(m)]$ and $var[Z(m)]=m2var[Y(m)]$. The variance to mean power law, demonstrated with the prime number deviations, implies the existence of long range correlations within the initial sequence of deviations. Long range correlations have been studied by Leland [17] in the context of self-similar processes, where the general form for the correlation function $r(k)$ takes the form,

$$r(k) = k^{-\beta} L(k), \quad k \to \infty \tag{10}$$

Here, k is the autocorrelation shift, β is a real valued constant such that $0 < \beta < 1$, and $L(k)$ is a slowly varying function for large values of k. For large values of k, this correlation approximates a power law with $r(k) \sim k^{-\beta}$.

Consider the case where the reproductive sequences $Y^{(m)}$ obey the property:

$$\mathrm{var}[Y^{(m)}] = \hat{\sigma}^2 m^{-\beta} \tag{11}$$

Tsybakov and Georganis [18] have shown that Equation (11) holds if and only if the autocorrelation for the sequence Y is:

$$r(k) = \frac{1}{2}[(k+1)^{2-\beta} - 2k^{2-\beta} + (k-1)^{2-\beta}] \tag{12}$$

In the limit this formula takes the form [18]:

$$\lim_{k \to \infty} \frac{r(k)}{k^{-\beta}} = \frac{1}{2}(2-\beta)(1-\beta)) \tag{13}$$

where the constant β is related to the Hurst exponent H, $\beta = 2(1-H)$ For additive sequences the variance takes the form:

$$\mathrm{var}[Z_i^{(m)}] = m^2\, \mathrm{var}[Y^{(m)}] = (\hat{\sigma}^2/\hat{\mu}^{2-\beta})E[Z_i^{(m)}]^{2-\beta} \tag{14}$$

Since $\hat{\mu}$ and $\hat{\sigma}^2$ are constant, this represents a variance to mean power law with exponent $p = 2 - \beta = 2H$. The power law found from the prime number deviations revealed that $p = 1.83$, and so the Hurst exponent was $H = 0.92$.

1/f Noise from the Prime Number Deviations

Autocorrelations of the form $r(k) \sim k-\beta$ imply, by virtue of the Wiener-Khintchine theorem [19], that the corresponding power spectral density will approximate $S(f) \sim f\beta-1$, where f is the frequency. In the case of the prime number deviations $|D(x)|$, this would imply $S(f) \sim f-0.83$, where the frequency is determined from the spacing of the consecutive values of the data sequence. Power spectra like this, with $S(f) \propto 1/f\gamma$ and $0 < \gamma < 2$, are the hallmark of $1/f$ noise.

Prime Number Deviations Described by a Tweedie Exponential Dispersion Model

In addition to $1/f$ noise, the variance-to-mean power law is associated with a class of probability distributions, known as the Tweedie exponential dispersion models, named after the man who first described them [20]. There are several types of Tweedie models, each distinguished by the values of their power law exponent p, including the Gaussian ($p = 0$), Poisson ($p = 1$), compound Poisson-gamma (PG) ($1 < p < 2$), gamma ($p = 2$) and inverse Gaussian ($p = 3$) distributions [13]. We employ here the class of PG distributions, characterized by gamma distributed jumps over the domain of the positive real numbers plus zero, to model the deviations $|D(x)|$. The additive form for the cumulant generating function for this PG distribution can be written as:

$$K^*(s) = \lambda \kappa(\theta) \left[\left(1 + \frac{s}{\theta} \right)^\alpha - 1 \right]$$

(15)

Here, s is the variable of the generating function, θ is the canonical parameter, λ is the index statistic, and the constant α and the cumulant function $k(\theta)$ are given by:

$$\alpha = (p-2)/(p-1) \text{ and}$$

(16)

$$\kappa(\theta) = \frac{\alpha-1}{\alpha} \left(\frac{\theta}{\alpha-1} \right)^\alpha$$

(17)

The PG distribution is thus specified by 3 independent adjustable parameters α, λ and θ. The parameter α relates to Taylor's power law exponent p, whereas λ and θ relate to the notions of shape and scale parameter employed in the conventional description of distributions.

The probability density function on the variable z, $p^*(z, \theta, \lambda, \alpha)$, that corresponds to this distribution does not exist in closed form. However, it can be expressed as an infinite series [13]:

$$p^*(z; \theta, \lambda, \alpha) = c^*(z; \lambda) \exp[\theta \cdot z - \lambda \kappa(\theta)] \text{, with}$$

(18)

$$c^*(z; \lambda) = \begin{cases} \dfrac{1}{z} \displaystyle\sum_{n=1}^{\infty} \lambda^n \kappa^n (-1/z)/\Gamma(-\alpha \cdot n)n! & \text{for } z > 0 \\ 1 & \text{for } z = 0 \end{cases}$$

(19)

Figure 2b provides a probability-probability plot of the respective CDF fitted to data derived from the sequence of $|D(x)|$ from the first 50,000 integers. The empirical and theoretical CDFs agreed well. Similarly Figure 3b provides the probability-probability plot corresponding to $|D(x)|$ derived from the integers $x = 10^6...10^7$, in steps of 10^3. The Tweedie PG CDF agreed closely with the empirical CDF. Figure 3d,f provides the probability-probability plots corresponding to the Tweedie PG distribution fitted from the prime number deviations corresponding to the integer sequences $x = 10^8...10^9$ in steps of 10^5 and $x = 10^{11}...10^{12}$ in steps of 10^8. The Tweedie PG distribution thus appeared to remain valid for over 12 orders of magnitude of data.

The Tweedie Convergence Theorem

For exponential dispersion models $ED(\mu, \sigma^2)$ with unit variance functions of the form $V(\mu) \propto \mu p$ Jørgensen *et al.* have shown that as $\mu \to 0$ or $\mu \to \infty$ then, within the constant factor c, $c - 1\ ED(c\mu, \sigma^2\ c^{2-p})$ will converge to the form of a Tweedie model as either $c \to 0$ or $c \to \infty$ [14]. Since the variance functions for many probability distributions approximate the form $V(\mu) \propto \mu p$, for small or large values of μ, the Tweedie distributions will act as the foci of convergence for a wide variety of data [14]. This convergence property appears related to stable generalizations of the CLT [13], suggesting that the Tweedie models have a role analogous to that of the Gaussian distribution in statistical theory. As will be shown below, this theorem has implications with respect to the distribution of prime numbers. Parenthetically, it should also be mentioned that the mean and variance of an additive random variable Z are given by the equations $E(Z) = \lambda\mu$ and $\text{var}(Z) = \lambda V(\mu)$.

The Theoretical Behavior of the Prime Number Deviations

We have described the empirical behavior of the statistic $|D(x)| = |R(x) - \pi(x)|$. The Tweedie convergence theorem requires for complex statistical systems with variance functions, which approximate $V(\mu) \propto \mu p$, that these systems be mathematically required to converge towards a Tweedie exponential dispersion model. The empirical demonstrations from the prime number deviations established that their variance function had this approximate behavior and the convergence requirements appeared justified.

Other Examples of Variance to Mean Power Laws and 1/f Noise

Similar variance to mean power laws have been extensively reported with nonlinear phenomena in connection with the clustering of species within their habitats (where it is known as Taylor's power law) [21], human sexual contacts in AIDS [22], pediatric leukemia cases [23] and murine experimental metastases [24]. As well, this relationship is evident with spatial and temporal data associated with long range correlations, such as temporal changes in measles incidence [25], regional blood flow heterogeneity (Bassingthwaighte's fractal scaling relation) [26], the genomic distributions of SNPs [27] and genes [28], auditory nerve spike trains [29], fluctuations in foreign exchange quotations [30], and with internet traffic [18] (where it has been called fluctuation scaling). With such diverse manifestations of this power law, the question would arise as to whether a more fundamental process could be implicated. We explain the wide manifestation of this power law by an asymptotic convergence of systems towards the Tweedie distributions [14].

The Gaussian and other stable distributions have their elementary role in probability theory consequent to the (generalized) central limit theorem, where they appear as the limiting forms for standardized sums of independent random variables with the Gaussian distribution appearing when the components have finite variances. The Tweedie convergence theorem [13] implies that a wide range of exponential dispersion models can be approximated by the Tweedie models according to a central limit-like effect. For this reason many types of non-Gaussian data will exhibit a power law relationship between the variance and the mean.

Recently, Taylor's power law and its implicit fluctuation scaling have been explained in terms of physical processes. Eisler *et al.* have provided a mean field framework that employs the summation of random variables to show that Taylor's law can be considered a consequence of a limiting theorem they relate to impact inhomogeneity [31]. Fronczak and Fronczak provided an alternative explanation that employed fluctuation dissipation in equilibrium and non-equilibrium systems to argue that Taylor's law is a consequence of the second law of thermodynamics and the action of a putative external field [32]. In the case of the prime number deviations, an explanation based on thermodynamic principles would not seem applicable.

The Tweedie convergence theorem has a more general scope than the limit theorem offered by Eisler *et al.* Indeed, the Tweedie models act as limiting distributions for a wide range of statistical models through a convergence effect that generalizes upon the central limit theorem [13,14,33,34]. Distributions within the domain of attraction of positive or extreme stable distributions can thus converge to manifest the variance to mean power law. One can hypothesize physical principles, or specific biological mechanisms, to explain the manifestation of Taylor's law, but it would seem that this mathematically demonstrable convergence property would provide a more robust and general explanation. Sequences like the Riemann deviations $D(x)$, that express a variance to mean power law and that also obey the Tweedie PG distribution have similarly been generated from random matrices of the Gaussian unitary and orthogonal ensembles [35], the Mertens function (Equation (6)) [36], Chebyshev's function [37], as well as within the genomic distribution of SNPs along chromosome 1 of the horse [36]. Fronczak and Fronczak's derivation of the variance to mean power law, mentioned above, [32] can be shown to exactly yield the Tweedie distributions (Equations (15)–(17)) [36], and Eisler *et al.* have shown their model for impact inhomogeneity to yield equations similar to Tweedie's [31]. The Tweedie convergence effect, which has as its focus these equations, can be further extended by theorem to multiplicative statistical models which, in turn, allow for a mechanistic explanation for the genesis of multifractality [38].

$1/f$ noise is another empirical phenomenon that has been widely observed in physical and biological systems [39,40]. There is no generally accepted explanation for the appearance of $1/f$ noise, though it notably has been explained in terms of the self-organized criticality of evolutionary behavior in extended dissipative systems [41]. As noted above, $1/f$ noise is directly related to long-range correlations and, in turn, to a variance to mean power law [18]. In this latter context, certain types of $1/f$ noise can thus be viewed as consequences of the statistical convergence behavior associated with the Tweedie PG model. In fact, Taylor's law and $1/f$ noise have been demonstrated to occur together within the genomic distribution of genes and SNPs along human chromosome 1 [42].

CONCLUSIONS

The variance to mean power law implies a statistical self-similarity that can underlie certain fractal patterns. Related fractal patterns have been demonstrated within the distributions of the prime numbers [43] and the Riemann zeta zeros [44]. Holdom showed that the deviations $b(i)$ of the offset logarithmic integral of the *i*th prime number p_i, b(i)≡Li(pi)−i, exhibited scale invariant correlations that were related to the Riemann hypothesis [45]. Indeed, it is possible to express the Riemann hypothesis in terms of the behavior of a random walk on a fractal lattice [46]. In this context it should not seem surprising that Taylor's scale invariant law would manifest with the deviations $|D(x)|$. It is the long range correlations in the distribution of the prime numbers that have been interpreted to indicate a fractal pattern [43]. The variance to mean power law is a manifestation of a fractal pattern where the power law exponent directly relates to the fractal dimension D=H−2 through the equation, p=4−2D. The assessments from Figure 2a and Figure 3, though, yielded values for p that ranged between 1 and 2. Some of this variation might reflect numerical artifact, but a component might also be attributable to the distribution of the prime numbers, itself. Indeed p can be shown to vary between 1 to 2 over different regions of the natural numbers and with different sample sizes (data not provided). When regions of a data sequence are found to have different fractal properties, as implied by the regional differences in p within the sequence of prime number deviations, that sequence would be said to be multifractal [47]. Multifractality of this nature has been described from the sequential deviations of ordered eigenvalues from the Gaussian unitary and orthogonal ensembles [38]. A current area of investigation deals with the apparent multifractality of the Riemann deviations $|D(x)|$.

The Riemann deviations $D(x)$ are intimately related to the positions of the prime numbers. Our examinations have revealed their absolute values to behave asymptotically as Tweedie probability distributions, perhaps due to a central limit-like effect, much like Erdös and Kac's demonstration of the asymptotic Gaussianity of the number of prime divisors of the integers [48]. Thus far, our evidence for a power variance to mean relationship with each of these functions is empirical, but it extends over many orders of magnitude and it adds to the growing evidence of a unifying description for disparate mathematical, physical and biological phenomena obtainable through the Tweedie exponential dispersion models. Additionally, any statistical model for sequential data designed to produce Taylor's law is mathematically required to converge towards a Tweedie model, and to yield $1/f$ noise [35].

Self-organized criticality was mentioned earlier in the introduction. This is a widely-held hypothesis used to explain the genesis of $1/f$ noise and other power law scaling behaviors [41]. Sandpile model simulations have generally been used to demonstrate $1/f$ noise and, thus, to support this hypothesis [41], yet the fluctuations evident to sandpile simulations also manifest a variance to mean power law and conform to the Tweedie compound Poisson-gamma distribution, raising the possibility that phenomena attributed to self-organized criticality can alternatively be attributed to the convergence behavior associated with the Tweedie models [15]. The irregular distribution of prime numbers thus appears to be one of many examples of Taylor's law and fluctuation scaling that are similarly implicated with this mathematical convergence behavior.

Self-organized criticality is postulated to occur within dynamical systems that naturally progress to borderline unstable and organized states, without any outside manipulation of the dynamical parameters. However, with the Riemann deviations examined here the manifestation of $1/f$ noise (the hallmark of self-organized criticality) would appear here to be possibly attributable to a mathematical convergence effect related to the central limit theorem of statistics. Further investigation into this matter could lead to a paradigm change in our understanding of processes that have conventionally been attributed to self-organized criticality.

ACKNOWLEDGMENTS

The authors gratefully acknowledge financial support from Patricia Rinaldo.

AUTHOR CONTRIBUTIONS

The authors were responsible for all aspects of this study.

CONFLICTS OF INTEREST

The authors declare no conflict of interest.

REFERENCES

1. Kotnik, T. The prime-counting function and its analytic approximations $\pi(x)$ and its approximations. *Adv. Comput. Math.* **2008**, *29*, 55–70.

2. Legendre, A.M. *Essai sur la Théorie des Nombres*, 2nd ed.; Courcier: Paris, France, 1808; p. 394. (In French)

3. Gauss, C.F. *Werke Band 2*, 1st ed.; Königliche Gesellschaft der Wissenschaften zu Göttingen: Göttingen, Germany, 1863; Volume 2, pp. 444–447. (In German)

4. Hadamard, J. Sur la distribution des zéros de la fonction et ses conséquences arithmétiques. *Bull. Soc. Math. Fr.* **1896**,*24*, 199–220. (In German).

5. De la Vallée Poussin, C.J. Recherches analytiques la théorie des nombres premiers. *Ann. Soc. Sci. Brux.* **1896**, *20*, 183–256. (In German).

6. Riemann, G.F.B. Über die anzahl der primzahlen unter einer gegebenen grösse. *Mon. Berl. Akad.* **1859**, *2*, 671–680. (In German).

7. Borwein, J.M.; Bradley, D.M.; Crandall, R.E. Computational strategies for the Riemann zeta function. *J. Comp. Appl. Math.* **2000**, *121*, 247–296.

8. Mertens, F. Über eine zahlentheoretische funktion. *Akad. Wiss. Math. Nat. Kleine Sitz.* **1897**, *106*, 761–830. (In German).

9. Titchmarsh, E.C. *The Theory of the Riemann Zeta-Function*, 2nd ed.; The Clarendon Oxford University Press: New York, NY, USA, 1986.

10. Gamba, Z.; Hernando, A.; Romanelli, L. Are prime numbers regularly ordered? *Phys. Lett. A* **1990**, *145*, 106–108.

11. Wolf, M. $1/f$ noise in the distribution of prime numbers. *Phys. A* **1997**,

241, 493–499.

12. Kriecherbauer, T.; Marklof, J.; Soshnikov, A. Random matrices and quantum chaos. *Proc. Natl. Acad. Sci. USA* **2001**, *98*, 10531–10532.

13. Jørgensen, B. *The Theory of Dispersion Models*; Chapman & Hall: London, UK, 1997.

14. Jørgensen, B.; Martínez, J.R.; Tsao, M. Asymptotic behaviour of the variance function. *Scand. J. Stat.* **1994**, *213*, 223–243.

15. Kendal, W.S. Self-organized criticality attributed to a central limit-like convergence effect. *Phys. A* **2015**, *412*, 141–150.

16. Riesel, H.; Göhl, G. Some calculations related to Riemann's prime number formula. *Math. Comput.* **1970**, *24*, 969–983.

17. Leland, W.E.; Taqqu, M.S.; Willinger, W.; Wilson, D.V. On the self-similar nature of ethernet traffic. *IEE/ACM Trans. Netw.* **1994**, *2*, 1–15.

18. Tsybakov, B.; Georganas, N.D. On self-similar traffic in atm queues: Definitions, overflow probability bound, and cell delay distribution. *IEEE/ACM Trans. Netw.* **1997**, *5*, 397–409.

19. McQuarrie, D.A. *Statistical Mechanics*; Harper & Row: New York, NY, USA, 1976; pp. 553–561.

20. Tweedie, M.C.K. An index which distinguishes between some important exponential families. In *Statistics: Applications and New Directions*, Proceedings of the Indian Statistical Institute Golden Jubilee International Conference, Calcutta, India, 27 September–1 October 1982; Ghosh, J.K., Roy, J., Eds.; Indian Statistical Institute: Calcutta, India, 1984; pp. 579–604.

21. Taylor, L.R. Aggregation, variance and the mean. *Nature* **1961**, *189*, 732–735.

22. Anderson, R.M.; May, R.M. Epidemiological parameters of HIV transmission. *Nature* **1988**, *333*, 514–519.

23. Philippe, P. The scale-invariant spatial clustering of leukemia in San Francisco. *J. Theor. Biol.* **1999**, *199*, 371–381.

24. Kendal, W.S.; Frost, P. Experimental metastasis: A novel application of the variance-to-mean power function. *J. Natl. Cancer Inst.* **1987**, *79*, 1113–1115.

25. Rhodes, C.J.; Anderson, R.M. A scaling analysis of measles epidemics in a small population. *Philos. Trans. R. Soc. Lond. B Biol. Sci.* **1996**, *351*, 1679–1688.

26. Kendal, W.S. A stochastic model for the self-similar heterogeneity of regional organ blood flow. *Proc. Natl. Acad. Sci. USA* **2001**, *98*, 837–841.

27. Kendal, W.S. An exponential dispersion model for the distribution of human single nucleotide polymorphisms. *Mol. Biol. Evol.* **2003**, *20*, 579–590.

28. Kendal, W.S. A scale invariant clustering of genes on human chromosome 7. *BMC Evol. Biol.* **2004**, *4*, 3.

29. Lowen, S.B.; Teich, M.C. The periodogram and Allan variance reveal fractal exponents greater than unitiy in auditory-nerve spike trains. *J. Acoust. Soc. Am.* **1996**, *99*, 3585–3591.

30. Sato, A.H.; Nishimura, M.; Holyst, J.A. Fluctuation scaling of quotation activities in the foreign exchange market.*Phys. A* **2010**, *389*, 2793–2804.

31. Eisler, Z.; Bartos, I.; Kertesz, J. Fluctuation scaling in complex systems: Taylor's law and beyond. *Adv. Phys.* **2008**, *57*, 89–142.

32. Fronczak, A.; Fronczak, P. Origins of Taylor's power law for fluctuation scaling in complex systems. *Phys. Rev. E* **2010**,*81*, 066112.

33. Jørgensen, B.; Vinogradov, V. Convergence to tweedie models and related topics. In *Advances on Theoretical and Methodological Aspects of Probability and Statistics*; Balakrishnan, N., Ed.; Taylor & Francis: New York, NY, USA, 2002; pp. 473–489.

34. Vinogradov, V.; Jørgensen, B.; Wentzell, B.D. From extreme stable laws to Tweedie exponential dispersion models. In*Stochastic Models*, Proceedings of the International Conference on Stochastic Models, Ottawa, ON, Canada, 10–13 June 1998; Dawson, D.A., Gorostiva, L., Ivanoff, G., Eds.; American Mathematical Society: Providence, RI, USA, 1999; pp. 435–443.

35. Kendal, W.S.; Jørgensen, B. Tweedie convergence: A mathematical basis for Taylor's power law, $1/f$ noise and multifractality. *Phys. Rev. E* **2011**, *84*, 066120.

36. Kendal, W.S.; Jørgensen, B. Taylor's power law and fluctuation scaling explained by a central-limit-like convergence.*Phys. Rev. E* **2011**, *83*, 066115.

37. Kendal, W.S. Fluctuation scaling and $1/f$ noise: Shared origins from the Tweedie family of statistical distributions. *J. Basic Appl. Phys.* **2013**, *2*, 40–49.

38. Kendal, W.S. Multifractality attributed to dual central limit-like convergence effects. *Phys. A* **2014**, *401*, 22–33.

39. Dutta, P.; Horn, P.M. Low-frequency fluctuations in solids: $1/f$ noise. *Rev. Mod. Phys.* **1981**, *53*, 497–516.

40. Musha, T. 1/f Fluctuations in Biological Systems. In Proceedings of

the Sixth International Conference on Noise in Physical Systems, Washington, DC, USA, 6–10 April 1981; pp. 143–146.

41. Bak, P.; Tang, C.; Wiesenfeld, K. Self-organized criticality: An explanation of the $1/f$ noise. *Phys. Rev. Lett.* **1987**, *59*, 381–384.

42. Kendal, W.S. Scale invariant correlations between genes and SNPs on human chromosome 1 reveal potential evolutionary mechanisms. *J. Theor. Biol.* **2007**, *245*, 329–340.

43. Cattani, C. Fractal patterns in prime numbers distribution. *Lect. Notes Comput. Sci.* **2010**, *6017*, 164–176.

44. Van Zyl, B.P.; Hutchinson, D.A.W. Riemann zeros, prime numbers, and fractal potentials. *Phys. Rev. E* **2003**, *67*, 066211.

45. Holdom, B. Scale-invariant correlations and the distribution of prime numbers. *J. Phys. A Math. Theor.* **2009**, *42*, 345102.

46. Shlesinger, M.F. On the Riemann hypothesis: A fractal random walk approach. *Phys. A* **1986**, *138*, 310–319.

47. Stanley, H.E.; Meakin, P. Multifractal phenomena in physics and chemistry. *Nature* **1988**, *335*, 405–409.

48. Erdös, P.; Kac, M. The Gaussian law of errors in the theory of additive number theoretic functions. *Am. J. Math.* **1940**, *62*, 738–742.

Chapter 6

ASYMMETRIC WAVE PROPAGATION THROUGH SATURABLE NONLINEAR OLIGOMERS

Daniel Law[1], Jennie D'Ambroise[1], Panayotis G. Kevrekidis[2] and Detlef Kip[3]

[1] Department of Mathematics and Statistics, Amherst College, Amherst, MA 01002, USA

[2] Department of Mathematics and Statistics, University of Massachusetts, Amherst, MA 01003, USA

[3] Faculty of Electrical Engineering, Helmut Schmidt University, Hamburg 22043, Germany

ABSTRACT

In the present paper we consider nonlinear dimers and trimers (more generally, oligomers) embedded within a linear Schrödinger lattice where the nonlinear sites are of saturable type. We examine the stationary states of such chains in the form of plane waves, and analytically compute their reflection and transmission coefficients through the nonlinear oligomer, as well as the corresponding rectification factors which clearly illustrate the asymmetry between left and right propagation in such systems. We examine not only the existence but also the dynamical stability of the plane wave states. Lastly, we generalize our numerical considerations to the more physically relevant case of Gaussian initial wavepackets and confirm that the asymmetry in the transmission properties also persists in the case of such wavepackets.

INTRODUCTION

In the last two decades, the subject of nonlinear dynamical lattices has gained considerable attraction and interest due to its emergence and relevance to a wide range of diverse applications. These include, among others, arrays of nonlinear-optical waveguides [1], Bose-Einstein condensates (BECs) in periodic potentials [2], micromechanical cantilever arrays [3], Josephson-junction ladders [4], granular crystals of beads interacting through Hertzian contacts [5], layered antiferromagnetic crystals [6], halide-bridged transition metal complexes [7], and dynamical models of the DNA double strand [8]. On the other hand, a specific phenomenon that has been intensely explored in a wide variety of recent studies is that of potentially asymmetric (i.e., non-reciprocal) wave propagation. This has been examined e.g., in asymmetric phonon transmission through a nonlinear interface layer between two very dissimilar crystals [9]. Another example is the proposed thermal diode [10] which, in turn, led to its experimental realization [11]. The optical diode was theoretically suggested in [12] (see also [13] for a setup of unidirectional transmission in photonics crystals) and experimentally achieved in [14]. Such rectification effects have also been proposed in left-handed metamaterials [15], granular crystals [16] and systems with gain-loss bearing so-called PT -symmetry [17,18], among others. In the present work, we focus on an, arguably simpler, implementation of the diode effect, which is fundamentally due to nonlinearity, as has been presented recently in the work of [19]. There, a linear chain was considered with a pair (or more) of nonlinear sites between the two linear ends of the chain. The nonlinear nature of the dynamics, coupled to a potential asymmetry between the characteristics of the nonlinear sites, was at the heart of the asymmetric propagation observed. The nonlinear sites were modeled as a prototypical system that has arisen in numerous applications, either as a direct model of relevance or as an envelope approximation in the form of the so-called discrete nonlinear Schrödinger (DNLS) equation [20]. While structurally simple, this model incorporates the fundamental characteristics of such lattices, namely diffraction (i.e., a discrete analogue of dispersion) and nonlinearity. The scenario of [19] explores the standard cubic nonlinearity associated with the Kerr effect [1]. However, often in applications, other types of nonlinearities are important as well. For instance, defocusing lithium niobate waveguide arrays exhibit a different type of nonlinearity, namely a saturable, defocusing one due to the photovoltaic effect [21]. In the latter context, dark solitons have been identified not only in regular homogeneous lattices, but also in higher gaps [22], as parts of multi-component soliton complexes (such as dark-bright solitary waves) [23], and even in heterogeneous chains with alternating couplings [24].

In the present work, we combine the experimentally accessible form of the saturable nonlinearity with the asymmetric propagation nonlinear phenomenology of [19]. The model setup is presented in Section 2. We consider in Section 3 exact plane wave solutions by solving the linear parts of the chain and gluing them through the nonlinear saturable "defect" sites. In this way, we identify the asymmetry between left and right transmittivities, due to the nonlinear propagation through the asymmetric dimer, trimer or more generally oligomer (i.e., few site) configuration. We explore the stability of these configurations and generically identify them as unstable. The dynamical evolution of the corresponding instabilities is explored through direct numerical simulations. Finally, more realistic (for experimental purposes) wavepackets of a Gaussian form are also considered in Section 4 and the manifestation of the asymmetry in propagation under such initial data is systematically quantified. Section 5 summarizes our findings and presents our conclusions including some possible directions of future work.

THE MODEL

Motivated by the above application of lithium niobate waveguide arrays, we consider a nonlinear Schrödinger type chain with governing equation

$$i\dot{\phi}_n(t) + \phi_{n+1}(t) + \phi_{n-1}(t) = \frac{\gamma_n \phi_n(t)}{1 + |\phi_n(t)|^2}$$

(1)

for $\gamma n \in R$ and $\varphi n \in C$. Here t plays the role of the (spatial) evolution variable. The saturable nonlinearity term is present in a finite region in the middle of the chain. That is, $\gamma n \neq 0$ only for $1 \leq n \leq N$. The wave propagates freely i.e., linearly outside of the finite region containing the nonlinearity. The system is Hamiltonian [25] with

$$\mathcal{H} = \sum_n \left[\left(\phi_n^* \phi_{n+1} + \phi_n \phi_{n+1}^* \right) - \gamma_n \ln(1 + |\phi_n|^2) \right]$$

(2)

We will examine the transmission properties of stationary solutions that take the form of plane waves on the linear portions of the lattice. We also explore the linear stability analysis for these solutions and for a number of dynamically unstable scenarios, we evolve the corresponding solutions through direct numerical simulations over the propagation parameter t. Finally, we examine the propagation of more physically applicable localized Gaussian wave packets and summarize the corresponding transmission properties in connection to the corresponding (potentially observable experimentally) asymmetries.

STATIONARY SOLUTIONS

Plane Waves

We seek standing wave solutions by setting $\phi_n = \psi_n e^{-i\omega t}$ for $\omega \in \mathbb{R}$. This gives a set of algebraic equations which can be written in the form of a backwards transfer map

$$\psi_{n-1} = -\psi_{n+1} + \left(-\omega + \frac{\gamma_n \psi_n}{1 + |\psi_n|^2}\right)\psi_n$$

(3)

for $\psi_n \in \mathbb{C}$ independent of t. Following the procedure in [18] and for now assuming k≥0, we begin by assuming solutions in the form of plane waves on the linear portions of the chain. That is,

$$\psi_n = \begin{cases} R_0 e^{ikn} + R e^{-ikn} & n \le 1 \\ T e^{ikn} & n \ge N \end{cases}$$

(4)

with R0,R,T∈C representing the incident, reflected and transmitted amplitudes, respectively. The solution Equation (4) solves Equation (3) for n∉{1,···,N} only if the wavenumber k satisfies ω=−2cos(k). Also directly from Equation (4) with n=0,1 we have

$$R_0 = \frac{e^{-ik}\psi_0 - \psi_1}{e^{-ik} - e^{ik}} \quad \text{and} \quad R = \frac{e^{ik}\psi_0 - \psi_1}{e^{ik} - e^{-ik}}$$

(5)

The stationary solution across the whole lattice is then known by the following procedure: given values for γn for n∈{1,···,N} we start by specifying values for k and T. Then we compute ψ0,ψ1,···ψN−1 via Equation (3) and then R,R0by Equation (5). Such a procedure of finding the input as a function of the output is referred to as a "fixed output problem" [26].

Stationary solutions where the amplitude R0 is incident from the right-hand-side and the wavenumber is taken as −k≤0can also be formulated in a similar way. In order to avoid swapping the format of Equation (4) (so that R0,R would apply on the right and T on left), it is more convenient to leave Equation (4) as-is with positive wave number k and instead flip left-to-right the configuration of the nonlinearities, i.e., {γ1,γ2,···,γN}→{γN,γN−1,···,γ$_1$}. In this way the computation of the solution for negative wavenumber is unchanged from the above outline aside from the swap of the order of the γ's. Plots of these plane wave stationary solutions are shown in the next section, where we also address the stability of their t-propagation. In practice we truncate the lattice and refer to its finite length as L.

For a solution that has been determined by the processes described above, we next compute the transmission coefficient $\tau \overset{def.}{=} |T|^2/|R_0|^2$ explicitly assuming that T is given. For this purpose it is convenient to write $\psi_n \overset{def.}{=} T e^{ikN} \Psi_n$ with n=N−lso that l=0 corresponds to n=N, and incrementing l corresponds to decreasing n. In this notation we have Ψ_N=1 for l=0 and the value of Ψ_n for each subsequent node towards the left is given by rewriting Equation (3) as

$$\Psi_{N-l} = -\Psi_{N-l+2} + \delta_{N-l+1}\Psi_{N-l+1}$$

(6)

for $\delta_j \overset{def.}{=} -\omega + \frac{\gamma_j}{1+|T|^2|\Psi_j|^2}$. See the Appendix where we record a few iterations of Equation (6). Then by Equation (5) with ψ_0, ψ_1 computed according to Equation (6) we have

$$R_0 = \frac{T}{e^{-ik}-e^{ik}}(\delta_1 - 2e^{ik})) \qquad \tau = \left|\frac{e^{ik}-e^{-ik}}{2e^{ik}-\delta_1}\right|^2 \qquad \text{for } N = 1$$

$$R_0 = \frac{Te^{ik}}{e^{-ik}-e^{ik}}(-1 + (\delta_1 - e^{ik})(\delta_2 - e^{ik})) \qquad \tau = \left|\frac{e^{ik}-e^{-ik}}{1+(\delta_1-e^{ik})(e^{ik}-\delta_2)}\right|^2 \qquad \text{for } N = 2$$

(7)

and

$$R_0 = \frac{Te^{2ik}}{e^{-ik}-e^{ik}}\left(\delta_1 - e^{ik} + (\delta_3 - e^{ik})(1 - \delta_2(\delta_1 - e^{ik}))\right)$$

$$\tau = \left|\frac{e^{ik}-e^{-ik}}{e^{ik}-\delta_1+(e^{ik}-\delta_3)(1-\delta_2(\delta_1-e^{ik}))}\right|^2 \qquad \text{for } N = 3$$

(8)

In the linear case ($\gamma_1 = \gamma_2 = 0$) and in the symmetric case ($\{\gamma_1, \gamma_2, \cdots, \gamma N\} = \{\gamma N, \gamma N-1, \cdots, \gamma_1\}$ as an ordered set) it is immediately seen that τ is the same for waves incoming from the left or right side. For N=1 the transmission is always symmetric.

We also define here a quantity to measure the asymmetric propagation. We will use the definition of a rectification factor f in the form of

$$f = \frac{\tau(k,T) - \tau_{flip}(k,T)}{\tau(k,T) + \tau_{flip}(k,T)}$$

(9)

where the quantity $\tau(k,T)$ corresponds to transmission of a left-incoming wave with positive wavenumber k≥0 and $\tau_{flip}(k,T)$ with k≥0 is equivalent to the transmission of a right-incoming wave with negative wavenumber (recall the process described above of keeping k positive while flipping the order of the γ's). This way nonzero values of f in the range [−1,1] measure the asymmetry of transmission in the system. Symmetry in transmission corresponds to f=0 and f>0 corresponds to greater transmission of incident waves originating from the left (transmitted on the right) as compared with incident waves originating from the right (transmitted on the left). Of course, f<0 corresponds to greater transmission of waves originating from the right.

Figure 1 shows plots of the transmission coefficient τ and the rectification factor f as a function of the amplitude T and the wavenumber k of the extended plane wave solutions. We find that whether more is transmitted for waves incoming from the right or left is variable as a function of k and T. Notice that values for γ's are chosen in Figure 1 to be such that $\gamma_1 < \gamma_2$ in the N=2 case and $\gamma_1 < \gamma_2 < \gamma_3$ in the N=3 case. In other words, with increasing γ's from left to right we observe that transmission properties vary with the choice of the parameters T,k. We find that in accordance with our above analysis the N=1 case is symmetric. Although we do not show an N=1 analogue of Figure 1, such plots look similar to Figure 1 but there is exact symmetry and f=0 for all T,k. It is interesting to point out here that the rectification factor appears to acquire its largest (absolute) values for k close to π i.e., at the edge of the Brillouin zone. Furthermore, both in the N=2 and in the N=3 case, the dependence of f near this value appears to be a non-sign-definite function of T (i.e., different ranges of T values appear to favor propagation in one or the other direction).

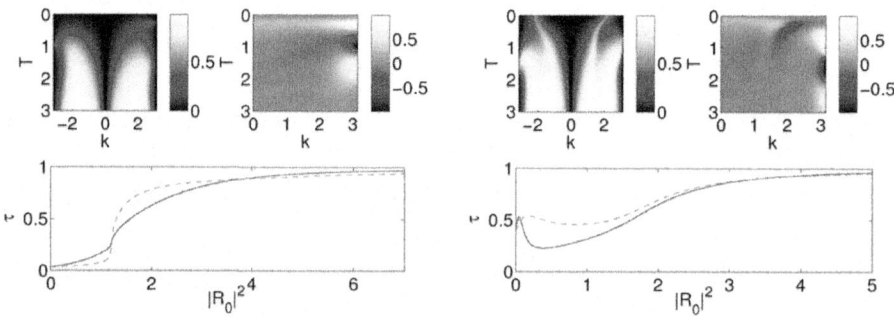

Figure 1: Each panel contains a contour plot of the transmission coefficient $\tau(k,T)$ (**top left**) and a contour plot of the rectification factor f(k,T) of Equation (9) (**top right**), plotted as a function of k and T. The panels also show a typical example of the dependence of τ for k=2 (solid lines) and k=−2 (dashed lines), so as to illustrate the asymmetry between the propagation for left and right incident waves (**bottom panel**). In the latter the dependence of τ is given as a function of $|R0|^2$. The left panel corresponds to N=2 and γ_1=2.5, γ_2=5 and the lattice size is L=200 in this case. The right panel is the N=3 case with γ_1=1.5, γ_2=2 and γ_3=3.5; here the lattice size is L=201.

Stability

In order to analyze the spectral stability of stationary states of the form discussed in the previous subsection we write

$$\phi_n(t) = e^{-i\omega t}\left(\psi_n + \varepsilon\left(a_n e^{i\nu t} + b_n e^{-i\nu^* t}\right)\right)$$

$$(10)$$

for $a_n, b_n, \nu \in \mathbb{C}$, ε_ε small, and with ψ_n being a stationary solution from the previous section. The resulting linear stability equations then read

$$\nu \begin{pmatrix} a_n \\ b_n^* \end{pmatrix} = \begin{pmatrix} F_1 & F_2 \\ F_3 & F_4 \end{pmatrix} \begin{pmatrix} a_n \\ b_n^* \end{pmatrix}$$

(11)

for

$$F_1 = diag\left(\omega - \frac{\gamma_n}{(1+|\psi_n|^2)^2}\right) + G, \qquad F_2 = diag\left(\frac{\gamma_n \psi_n^2}{(1+|\psi_n|^2)^2}\right)$$

$$F_3 = diag\left(\frac{-\gamma_n \psi_n^2}{(1+|\psi_n|^2)^2}\right), \qquad F_4 = diag\left(-\omega + \frac{\gamma_n}{(1+|\psi_n|^2)^2}\right) - G$$

(12)

where G is a sparse matrix with ones on both the super- and sub-diagonals. Given a stationary plane wave solution ψ_n and values of γ_n which encode the nonlinearity for $1 \le n \le N$, one then calculates the eigenvalues ν in Equation (11). If ν has a negative imaginary part this indicates that the perturbed solution $\varphi_n(t)$ is unstable, as is easily seen by Equation (10). In practice, one diagonalizes a finite truncation of the matrix in Equation (11), ensuring that the relevant eigenvalues are not affected by the truncation error. In other words, F_1, F_2, F_3, F_4 and G are all L×L matrices and in the matrix Equation (11) it is now convenient to think of an and bn as length L column vectors. Furthermore, the Hamiltonian symmetry of the solution ensures that the relevant instability eigenvalues come either in pairs (if ν is imaginary) or in quartets (if ν is genuinely complex).

In Figure 2 and Figure 3 we show a plot of min(Im(ν)) as a function of T and γ. We find that an increase in the magnitude of a γi parameter (with other nearby γ's held fixed) leads to min(Im(ν)) of larger magnitude indicating greater instability. Figure, Figure 4, Figure 5 and Figure 6 show eigenvector and eigenvalue plots alongside snapshots of $\varphi n(t)$ to show the behaviour of typical propagation in the t variable of the unstable plane waves. The boundary conditions are calculated according to Equation (4) at t=0 and evolved by multiplying by $e^{-i\omega t}$ for t>0 so as to conform with Equation (10). The unstable plane wave solution, when propagated in the evolution variable, exhibits a few effects: if k>0 then amplitude leaks over to the right-hand side (to the left if k<0), and due to the localized instability a peak appears in the center of the lattice. Of course, given the conservation laws of the system, the power $\sum_n |\phi_n(t)|^2$ and the Hamiltonian H(t) are preserved over t. The figures also show a transition in the eigenvalue plots for unstable solutions. A weak instability (corresponding to dim but nonzero regions of the min(Im(ν)) plots of Figure 2 and Figure 3) results in eigenvalue plots in the complex plane where a quartet appears off the real axis; see Figure 2 and Figure 5. As the instability is enhanced for larger values of γ (comparably brighter regions of the min(Im(v)) plots), the two pairs constituting the quartet merge on the imaginary axis and subsequently split with one pair headed towards zero; see Figure 4 and Figure 5. For the highest magnitude of instability (brightest regions on the plots of min(Im(v))) the eigenvalues indicating the instability are in the form of a pair on the imaginary axis; see Figure 4 and Figure 6. In the examples shown, the instability generically appears to transport power to the right part of the lattice, deforming (decreasing the power of) the corresponding n<0 portion of the plane wave. On the other hand, critically (per the localized eigenvector of the instability), a localized mode appears to form at the central nonlinear nodes within the domain.

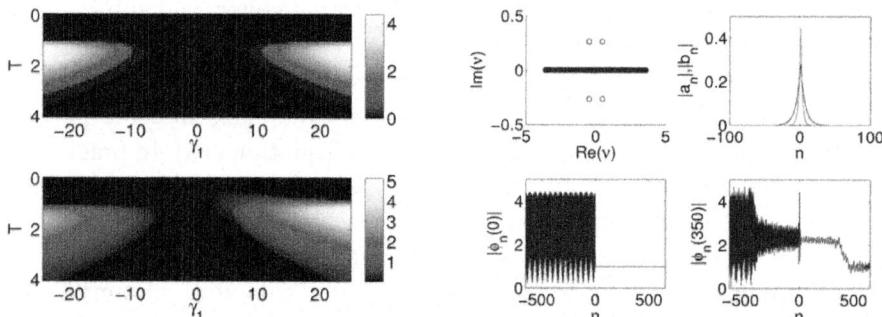

Figure 2: The left panel depicts the value of |min(Im(v))|, for N=1, plotted as a function of γ_1 and T; the top graph shows k=π/2 and in the bottom graph k=2.5. The lattice length in the two left panel plots is L=99. These two plots show that the magnitude of the minimum imaginary part of the calculated eigenvalue, i.e., the strength of the instability, increases as the magnitude of γ_1 increases. On the other hand for a fixed γ_1 value an extended solution of the form shown in Equation (4) is tending toward stabilization for large T. In the right panel we show four plots that correspond to a dim but nonzero region on the plot of |min(Im(v)| in the left panel. That is, the right-hand four plots correspond to k=2.5, γ_1=5 and T=1. The four plots show the eigenvalues in the complex plane (top left), eigenvector magnitude (top right with |an| blue and |bn| green), initial profile of the plane wave at t=0 (bottom left) and a later profile at t=350 of the plane wave (bottom right). For this instability the eigenvalues are in the form of a quartet.

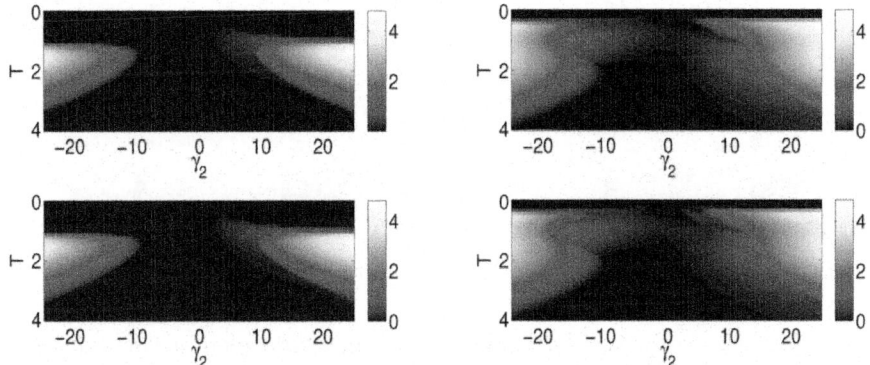

Figure 3: The plots are similar to the left panel in Figure 2. Here the left panel corresponds to N=2, L=100 and we plot |min(Im(v)| as a function of γ2 and Twhile the value of γ1 is fixed: γ_1=1 in the top graph and γ_1=−1 in the bottom graph. Here the right panel corresponds to N=3, L=101 and we plot |min (Im(v)| as a function of γ_2 and T while the values of γ_1 and γ_3 are fixed: γ_1=1, γ_3=5 in the top graph and γ_1=−1,γ_3=5 in the bottom graph.

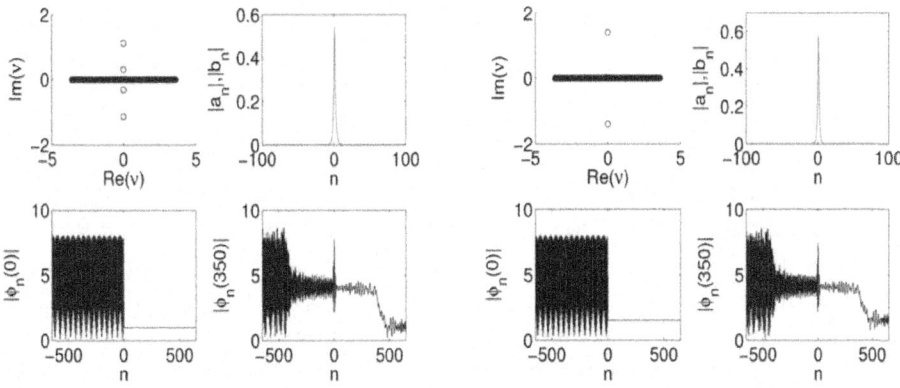

Figure 4: Here we focus on parameter values that correspond to a bright region in the left panel of Figure 2. We show four plots similar to the right panel of Figure 2. Here we have N=1, L=1299 and k=2.5. The left panel of four plots corresponds to γ_1=9.5, T=1, and the right panel corresponds to γ_1=10, T=1.5. Comparing the three sets of four plots in the present figure and in Figure 2 shows the transition in the eigenvalue plots as we move towards brighter regions of the |min(Im(v)| diagram.

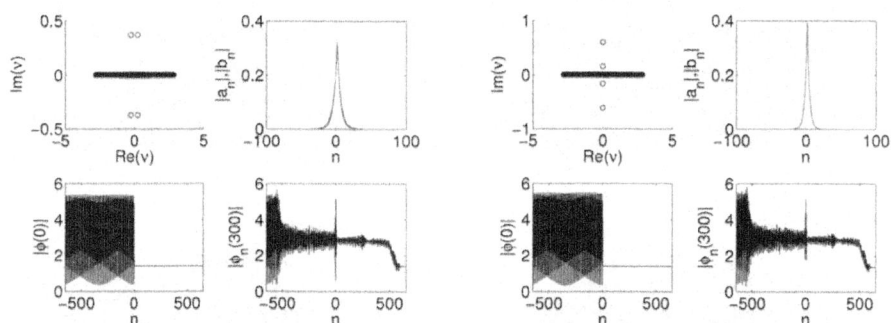

Figure 5: Here we focus on parameter values that correspond to the left panel of Figure 3. Again we show four plots similar to the right panel of Figure 2. Here we have N=2,L=1300 and k=2. The left panel of four plots corresponds to $\gamma_1=-1, \gamma_2=9.25$, T=1.4, and the right panel corresponds to $\gamma_1=-1, \gamma2=9.5$, T=1.4. Comparing these two sets of four plots shows the transition in the eigenvalue plots as we move towards brighter regions of the appropriate |min(Im(v))| diagram in Figure 3.

In comparing our results with those of [19], we find that the asymmetry associated with the saturable nonlinearity (presented here) is less pronounced than that of a system with a cubic nonlinearity and a linear potential term (presented in [19]). In the t propagation of extended solutions we can also compare the top plot in Figure 3 of [19] with our Figure 7 in which we show plots over space and the propagation parameter. The two systems both experience a concentration of amplitude at the center of the lattice as t moves forward. In the case of the cubic nonlinearity in [19] there are three concentrations of amplitude (two of which are moving). Here we see predominantly a concentration of amplitude at the center, while the large amplitude sites nearby decrease in amplitude in comparison to their respective values at the initialized state at t = 0. Also, in contrast to [19] where the amplitude concentrations more dramatically rise above the background, here the central concentration of amplitude is more similar to the maximal amplitude of the initialized state.

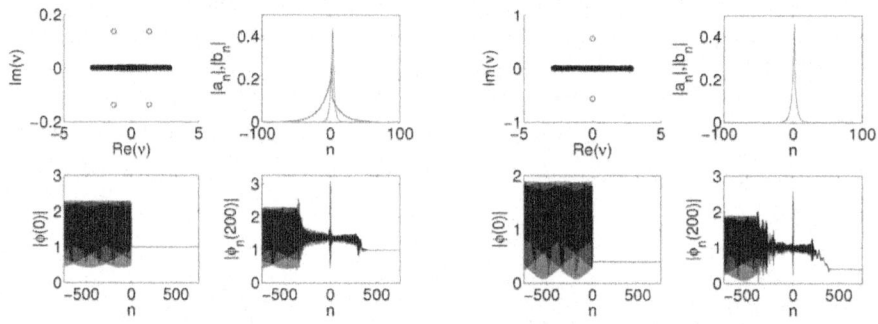

Figure 6: Here we focus on parameter values that correspond to the right pan-

el ofFigure 3. Again we show four plots similar to the right panel of Figure 2. Here we haveN=3, L=1301 and k=2. The left panel of four plots corresponds to γ_1=1,γ_2=3.25,γ_3=4, T=1, and the right panel corresponds to γ_1=1,γ_2=5.5,γ_3=4, T=0.4. Comparing these two sets of four plots shows the transition in the eigenvalue plots as we move towards brighter regions of the appropriate |min(Im(v))| diagram in Figure 3.

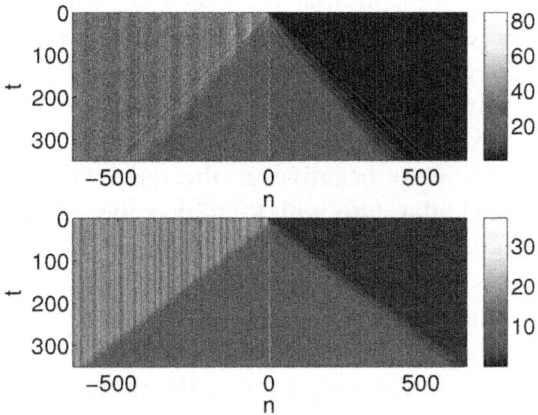

Figure 7: The plots show $||\varphi n(t)|^2$ as a function of n and t. The top plot corresponds to parameters the same as in the right four plots in Figure 4. The bottom plot corresponds to parameters the same as in the left four plots in Figure 5.

PROPAGATION OF A GAUSSIAN

Finally, in this section we look at the propagation of a Gaussian wavepacket through the lattice for each of the cases N=1,2,3. While it is less straightforward to prepare the delocalized initial conditions needed for the plane wave solutions of the previous section (whose asymmetric propagation, however, can be analytically quantified), preparing the Gaussian initial data of the present section appears to be considerably more tractable in optical experiments e.g., with lithium niobate waveguide arrays. On the other hand, in this latter setting, we will have to rely on detailed numerical computations of the rectification factor, as this set of initial conditions is less amenable to detailed analytical considerations. The wavepacket considered is given by the equation

$$\phi_n(0) = I e^{-(n-n_0)^2/s^2}$$

(13)

for starting position index n_0 and width parameter s. We measure the transmission at some value t=t0 sufficiently large so that the wavepacket has interacted with the nonlinear region and, as a result, some portion of it has been accordingly transmitted through and reflected from the relevant interval. The transmission is then measured by

$$\tau_+ = \frac{\sum_{n>N} |\phi_n(t_0)|^2}{\sum_n |\phi_n(t_0)|^2} \qquad \tau_- = \frac{\sum_{n<1} |\phi_n(t_0)|^2}{\sum_n |\phi_n(t_0)|^2}$$

(14)

for k>0 and k<0, respectively. Then the rectification factor takes the form f=$(\tau_+-\tau_-)/(\tau_++\tau_-)$. The transmission is, of course, equal in both directions (f=0) in the N=1 case. In Figure 8 we plot f as a function of γ_1,γ_2 in the case of N=2, and as a function of γ_1,γ_3 with γ_2 fixed in the case of N=3. In Figure 8 we also show some typical space-t propagation plots for the Gaussian initial data case. Similar to the plane wave solutions case, the rectification factor may be positive or negative as the parameters change. Here the Gaussian in this particular case with k=$\pi/2$ has the following property. For parameter values concentrated in the vertical and horizontal bands in the right-side four plots of Figure 8 more is transmitted if the wave hits the lower γ-value first, in comparison to the wave hitting the larger γ-value first, i.e., encountering the region which is closer to linear is more conducive towards transmission, while encountering the more nonlinear sites at first is more prone to reflection, a feature that seems to be intuitively justified.

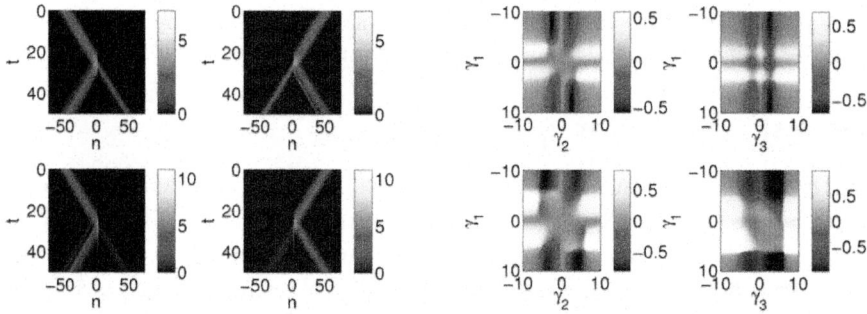

Figure 8: In the left four graphs, we plot $||\varphi n(t)|^2$ as a function of the lattice n and the propagation variable t for initial profile according to Equation (13) with $|I|2$=2.5. The left columns have wavenumber k=$\pi/2$ and starting position n0=−50 and the right columns have wavenumber k=−$\pi/2$ and starting position n0=50. The top two plots correspond to N=2 with γ_1=2.75 and γ_2=5; we calculate that the rectification factor in this case is f≈−0.1767. The bottom two plots correspond to N=3 with γ_1=3, γ_2=3.25, γ_3=4; we calculate that f≈−0.1525. In the right four graphs, we plot the rectification factor of the Gaussian waves for N=2 (left column) and N=3(right column) as a function of variable γ parameters. The top row corresponds to $|I|^2$=1 in Equation (13) and the bottom row to $|I|^2$=2.5. For the N=3 plots, the value γ_2=5 (top) and γ_2=3.25 (bottom) is fixed.

CONCLUSIONS

In the present work, we considered a lattice setting where embedded in a linear Schrödinger chain was a nonlinear "segment" of the saturable type. Our analytical considerations were focused around plane waves enabling us to analytically compute both the transmittivity and the rectification factor between left- and right-propagating such waves. These features evidence the asymmetric nature of the propagation in a way that is analytically tractable. This asymmetry can be explicitly traced in the nonlinearity of the relevant setup. We also considered the spectral stability of such states in which we observe some effects of low versus high rates of instability. Finally, we considered the asymmetry of propagation of a Gaussian wavepacket. The latter was also clearly evidenced both in the case of a dimer, as well as in that of a trimer, paving the way for the experimental observation of relevant phenomena, such as the enhanced transmission of a wavepacket when encountering a region of increasing, rather than that of a decreasing nonlinear index profile.

There are numerous aspects that may be worthwhile to further explore. In the context of lithium niobate waveguide arrays, it may be relevant to examine settings that involve a genuinely nonlinear lattice but with its central sites bearing a different nonlinearity than the background. This type of "spatial profile" of the nonlinearity coefficient has attracted considerable interest in numerous recent studies as evidenced by the review of [27] and may also be quite experimentally tractable. On the other hand, the vast majority of the present studies on the nonlinearity-induced asymmetry that we are aware of have focused chiefly on one-dimensional configurations. However, it would be both more numerically challenging and also theoretically intriguing to explore scattering of two-dimensional wavepackets from a central (two-dimensional) segment of a lattice which is genuinely nonlinear. Such aspects are currently under investigation and will be reported in future publications.

Finally, we should note that after submission of this manuscript, we were notified of a related work in [28]. While the models analyzed in these works are fairly similar, distinctive features of the present work are (a) that we explored systematically the stability of the extended waves and we studied the outcomes of their dynamical instabilities; and (b) we also considered the effect of the incidence of a Gaussian wave packet. Lastly, (c) although the latter work is restricted to dimers, our considerations here have been provided for general N and examined for N=1,2,3.

ACKNOWLEDGMENTS

The authors acknowledge useful discussions with Stefano Lepri on the subject.

P.G.K acknowledges support from the National Science Foundation under grants CMMI-1000337, DMS-1312856, from ERC and FP7-People under the grant IRSES-606096 and from the US-AFOSR under grant FA9550-12-10332, as well as from the Binational Science Foundation under grant 2010239.

AUTHORS CONTRIBUTION

P.G.K. and D.K conceived of the general research direction. P.G.K. and J.D.A. designed the research plan. D.L. and J.D.A. obtained the results. All authors contributed to the writing of the manuscript.

CONFLICTS OF INTEREST

The authors declare no conflict of interest.

A. APPENDIX

We show a few iterations of the backwards transfer map in Equation (6).

$$
\begin{aligned}
\delta_N &= -\omega + \frac{\gamma_j}{1 + |T|^2} \\
\Psi_{N-1} &= -e^{ik} + \delta_N \\
\delta_{N-1} &= -\omega + \frac{\gamma_{N-1}}{1 + |T|^2|\delta_N - e^{ik}|^2} \\
\Psi_{N-2} &= -1 + \delta_{N-1}(\delta_N - e^{ik}) \\
\delta_{N-2} &= -\omega + \frac{\gamma_{N-2}}{1 + |T|^2|1 + \delta_{N-1}(e^{ik} - \delta_N)|^2} \\
\Psi_{N-3} &= -\delta_{N-2} + (e^{ik} - \delta_N)(1 - \delta_{N-2}\delta_{N-1}) \\
\delta_{N-3} &= -\omega + \frac{\gamma_{N-3}}{1 + |T|^2|\delta_{N-2} + (\delta_N - e^{ik})(1 - \delta_{N-2}\delta_{N-1})|^2} \\
\Psi_{N-4} &= 1 - \delta_{N-3}\delta_{N-2} + (e^{ik} - \delta_N)(\delta_{N-1} + \delta_{N-3}(1 - \delta_{N-2}\delta_{N-1}))
\end{aligned}
\tag{15}
$$

REFERENCES

1. Lederer, F.; Stegeman, G.I.; Christodoulides, D.N.; Assanto, G.; Segev, M.; Silberberg, Y. Discrete solitons in optics.Phys. Rep. **2008**, 463, 1–126.

2. Morsch, O.; Oberthaler, M. Dynamics of Bose-Einstein condensates in optical lattices. Rev. Mod. Phys. **2006**, 78, 179.

3. Sato, M.; Hubbard, B.E.; Sievers, A.J. Colloquium: Nonlinear energy localization and its manipulation in micromechanical oscillator arrays.

Rev. Mod. Phys. **2006**, 78, 137.

4. Binder, P.; Abraimov, D.; Ustinov, A.V.; Flach, S.; Zolotaryuk, Y. Observation of Breathers in Josephson Ladders. Phys. Rev. Lett. **2000**, 84, 745.

5. Boechler, N.; Theocharis, G.; Job, S.; Kevrekidis, P.G.; Porter, M.A.; Daraio, C. Discrete Breathers in One-Dimensional Diatomic Granular Crystals. Phys. Rev. Lett. **2010**, 104, 244302.

6. English, L.Q.; Sato, M.; Sievers, A.J. Modulational instability of nonlinear spin waves in easy-axis antiferromagnetic chains. II. Influence of sample shape on intrinsic localized modes and dynamic spin defects. Phys. Rev. B **2003**, 67, 024403.

7. Swanson, B.I.; Brozik, J.A.; Love, S.P.; Strouse, G.F.; Shreve, A.P.; Bishop, A.R.; Wang, W.-Z.; Salkola, M.I. Observation of Intrinsically Localized Modes in a Discrete Low-Dimensional Material. Phys. Rev. Lett. **1999**, 82, 3288.

8. Peyrard, M. Nonlinear dynamics and statistical physics of DNA. Nonlinearity. **2004**, 17, R1.

9. Kosevich, Y.A. Fluctuation subharmonic and multiharmonic phonon transmission and Kapitza conductance between crystals with very different vibrational spectra. Phys. Rev. B **1995**, 52, 1017.

10. Terraneo, M.; Peyrard, M.; Casati, G. Controlling the Energy Flow in Nonlinear Lattices: A Model for a Thermal Rectifier. Phys. Rev. Lett. **2002**, 88, 094302.

11. Chang, C.W.; Okawa, D.; Majumdar, A.; Zettl, A. Solid-state thermal rectifier. Science **2009**, 314, 1121.

12. Scalora, M.; Dowling, J.P.; Bowden, C.M.; Bloemer, M.J. The photonic band edge optical diode. J. Appl. Phys. **1994**, 76, 2023.

13. Konotop, V.V.; Kuzmiak, V. Nonreciprocal frequency doubler of electromagnetic waves based on a photonic crystal.Phys. Rev. B **2002**, 66, 235208.

14. Gallo, K.; Assanto, G.; Parameswaran, K.; Fejer, M. All-optical diode in a periodically poled lithium niobate waveguide. Appl. Phys. Lett. **2001**, 79, 314.

15. Feise, M.W.; Shadrivov, I.V.; Kivshar, Y.S. Bistable diode action in left-handed periodic structures. Phys. Rev. E **2005**,71, 037602.

16. Boechler, N.; Theocharis, G.; Daraio, C. Bifurcation-based acoustic switching and rectification. Nature Mater. **2011**, 10, 665.

17. Lin, Z.; Ramezani, H.; Eichelkraut, T.; Kottos, T.; Cao, H.; Christodoulides,

D.N. Unidirectional Invisibility Induced by PT-Symmetric Periodic Structures. Phys. Rev. Lett. **2011**, 106, 213901.

18. D'Ambroise, J.; Kevrekidis, P.G.; Lepri, S. Asymmetric wave propagation through nonlinear PT-symmetric oligomers.J. Phys. A: Math. Theor. **2012**, 45, 444012.

19. Lepri, S.; Casati, G. Asymmetric wave propagation in nonlinear systems. Phys. Rev. Lett. **2011**, 106, 164101.

20. Kevrekidis, P.G. The Discrete Nonlinear Schrödinger Equation; Springer-Verlag: Heidelberg, Germany, 2009.

21. Smirnov, E.; Rüter, C.E.; Stepić, M.; Kip, D.; Shandarov, V. Formation and light guiding properties of dark solitons in one-dimensional waveguide arrays. Phys. Rev. E **2006**, 74, 065601.

22. Dong, R.; Rüter, C.E.; Song, D.; Xu, J.; Kip, D. Formation of higher-band dark gap solitons in one dimensional waveguide arrays. Opt. Express **2010**, 18, 27493.

23. Dong, R.; Rüter, C.E.; Kip, D.; Cuevas, J.; Kevrekidis, P.G.; Song, D.; Xu, J. Dark-bright gap solitons in coupled-mode one-dimensional saturable waveguide arrays. Phys. Rev. A **2011**, 83, 063816.

24. Kanshu, A.; Rüter, C.E.; Kip, D.; Cuevas, J.; Kevrekidis, P.G. Dark lattice solitons in one-dimensional waveguide arrays with defocusing saturable nonlinearities and alternating couplings. Eur. Phys. J. D **2012**, 66, 182.

25. Samuelsen, M.R.; Khare, A.; Saxena, A.; Rasmussen, K.O. Statistical mechanics of a discrete Schrödinger equation with saturable nonlinearity. Phys. Rev. E **2013**, 87, 044901.

26. Knapp, R.; Papanicolaou, G.; White, B. Transmission of waves by a nonlinear random medium. J. Stat. Phys. **1991**, 63, 567.

27. Kartashov, Y.V.; Malomed, B.A.; Torner, L. Solitons in nonlinear lattices. Rev. Mod. Phys. **2011**, 83, 247. [

28. Assunçao, T.F.; Nascimento, E.M.; Lyra, M.L. Nonreciprocal transmission through a saturable nonlinear asymmetric dimer. Phys. Rev. E **2014**, 90, 022901.

Chapter 7

MATHEMATICAL ANALYSIS FOR RESPONSE SURFACE PARAMETER IDENTIFICATION OF MOTOR DYNAMICS IN ELECTRIC VEHICLE PROPULSION CONTROL

Richard A. Guinee[1]

[1]Department of Electrical and Electronic Engineering, Cork Institute of Technology, Cork, Ireland

INTRODUCTION

This chapter addresses the topographical examination of various mean squared error (MSE) cost surface structures and selecting the most suitable MSE fitness function for accurate brushless motor drive (BLMD) dynamical parameter system identification (SI) of BLMD shaft load inertia and viscous damping for electric vehicle controlled propulsion. The parameter extraction procedure employed here is in the offline mode for optimal drive tuning purposes during the installation and commissioning phase of embedded BLMD systems in high performance electric vehicle torque, speed and position control scenarios. Two types of penalty function, based on the transient step response of the permanent magnet (PM) motor shaft velocity and its stator winding current feedback in torque control mode [1,2], are examined here for arbitration of a suitable choice of cost objective function as the response surface in the accurate extraction of the BLMD dynamics. The choice of a particular MSE cost surface as an objective function in BLMD load parameter identification is motivated by the need for reliable tuning of the proportional and integral (PI) term settings during the drive installation phase for controller robustness and optimal performance in adjustable speed drive (ASD) or torque controlled embedded PM motor applications for electric vehicle propulsion. This chapter will focus on the mathematical analysis of embedded motor drive dynamical parameter identification over an MSE multiminima response surface with the following key results obtained:

- the development of a novel quadratic mathematical model approximation for the investigation of the (i) nature of the MSE objective function and (ii) existence of a bounded MSE global minimum stationary region, based on transient step response motor current feedback signals, for mechanical parameter identification in sensorless drive torque control of electric vehicles.

- the examination of the phenomenon of multiminima proliferation in the MSE cost formulation due to target data choice and 'noisiness' arising from evaluation of pulse width modulated (PWM) edge transition times during BLMD simulation [1,2].

- the measurement of cost surface selectivity based on shaft velocity and current feedback target data and the decision favouring the choice of the latter data training record as the target function for dynamical parameter identification

- the development of a novel parameter quantization metric to overcome cost surface 'noisiness', arising from computational uncertainty in the simulated PWM edge transitions [1], for avoidance of local minima trapping in the MSE cost surface during identification of the BLMD dynamics.

- the development of a novel parameter convergence radius measure of encirclement of the cost surface stationary region global minimum, arising from the parameter quantization metric in (d), for determination of the bounds of accuracy that can be imposed on the returned estimates of the global optimum dynamical parameter vector during BLMD identification.

Motivation

BLMD control tuning is necessary during the commissioning phase of embedded drives applications, for accurate torque and speed control in electric vehicle propulsion systems setup [3,4], accurate robotic end effector [5] or CNC tool positioning [6], where detailed apriori knowledge of expected drive load inertia and friction parameters are unknown to the electric motor drive supplier/ manufacturer in the intended application beforehand. The choice of ASD [7] in high performance industrial applications, such as a small electric vehicle [4], robot manipulator [8, 9, 10] or machine tool feed drive [6,4], is usually based on consideration of a BLMD manufacturer's catalogued specifications, relating to drive performance capabilities and limitations, by the customer or embedded drive equipment designer/manufacturer. The BLMD selection is often done independently of the motor drive manufacturer by the equipment designer for reasons of embedded systems design confidentiality and second sourcing of matching drive equivalents from different manufacturers for the purpose of cost reduction and embedded product protection from obsolescence via alternative drive substitution. The range of motor sizes available and spread of possible BLMD embedded applications has resulted in the provision of flexible drive tuning facilities with either manual or autotuning features [11] by motor drive manufacturers as a sales and marketing expedient to embedded equipment designers. This flexible approach to drive tuning policy eliminates the need for the BLMD manufacturer to participate in the detailed design of embedded drive applications except in the provision of motor drive systems with high output torque and speed ranges to cater for a range of anticipated high performance applications [12,8,13]. BLMD systems with high peak current capability and fast response times due to low PM rotor mass are designed [6, 7] to handle large inertial load torques [14] experienced in robotic applications [4, 5] and electric vehicle propulsion systems, with a no-load to full-load inertia variation [9] of 10 is to 1. It is in response to this background of applications diversity, regarding the particular design details of embedded drive products about the size of inertial loads and friction coefficients encountered [4,6,9], that the present work on cost surface analysis for parameter extraction in electric vehicle control is directed from a motor manufacturer's perspective.

Since the possible variation in the load dynamics of an intended BLMD application is unknown at the outset the initial task here for an end user is to identify the actual load inertia and friction coefficients experienced during startup of a given embedded drive in the offline mode for robust PI controller tuning [15]. In this scenario the customer has the flexibility of manually tuning the BLMD speed loop, which is provided as a PI adjustment option along with procedural details for tuning by the motor manufacturer, during the setup and commissioning phase for a particular ASD application. The challenge then posed for the motor drive designer in this instance is the provision of an automated tuning facility for the velocity or torque loop during the commissioning stage thus eliminating the need for any manual input by the customer. This feature requires the identification of a fixed embedded load configuration during setup and subsequent automated optimal configuration of the velocity controller PI terms [15]. In the absence of embedded load information the cost surfaces and identification methods investigated here focused on inertial load spreads for vehicular and robotic applications [9] of up to ten times the inherent motor shaft inertia as recommended by the BLMD manufacturer for the drive [16] modelled in [1].

The concept of a simulated cost surface is developed here [17] as an objective function to facilitate parameter extraction of the installed drive dynamics, during offline BLMD system identification, with MSE minimization. This methodology provides useful insight into the nature and formulation of the most suitable MSE objective function to be minimized, based on actual drive experimental test data available and BLMD model simulation, coupled with an effective system identification (SI) strategy for accurate motor parameter extraction [18]. This approach can also be used as an alternative means of providing the optimal set of extracted parameter estimates from inspection of the global minimum location on the simulated cost surface with embedded local minima. Furthermore it can be used as a basis for comparison of the effectiveness of the actual identification search strategy deployed in terms of the accuracy of returned parameter estimates. The problem of inertia (J) and friction (B) parameter extraction of an actual BLMD system over a sinc function shaped multiminima cost surface [19], based on step response feedback current (FC) target data which has a constant amplitude swept frequency characteristic, is investigated as a test case using response surface simulation.

The global minimum estimation, from response surface simulation discussed in section 2.0 below, is targeted towards offline identification of the fixed dynamical load possibly encountered by an embedded BLMD system during the setup and commissioning phase. This is necessary for optimal tuning of the installed BLMD velocity and position loops in any high performance electric vehicle and industrial application. The present work on optimal parameter estimation is mainly concerned with the offline identification of the worst case inertial load that could possibly be experienced by an installed embedded BLMD. This is articulated here through BLMD simulation in torque control mode, using the full reference model developed in [1, 2], and drive experimental step response measurements with three known test cases of shaft load inertia, for validation of the accuracy of the parameter identification strategy, corresponding to:

- the no-load rotor inertial value $J_T = J_m$,
- medium shaft load inertia $J_T \sim 4J_m$ and
- large shaft load inertia $J_T \sim 7J_m$.

where J_T is the total inertia consisting of rotor J_m and additional shaft load J_l with

$$J_T = J_l + J_m.$$

(1)

The problem of a numerically 'noisy' multiminima cost function resulting in non optimal parameter convergence because of local minimum trapping, associated with the adoption of the BLMD reference model in [1, 2] during simulation, in motor parameter identification is examined [20, 21]. An explanation is provided as to the existence of 'false' local minima plurality with inaccurate resolution of PWM edge transition times, associated with the choice of fixed step sizes Δt in BLMD model simulation, in both the current feedback I_{fj} and shaft velocity $V_{\omega r}$ MSE objective functions. An explanation is also furnished as to the existence and proliferation of genuine local minima with the observed feedback current (FC) target data used in penalty cost surface generation, which will be shown to posses an inverted sinc function-like shape. Details are presented, through MSE response surface simulation with coarse step sizes chosen initially for the inertia J and friction B parameters employing shaft velocity (parabolic cost surface) and feedback current (sinc-like surface) experimental target data respectively, to shed light on the numerical noise problem for SI purposes. Both simulated MSE response surfaces reveal on a macro-scale the presence of a 'line minimum' of possible feasible solutions in a stationary region, enveloping a global extremum within the central surface fold, principally in the B-parameter direction. A novel mathematical approximation [17], which provides verification of the cost surface shape in both cases, is given and is used to provide information on the existence of a unique global minimum with an accompanying optimal parameter set

$\overline{X}_{opt} = \{\overline{J}_{opt}, \overline{B}_{opt}\}^T$ Instead of a multiplicity of candidate options, $\overline{X}_{opt}^j = \{\overline{J}_{opt}, \overline{B}_{opt}^j\}^T$, along a 'B - line minimum', for j = 1,2 .... Details of BLMD model simulation at a finer parameter step size δX, which illuminates the problem of a noisy cost surface, are also provided for both objective functions. An independent statistical analysis appraisal of the computation 'noise' voltage engendered in the search for accurate PWM transitions, based on a novel theoretical estimation [18] for the random error pulse energy expectation associated with PWM replication with chosen simulation step size Δt, is also provided. This probability analysis in itself provides a useful insight into the induced noise mechanism with chosen time step size and highlights the magnitude contribution of the random error 'noise' voltage with PWM resolution to the overall accuracy in the BLMD model simulation exercise. The effect of inherent 'noisy' evaluation of the PWM edge transition times during BLMD simulation is transferred as a lack of smoothness in the simulated construction of the MSE cost surface at the micro-scale for very low step changes δX in the BLMD dynamical parameters J and B.

A novel mathematical analysis [21] is presented, via embedded quadratic curve fitting in the MSE cost surface, to establish the worst case parameter quantization step size δX^L necessary to overcome cost function 'noisiness'. This analysis also provides a radius of convergence rX in parameter space about the global minimum for any parameter identification search strategy and establishes a bound on the limits of accuracy for the returned optimal parameter estimate $\hat{X}_{opt} = \{\hat{J}_{opt}, \hat{B}_{opt}\}^T$. Furthermore this methodology provides a sensitivity measure of the MSE cost surface selectivity for both the step response shaft velocity and current feedback response surfaces in the neighbourhood of the global extremum $\iota \overline{X}_{opt}$. This surface variability metric dependency on elemental parameter variation δX can then be used to decide on the best objective function for parameter extraction purposes based on the accuracy of the returned estimate. The choice of the FC target data is explained for its excellent coherence properties, based on frequency and phase attributes from step response tests, in checking BLMD model fidelity and accuracy and also for its high selectivity in penalty cost function formulation for accurate parameter identification. Furthermore it will be shown that there is an improvement in FC cost surface selectivity with longer data training records while the converse effect is manifested for shaft velocity target data with measurement data length in the reduction of cost surface curvature in the vicinity of the global minimum. These current feedback step response attributes arbitrate in its favour as the most suitable choice of target data in MSE cost function formulation.

In the absence of embedded drive application details from the BLMD manufacturer [17] no precise limits on the desired accuracy of the returned J and B parameter estimates could be affixed to the parameter identification strategy for velocity controller tuning purposes in the commissioning phase. However the use of a quantized metric $δX^L$, as mentioned previously, in parameter space puts a limit on the parameter resolution accuracy possible during identification of the BLMD dynamics in electric vehicles. It should be noted that without the imposition of this parameter quantization strategy there is a risk of false minimum trapping of the identification search algorithm in a 'noisy crevice' [18] in a side-wall of a cost surface, besides local minimum capture, well away from the global minimum estimate. This novel quantization procedure in parameter space, which eliminates the effect of simulation step size related computation induced noise, results in the availability of a smooth cost surface over which a parameter identification search algorithm will work and converge to an optimal estimate [17,18,21]. One other benefit of the parameter quantization process is that it divides up parameter space and restricts the identification strategy to a countable number of parameter lattice points [18] and thus minimizes the search time to global optimality.

A further aspect of concern besides false minimum trapping is that all optimization algorithms for BLMD parameter identification proceed in a continuous search of parameter space to a convergence estimate of the parameter vector sought with an end stopping criterion [22,23]. The norm of cessation of the optimal parameter search strategy is generally based on the smallness of cost reduction over successive iterations within a specified error bound ε at termination. The termination criteria are generally not focused on the smallest percentage variation of the parameter estimates acceptable. However with the quantization δX^L of parameter space for response surface smoothness, limits for parameter resolvability can be imposed by restricting the identification search process to an integral number k of quantum steps $k\delta X$ commensurate with the percentage accuracy %X required in absolute terms such that %X = $k\delta X$. This restricted step approach, in terms of the specified parameter accuracy sought for BLMD tuning purposes during the setup and commissioning phase, can reduce the SI computation time to optimality [18].

RESPONSE SURFACE SIMULATION AND ANALYSIS [17]

The concept of a simulated response surface (RS) is presented as an aid to motor dynamical parameter optimization in high performance Brushless Motor Drive (BLMD) identification with a multiminima objective function. This methodology provides useful information concerning the formulation and nature of the most suitable objective function to be minimized, based on actual drive experimental test data available and BLMD model simulation, coupled with an effective system identification (SI) strategy for accurate motor parameter extraction. This simple approach, although computationally intensive, can also be used as an alternative means of providing the optimal set of parameter estimates from inspection of the global minimum location on the simulated cost surface with embedded local minima. Furthermore it can be used as a basis for comparison of the effectiveness of other identification search strategies deployed, such as the Powell Conjugate Direction search method [18] and Fast Simulated Diffusion algorithm [20,21], in terms of the accuracy of returned parameter estimates. The problem of inertia J and viscous friction B parameter extraction of an actual BLMD system over a sinc-function (sinx/x) shaped multiminima cost surface, based on step response feedback current (FC) target data which has a constant amplitude swept frequency characteristic, is investigated using response surface simulation. The choice of the FC target data is based on its excellent coherence properties [24] from step response testing, for checking BLMD model fidelity and accuracy and for the penalty cost function formulation in SI. This difficulty with a multiminima objective function converging to a non optimal parameter estimate, associated with the adoption of the FC target data for motor parameter identification, is examined in the FSD method [20, 21]. An explanation is provided as to the existence of local minima plurality with the observed FC target data used in the sinc-like penalty cost surface generation. All classical optimization techniques [22], with the exception of modern statistical methods [21], are known to have difficulty with this type of cost surface in identifying the optimal parameter vector. The problem arises with initialization of the search strategy far from the global minimum resulting in possible local minimum trapping and non optimal convergence of the cost minimization algorithm during the parameter extraction process. This response surface [RS] methodology, however, provides a simple and effective alternative to classical methods in acquiring an accurate estimate of the global minimum. Results are presented, which demonstrate the efficacy and reliability of the RS method in returning accurate estimates for 'known' values of the BLMD shaft dynamics. The application of this FC step response related multiminima cost function in parameter extraction is compared with the alternative parabolic shaped shaft velocity objective function for cost surface selectivity in the vicinity of the global minimum and for accuracy of the returned identified parameter estimates. A mathematical approximation analysis is provided for verification of the cost surface shapes resulting from the deployment of step response FC and shaft velocity as target data in objective function formulation.

Cost Function Formulation [18]

Response surface simulation is a useful graphical tool [25] in system identification and can easily be applied to motor parameter extraction and BLMD model validation. This visual concept, which has been used in process control optimization [25], provides an intuitive insight into the topographical structure of the cost function to be minimized and the rapid location of the global minimum. It also provides information on the most suitable identification search strategy that should be adopted in parameter space to obtain an accurate estimate $\hat{X}_{opt} = \{\hat{x}_{1\,opt}, \hat{x}_{2\,opt}\}^T$ of the motor dynamics where the inertia $J \equiv x1$ and viscous friction $B \equiv x2$ are the coded variables. The location of the global minimum stationary point can be obtained by inspection from the simulated cost surface. This approach, although computationally expensive, can be used to secure an independent alternative optimal estimate $\overline{X}_{opt} = \{\overline{J}_{opt}, \overline{B}_{opt}\}^T$ as a reference against which the accuracy of other parameter identification search schemes such as the Fast Simulated Diffusion [26] can be judged.

BLMD parameter extraction is generally based on the minimization of the errors of fit ek between the observed motor drive target data and BLMD model responses in terms of the controlled parameter vector X. This identification process results in the adjustment of the J and B parameters towards global optimality. The search strategy is performed in the neighbourhood of the global extremum using the least squares error criterion in the cost function formulation between each value of a time series

$$t = \left\{ t_k \middle| t_k = kT, \ T = \Delta t, \ 1 \le k \le n, \ k \in N \right\}$$
(2)

of n experimental sample points of the actual motor drive response g(tk) as the target data reference and the corresponding simulated model response f(X, tk). The objective function E(X) is defined as the mean square error (MSE) from the residual vector as

$$e^T = \left[e_1, e_2, \ldots, e_n \right]$$
(3)

as

$$E(X) = \tfrac{1}{n} e^T e = \tfrac{1}{n} \sum_{k=1}^{n} e_k^2$$
(4)

where

$$e_k = g(t_k) - f(X, t_k)$$
(5)

The MSE generates an error response cost surface in parameter X space based on target data from one of the internal test points in [1]. The Powell Conjugate Direction (PCD) [23] and Fast Simulated Diffusion optimization techniques [27] can be applied in conjunction with the BLMD model in [1] to the response surfaces corresponding to motor shaft velocity Vωr and winding FC Ifa target data respectively, obtained in torque control mode for different shaft load inertia listed in [1], for optimal parameter Xopt extraction.

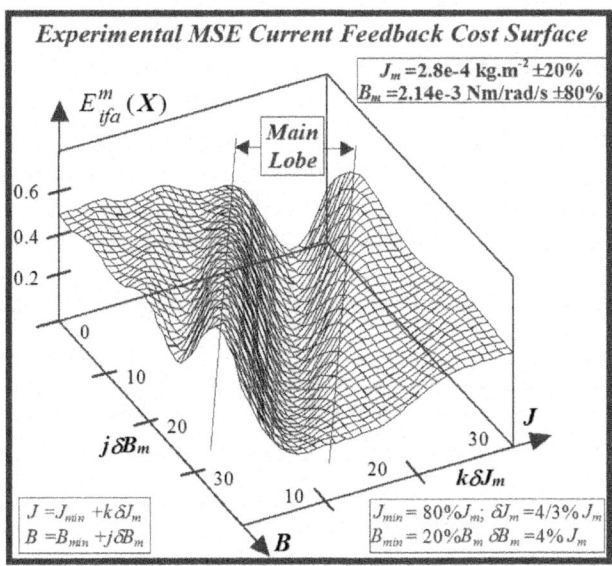

Figure 1: Experimental FC Cost Surface

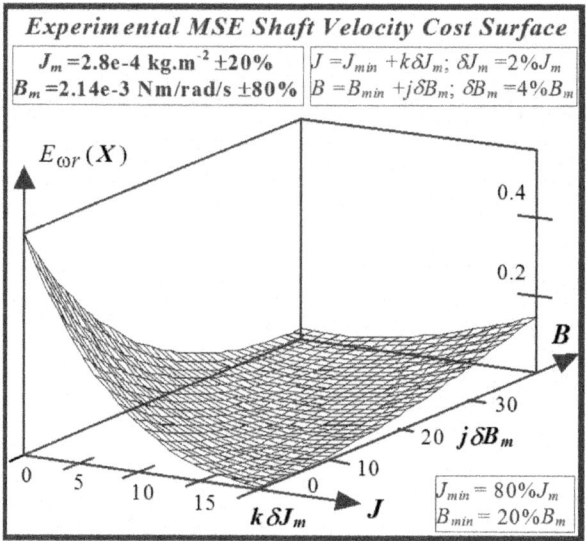

Figure 2: Shaft Velocity Cost Surface

The simulated response surfaces E(X) are derived from BLMD simulation, with a fixed time step of 1 μs and appropriate decimation factor, using the model test point o/p f(X,tk) in conjunction with the sampled experimental target data

$$g(t_k) \in \{V_{\omega_r}(kT), I_{fa}(kT)\}$$

(6)

as the target reference. These penalty cost functions are depicted in Figs.1 and 2 for zero shaft load conditions over parameter space $X=[J,B]^T$ with a crude mesh size δX chosen as per Table 1 to initially determine surface shape, according to the rotor inertia and friction tolerances likely to be encountered in practice. The experimental test data training records used in the MSE formulation for each objective function are displayed in Figs.1 and 2.

Table 1: Experimental Cost Surface Formulation for Zero Shaft Load (NSL) Conditions

MSE Cost Surface Type E(X)E(X)	Current Feedback: Eifa(J,B)Eifa(J,B)	Shaft Velocity: Eωr(J,B)Eωr(J,B)
Data Training Record g(tk)g(tk)	Current Feedback: Ifa(tk)Ifa(tk)	Shaft Velocity: Vωr(tk)Vωr(tk)

No. of Data Points N_d @ 20μs	4095		4095	
BLMD Parameter varied x	J_m(kg.m²)	B_m(Nm/rad/sec)	J_m (kg.m²)	B_m(Nm/rad/sec)
Nominal Parameter Value $_x$m	2.8×10^{-4}	2.14×10^{-3}	2.8×10^{-4}	2.14×10^{-3}
Parameter Tolerance Band Δx	±20%	±80%	±20%	±80%
Crude Parameter Step size δx	1.33%	4%	2%	4%
No. of Parameter Steps $_N x$	30	40	20	40
Parameter Value Returned	2.99×10^{-4}	$\sim 1.54 \times 10^{-3}$	3.024×10^{-4}	$\sim 1.626 \times 10^{-3}$
Assumed Optimal Parameter Vector X$_o$ for Response Surface Analysis				
$_x 0$	3.0×10^{-4}	2.14×10^{-3}	3.0×10^{-4}	2.14×10^{-3}

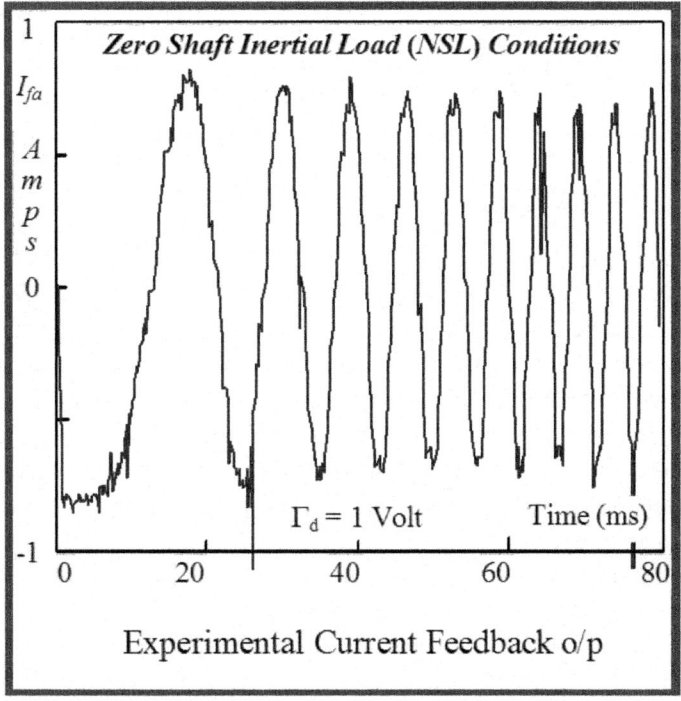

Figure 3: BLMD Current Feedback

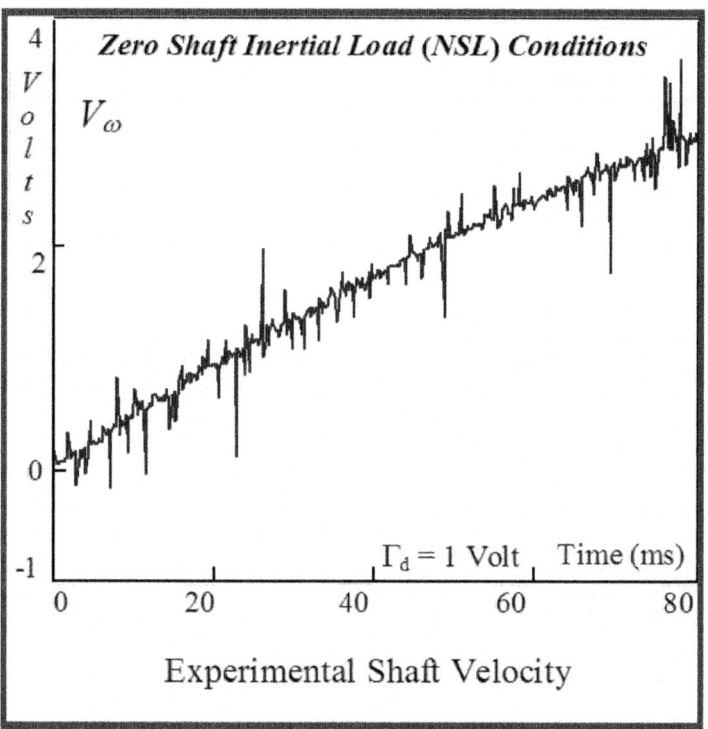

Figure 4: Rotor Shaft Velocity Vω

The anticipated variation in the search cost, likely to be encountered during BLMD system identification (SI) over the parameter tolerance band of interest, can be gauged from cross sections through the chosen response surface at nominal values of the rotor parameters $[Jm,Bm]^T$. The cost variations associated with specific dynamic parameters are illustrated in Figs.5 and 6 for motor current feedback and in Figs.7 and 8 for shaft velocity target data. These cross sections provide important information regarding the surface shape and curvature and consequently about the nature of the stationary points found and type of SI search algorithm that should be deployed over such hitherto surface 'terra incognita'.

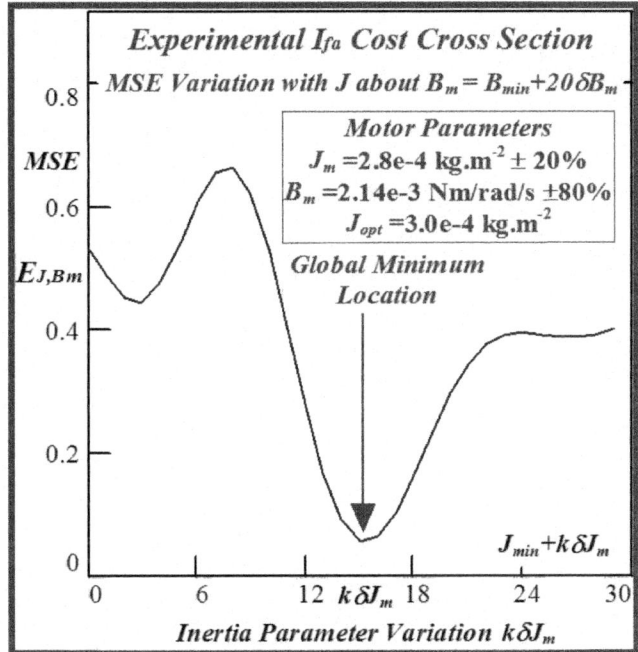

Figure 5: MSE-Ifa Cross section at Bm

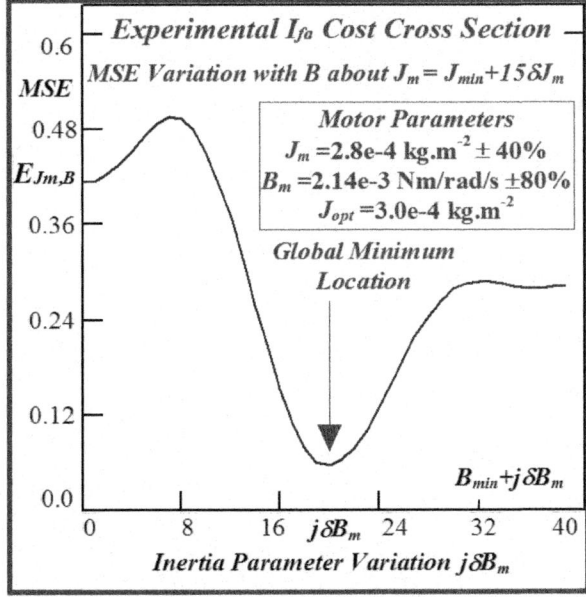

Figure 6: MSE-Ifa Cross section at Jm

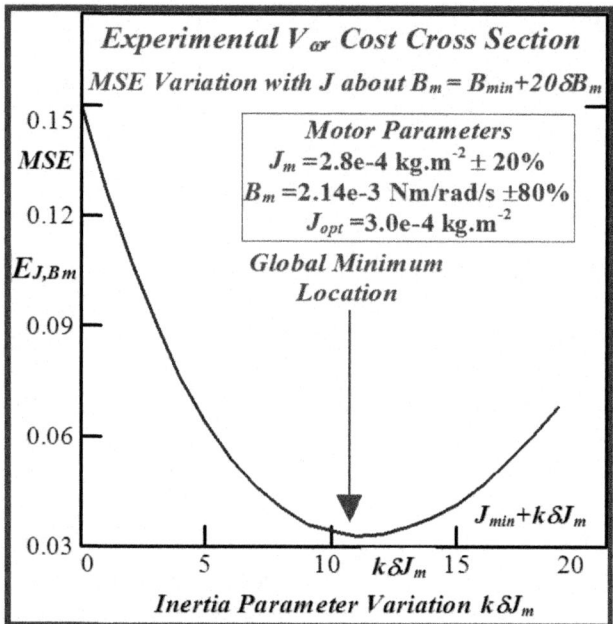

Figure 7: MSE-Vωr Cross section at Bm

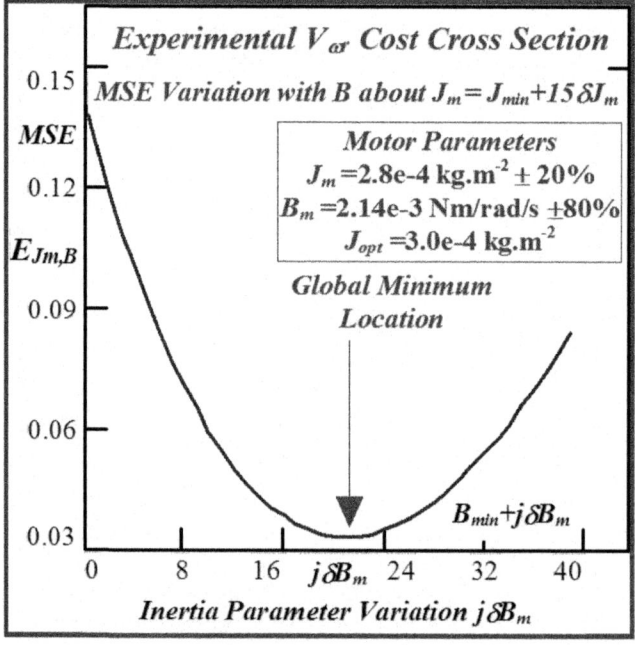

Figure 8: MSE-Vωr Cross section at Jm

The FC cost 'landscape' highlights the existence of several parabolic shaped ridges, interspersed with embedded synclines within its sinc-like folded topography, with a consequent plurality of local minima. The cost terrain also shows the presence of a stationary elliptical shaped ridge system centrally located in the contour map of Fig. 9 with the possible existence of a 'line minimum' [25] along the principal/major axis. These multiminima folds are manifested in the constructive and destructive interference patterns encountered in the frequency ramp up of the FC sinusoid, when compared with the optimal parameter reference or test data waveform, during the transient phase of motor acceleration. The shaft velocity cost surface is parabolic shaped as seen from the contour map in Fig. 10 but is less selective than its FC equivalent in the vicinity of the global minimum when the respective cost surface cross sections with equivalent parameter grid sizes are compared.

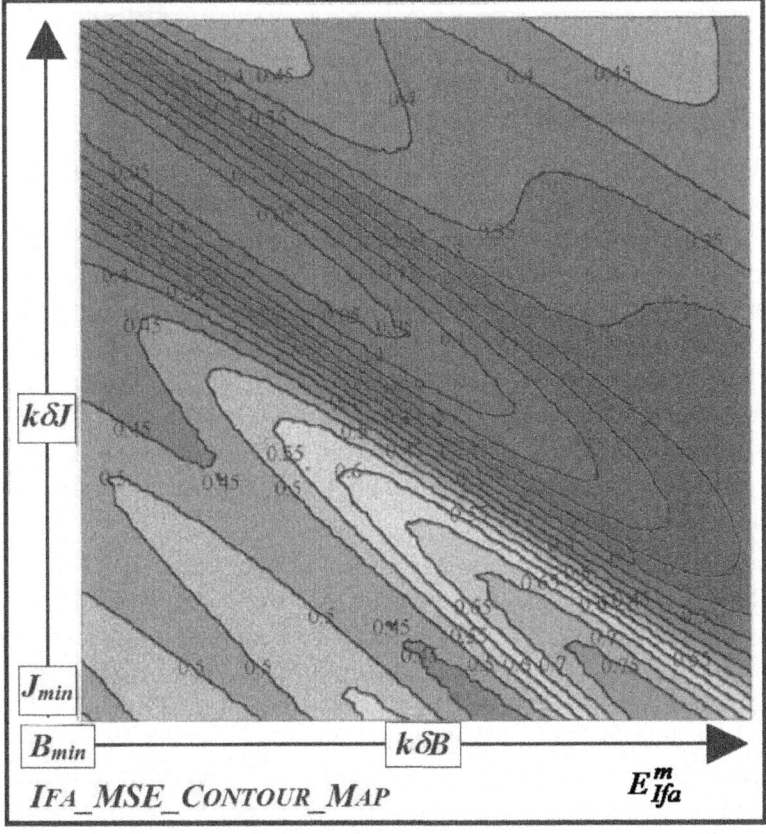

Figure 9: Experimental MSE-I_{fa} Contour Map

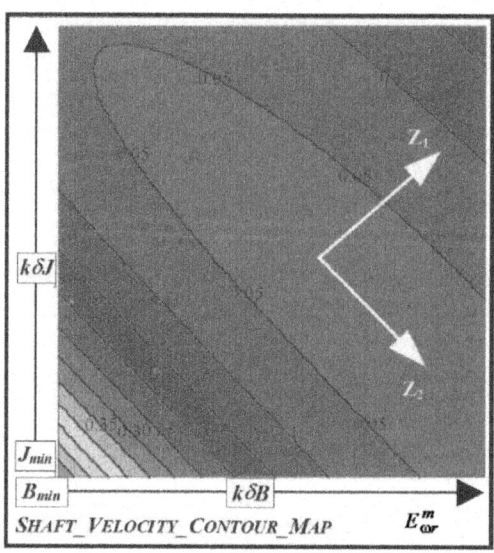

Figure 10: Experimental MSE-V$_{\omega r}$ Contour Map

It is evident from Figs. 9 and 10 that both objective functions possess long wedge shaped stationary valleys in the response surfaces with no 'apparent' clearly defined global minimizer. The observed near linear dependence of the surface shape on the parameters in a 'line minimum' along the valley floor indicates that B is commensurate with J in the ratio J/B which is the dynamical time constant τm of the motor. The B parameter, which is the least likely of the two to vary in the electromechanical drive applications [8,9] can to be acquired from dynamic testing as per [1] to free the other parameter for identification purposes. This reduces the identification problem to single parameter extraction in J or alternatively in τm, where parameter decoupling is non essential, for controller design purpose.

Response surface simulation provides an alternative route of accurately estimating the optimal parameter vector X_{opt} by means of inspection of the surface minimum cost. This method, although computationally expensive, can be used as a yardstick by which the overall convergence performance of other identifications schemes [21] can be contrasted, such as FSD, over a range of motor shaft inertial loads. The response surfaces can be simulated initially using a coarse parameter mesh size, for a range of supposedly 'unknown' motor inertial load test cases for shaft velocity and current feedback MSE objective functions, for rapid location of the global minimum. Further refinement in mesh size can be made down to the parameter step sizes necessary in the vicinity of the global minimum for accurate resolution of the optimal parameter set. Results, which demonstrate the accuracy and effectiveness of RS simulation, are presented for global minimum estimates of motor shaft inertia which are in close agreement with known test inertial load values.

Novel Mathematical Analysis of Response Surface [18] – Modelling and Simulation

Response surfaces can be generated for the BLMD shaft velocity and current feedback step responses, as the mean squared error cost function between an actual drive experimental target data record and simulated model responses, by varying J and B over the two dimensional dynamical parameter space of interest. This graphical procedure is then used to shed light on the shape of the respective cost surfaces and to make a decision as to the most efficient parameter identification strategy to be deployed in each case. Inspection of each of the 2-D MSE response surfaces reveal the existence of 'open' wedge shaped stationary regions principally in the B-parameter direction containing what appears to be a global 'line' minimum in both cases. From a parameter identification perspective such open stationary regions would mean an infinite number of admissible solutions and thus uncertainty in the parameters extracted. The presence of such a difficulty would require careful measurement of one the parameters, in this case the friction as this is the principal direction that the line minimum appears to exists, in order to free the other (J) for identification. A novel mathematical analysis is presented in this chapter to determine whether or not these embedded stationary regions are open. This approach is articulated by formulating a simple quadratic model approximation of the cost surface stationary regions over a small neighbourhood of parameter space, with interacting J and B terms, for proposed model accuracy. The BLMD model step responses are also approximated by simple analytical expressions over response time spans that are very short by comparison with the dynamical time constant τ_m for validation and accuracy of the response surface quadratic model approximation. These simple step response representations, in which the parameters J and B can be adjusted over the space of interest for local cost surface generation and analysis of the stationary region, are included along with the relevant experimental target data in the cost surface quadratic model approximation. This mathematical analysis, employing the simplified quadratic model for both cost surfaces, can be used to show:

- that the stationary regions for the current feedback and shaft velocity objective functions are closed and bounded indicating the presence of a trapped global minimum,

- how closely the dynamical J and B parameters are coupled by making a comparison of the extracted quadratic model eigenvalues,

- that a line minimum exists principally in the B parameter direction and quantifies the extent of thisB-line minimum by the eigenvalue ratio

- establishes the degree of ill conditioning for the global minimum solution parameter vector estimate XS extracted from the minimized quadratic model.

Furthermore this analysis also demonstrates that the current feedback response surface has better selectivity in the global stationary region than the shaft velocity equivalent with increasing data record lengths. This outcome helps in the decision analysis that favours the use of current feedback target data in cost function formulation for dynamical parameter identification.

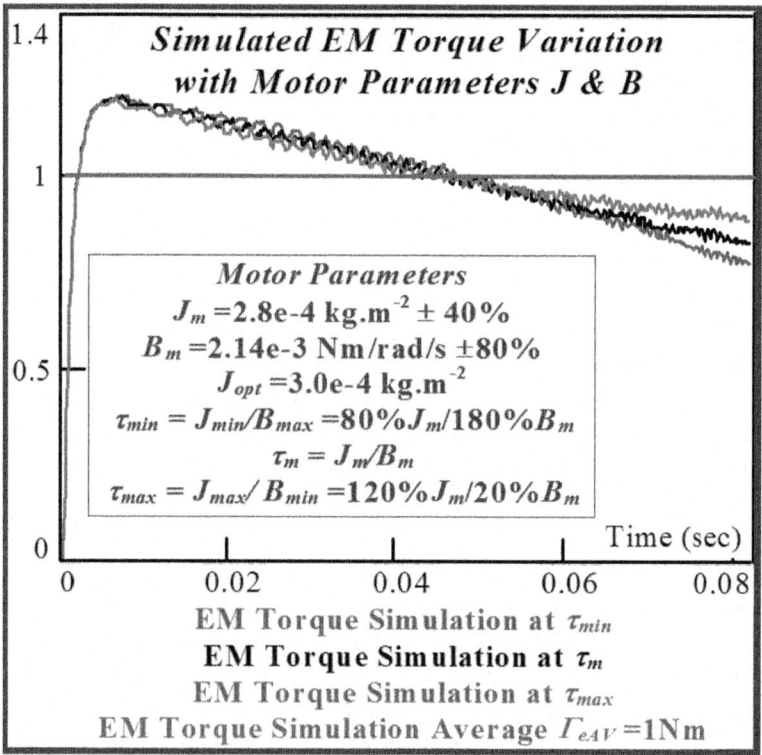

Figure 11: EM Torque Variation with Bm & Jm

The observed topographical features in the above penalty response surfaces can be anticipated from the following approximation analysis. Initially the developed electromagnetic torque Γ_e is at a maximum for unit torque demand step input Γ_d and remains so for a very short time as per the BLMD model simulation in Fig.11 until the shaft speed starts to build up exponentially as in Fig. 12 with time constant τ_m.

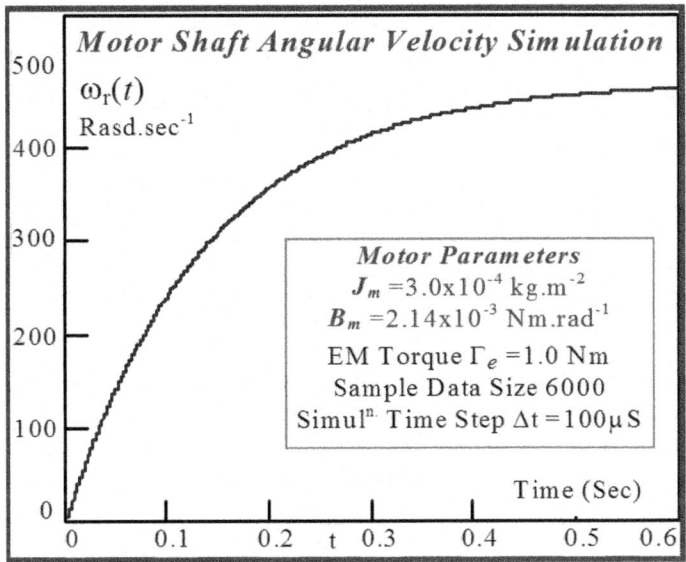

Figure 12: Simulated Shaft Velocity

The back-emf term v_{ej} in [1] becomes substantial causing a decrease in winding current i_{js} which reduces the applied torque. Furthermore the increased rotor angular velocity ω_r causes the machine impedance angle φ_z in [1] to approach $\pi/2$ and forces the winding currents into quadrature with the current command signals i_{dj} with subsequent torque reduction as in [1]. The variation in applied motor torque with the worst case spread of dynamical time constant τ_m values, observed for the parameter tolerance ranges in Table 1 with zero shaft load conditions, is small over the motor acceleration period $(\hat{t}=0.08\text{sec}\approx 60\%\,\tau_m)$ shown in Fig.11. The average value of applied mechanical torque Γ_{em} is 1Nm and is assumed constant over the period $\hat{t}$ for tractability reasons in the following analysis of the cost surfaces used in the PCD and FSD methods of parameter extraction. However this value deteriorates over longer time spans as the winding current moves out of phase alignment with current demand as motor speed increases and thus with the back EMF. The simulated shaft speed variation with time, based on the nominal parameter vector X_m in Table 1 and displayed inFig.12 for a step i/p torque demand Γ_d (~1v) is given by (7)-(a)

$$\omega_r^m(t) = K_m(1-e^{-t/\tau_m}) \quad \text{with} \quad K_m = \Gamma_{em}/B_m \quad \text{and} \quad 0 \le t \le \hat{t} \qquad (a)$$

$$\omega_r^0(t) = K_0(1-e^{-t/\tau_0}) \quad \text{with} \quad K_0 = \frac{\Gamma_{em}}{B_0}; \quad \tau_0 = \frac{J_0}{B_0}. \qquad (b)$$

$$(7)$$

Similarly the corresponding shaft speed variation with time at the assumed optimum parameters $\{J_o, B_o\}$ in Table 1, which are be identified from cost surface trial analysis, is given by (7)-(b) The sampled motor speed 'test' data $\omega_r^o(t_k)$ generated via (7)-(b) can now be used as target reference 'test' data in the simulated trial cost function $E_{\omega r}^O$ for analytical purposes. The optimal parameter set X is supposedly unknown and the task here is to obtain a good estimate $[\bar{J}_o, \bar{B}_o]^T$ of this vector in the following cost surface analysis for verification of the RS strategy. The variation in the time constant τ_m over the permitted parameter tolerance ranges employed in the response surface generation, such as those in Figs 1 and 2 relying on experimental test data, is insufficient to cause departure from nominal applied torque Γ_{em} for the short time span shown in Fig.11. The shaft speed variation in this instance is approximated by

$$\omega_r(t) = K(1 - e^{-t/\tau}) \quad \text{with} \quad K = \frac{\Gamma_{em}}{B}; \quad \tau = \frac{J}{B}.$$

$$(8)$$

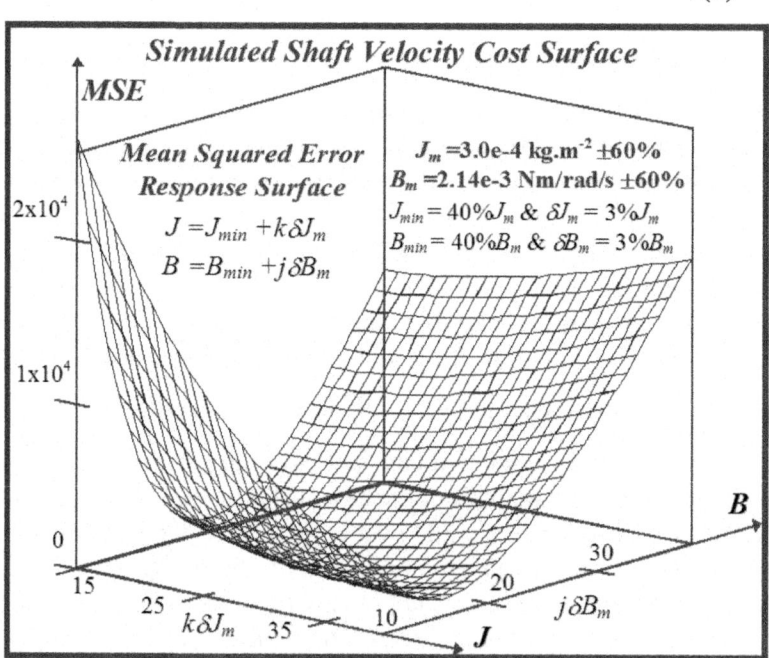

Figure 13: Simulated Velocity Cost Surface

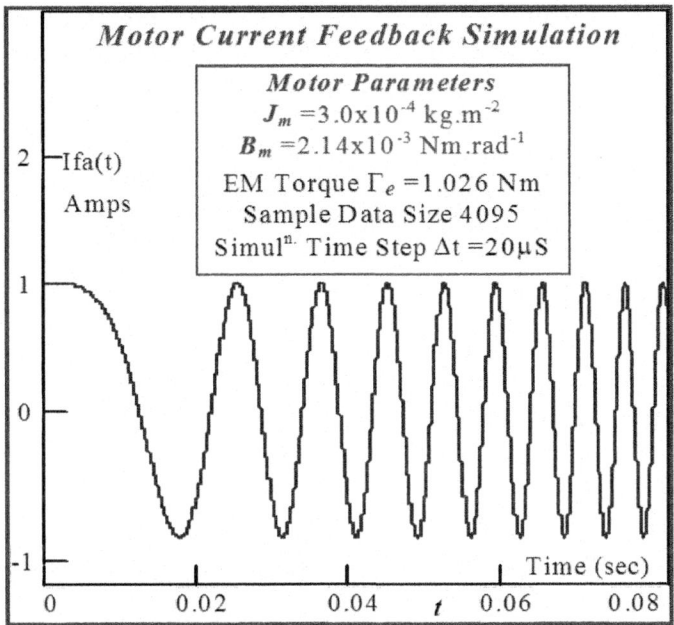

Figure 14: BLMD Current Feedback

The resulting MSE cost function construct, illustrated in Fig.13 with details in Table 2, is for simulation purposes given by

$$E_{\omega r}^m(\mathbf{X}) = \frac{1}{N_d}\sum(\omega_r - \omega_r^m)^2$$

(9)

with target data ω_r^m. The parabolic cost variations associated with specific dynamic parameters for shaft velocity target data are illustrated in Figs.15 and 16. The corresponding winding current feedback $i_{fa}(t)$ has the characteristics of a frequency modulated sinusoid during the exponential buildup of motor shaft speed in that it exhibits the features of a constant amplitude swept frequency waveform as shown in Fig.14. The effect of shaft speed increase on the phase angle φ of the FC response is determined from (8) as

$$\varphi = \int_0^t \omega_r(x)dx = K\left\{t + \tau(e^{-t/\tau} - 1)\right\} = Kt - \tau\omega_r(t)$$

(10)

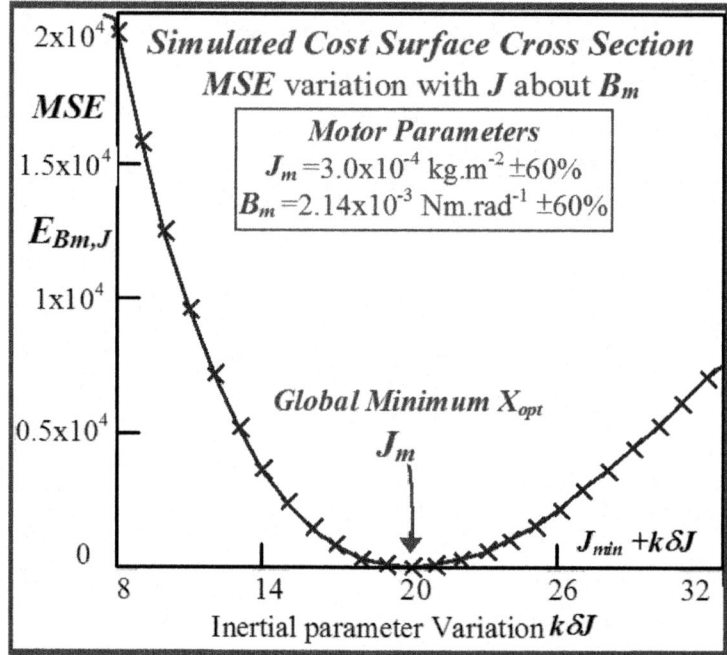

Figure 15: MSE-V$_{or}$ Cross section at B$_m$

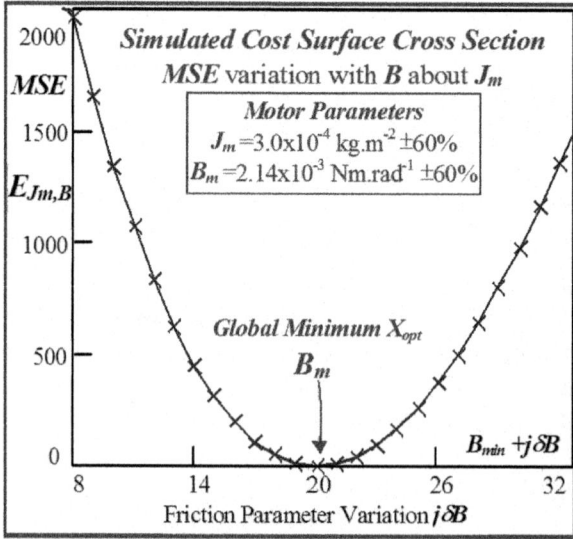

Figure 16: MSE-V$_{or}$ Cross section at J$_m$

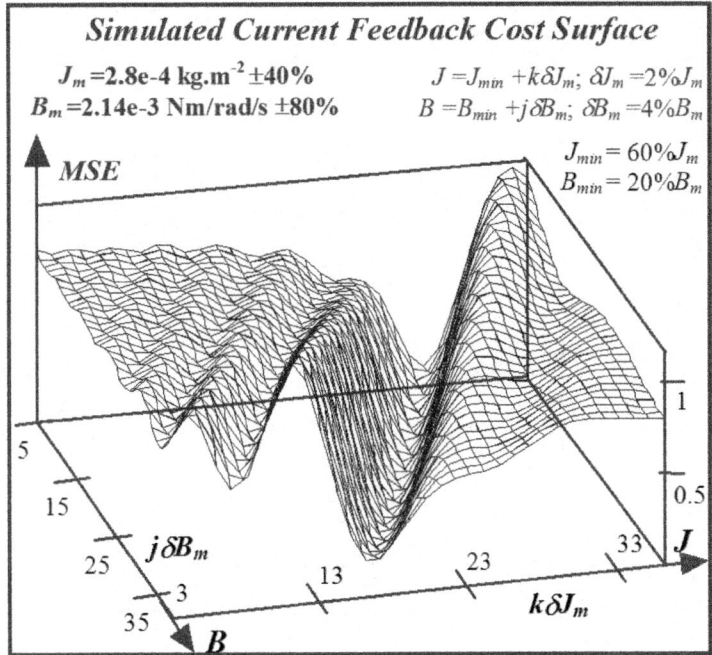

Figure 17: Simulated FC Cost Surface

The frequency modulated FC, which is current regulated by G_I in [1], is given by

$$I_{fa}(t) = I_f \cos p\varphi = I_f \cos p\left(Kt - \tau\omega_r(t)\right)$$

(11)

with $I_f \approx 1$ amp for a unit step torque demand i/p. The resultant FC cost surface generated from simulation in Fig.17, with parameter grid sizes in Table 1, is based on the target shaft velocity $\omega_r^m(t)$ in (7)-(a) for nominal values of the dynamical parameters X_m with

$$E_{Ifa}^m(\mathbf{X}) = \frac{1}{N_d}\sum\left(I_{fa} - I_{fa}^m\right)^2$$

(12)

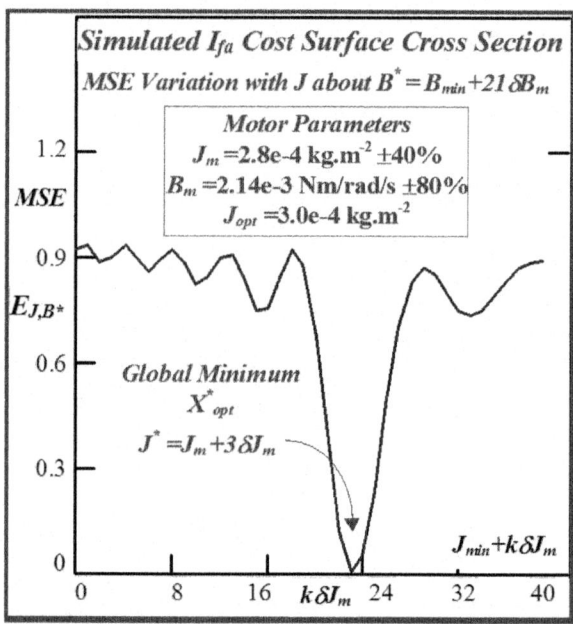

Figure 18: MSE-I_{fa} Cross section at B^*

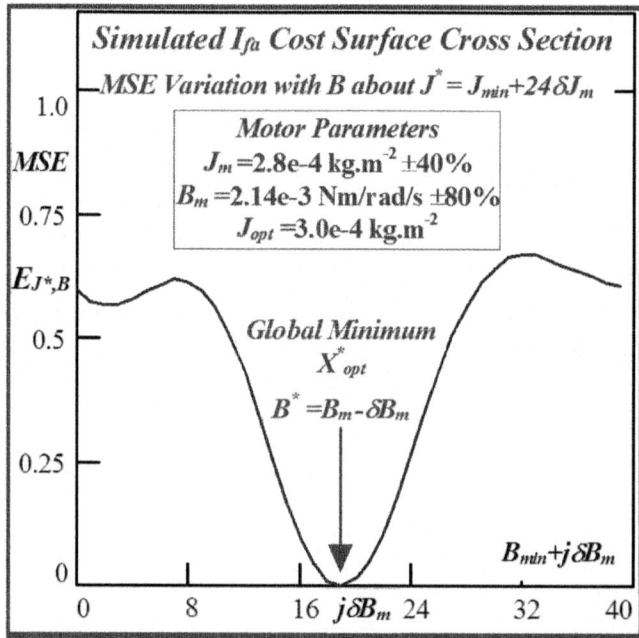

Figure 19: MSE-I_{fa} Cross section at J^*

The sinc-profile cost variations associated with specific dynamic parameters for motor current feedback target data at nominal values of the BLMD parameters $[J_m, B_m]^T$ are illustrated in Figs.18 and 19. TheMSE penalty cost function can described in a more general form about X_m as

$$E_f^m(\mathbf{X}) = \frac{1}{N_d} \sum_k \left(f(t_k) - f^m(t_k) \right)^2 \tag{13}$$

with either target data training record deployed using the representation

$$f(\mathbf{X}, t_k) \in \left\{ \omega_r(t_k), I_{fa}(t_k) \right\} \tag{14}$$

The nature of the global stationary region embedded in either cost surface, described by (9) or (12), can be explored in canonical form [25] by fitting a quadratic model using a Taylor series. This two dimensional truncated series expansion, with quadratic terms measuring the surface curvature, is anchored at the nominal value X_m to establish the principal axes/directions in parameter space for global minimum search. It is assumed that the expansion pivot X_m is in proximity to the supposed global optimum X_0 in the case of the FC objective function as this consists of parallel ridges interlaced with folds enveloping local minima regions which obscure the global extremum position. The response surface model $\mathfrak{R}^f$ can be expressed, in either case with (14), in terms of the variables $J \equiv x_1$ & $B \equiv x_2$ and low order interactive terms β_{ij} about X_m as

$$\mathfrak{R}^f = \beta_0 + \beta_1(x_1 - x_{1m}) + \beta_2(x_2 - x_{2m}) + \frac{\beta_{11}}{2!}(x_1 - x_{1m})^2$$
$$+ \frac{\beta_{22}}{2!}(x_2 - x_{2m})^2 + \beta_{12}(x_1 - x_{1m})(x_2 - x_{2m}) + \varepsilon \tag{15}$$

with random modelling error ε. The surface model can alternatively be approximated in compact matrix form as

$$\hat{\mathfrak{R}}^f = \beta_0 + \mathbf{B}^T(\mathbf{X} - \mathbf{X}_m) + \frac{1}{2}(\mathbf{X} - \mathbf{X}_m)^T \hat{\mathbf{G}}(\mathbf{X} - \mathbf{X}_m) \tag{16}$$

with constant coefficient matrices determined from the cost at X_m, based on target data

$$f^0(t_k) = f(\mathbf{X}_0, t_k) \in \left\{ \omega_r^0(t_k), I_{fa}^0(t_k) \right\} \tag{17}$$

by the gradient vector B given by

$$\begin{bmatrix} \beta_1 \\ \beta_2 \end{bmatrix} = \nabla E_f^0(\mathbf{X}_m) = \begin{bmatrix} \dfrac{\partial E_f^0}{\partial x_1} \\ \dfrac{\partial E_f^0}{\partial x_2} \end{bmatrix}\Bigg|_{\mathbf{X}_m}$$

(18)

and the symmetric Hessian matrix $\hat{G}$

$$\begin{bmatrix} \beta_{11} & \beta_{12} \\ \beta_{12} & \beta_{22} \end{bmatrix} = \nabla^2 E_f^0(\mathbf{X}_m) = \begin{bmatrix} \dfrac{\partial^2 E_f^0}{\partial x_1^2} & \dfrac{\partial^2 E_f^0}{\partial x_1 \partial x_2} \\ \dfrac{\partial^2 E_f^0}{\partial x_1 \partial x_2} & \dfrac{\partial^2 E_f^0}{\partial x_2^2} \end{bmatrix}\Bigg|_{\mathbf{X}_m}$$

(19)

which determines the curvature in the vicinity of a local minimum via

$$E_f^0(\mathbf{X}_m) = \tfrac{1}{N_d} \sum_k \left(f^m(t_k) - f^0(t_k) \right)^2$$

(20)

The set of constant coefficient differential equations pertaining to (15) are obtained via (13), using either target data record (7)-(b) or (11) with $I_{fa}(t)\big|_{\omega_r(t)=\omega_r^0(t)}$, as

$$\beta_0 = E_f^0(\mathbf{X}_m) = \tfrac{1}{N_d} \sum (f^m - f^0)^2$$

(21)

$$\beta_1 = \frac{\partial E_f^0}{\partial x_1}(\mathbf{X}_m) = \tfrac{2}{N_d} \sum (f^m - f^0)\frac{\partial f}{\partial J}\Big|_{\mathbf{X}_m}$$

(22)

$$\beta_2 = \frac{\partial E_f^0}{\partial x_2}(\mathbf{X}_m) = \tfrac{2}{N_d} \sum (f^m - f^0)\frac{\partial f}{\partial B}\Big|_{\mathbf{X}_m}$$

(23)

$$\beta_{11} = \frac{\partial^2 E_f^0}{\partial x_1^2}(\mathbf{X}_m) = \tfrac{2}{N_d} \sum \left[\left(\frac{\partial f}{\partial J}\right)^2 + (f^m - f^0)\frac{\partial^2 f}{\partial J^2} \right]_{\mathbf{X}_m}$$

(24)

$$\beta_{22} = \frac{\partial^2 E_f^0}{\partial x_2^2}(\mathbf{X}_m) = \tfrac{2}{N_d} \sum \left[\left(\frac{\partial f}{\partial B}\right)^2 + (f^m - f^0)\frac{\partial^2 f}{\partial B^2} \right]_{\mathbf{X}_m}$$

(25)

$$\beta_{12} = \beta_{21} = \frac{\partial^2 E_f^0}{\partial x_2^2}(\mathbf{X}_m) = \tfrac{2}{N_d} \sum \left[\frac{\partial f}{\partial J}\cdot\frac{\partial f}{\partial B} + (f^m - f^0)\frac{\partial^2 f}{\partial J \partial B} \right]_{\mathbf{X}_m}$$

(26)

The required first and second order partial differential equations, based on the shaft velocity ω_r, to substantiate expressions (22) to (26) are given by

$$\frac{\partial \omega_r}{\partial J} = \left(\frac{t}{\tau}\right)\left(\frac{\omega_r - K}{J}\right)$$

(27)

$$\frac{\partial \omega_r}{\partial B} = -\left(\frac{\omega_r}{B} + \tau \frac{\partial \omega_r}{\partial J}\right)$$

(28)

$$\frac{\partial^2 \omega_r}{\partial J^2} = -\left(\frac{1}{J}\right)\left(2 - \frac{t}{\tau}\right)\frac{\partial \omega_r}{\partial J}$$

(29)

$$\frac{\partial^2 \omega_r}{\partial B^2} = \left(\frac{2}{B^2}\right)\omega_r + \left(\frac{\tau}{B}\right)\left(2 + \frac{t}{\tau}\right)\frac{\partial \omega_r}{\partial J}$$

(30)

$$\frac{\partial^2 \omega_r}{\partial B \partial J} = -\left(\frac{t}{J}\right)\frac{\partial \omega_r}{\partial J}$$

(31)

Table 2: Summary of Cost Surface Quadratic Modelling Details at X_m

	2000 Points - $\omega_r^0(t_k)$	2000 Points - $I_{fa}^0(t_k)$
Target Data Record Length N_d with Time Step 20µs	t=0.04sec ~31%τ_m	t=0.04sec ~31%τ_m
Target Data Parameters $X_0 = [J_0, B_0]^T$ "To be identified"	Shaft Velocity Reference Data $[3.0\times10^{-4}, 2.14\times10^{-3}]^T$	Current Feedback Reference Data $[3.0\times10^{-4}, 2.14\times10^{-3}]^T$
Quadratic Model Fulcrum $X_m = [J_m, B_m]^T$	Model Surface $\hat{R}^{\omega r}$: $[2.8\times10^{-4}, 2.14\times10^{-3}]^T$	Model Surface $\hat{R}^{Ifa}$: $[2.8\times10^{-4}, 2.14\times10^{-3}]T$
Model Cost $\hat{R}_m^f$ at X_m	19.553	0.098
Constant β_0 via (21)	19.553	0.098
Gradient Vector B $[\beta_1, \beta_2]^T$ via (22/3)	$[-2.079\times10^6, -3.279\times10^4]^T$	$[-9.827\times10^3, -113.345]^T$
Hessian Matrix $\hat{G} \begin{bmatrix} \beta_{11} & \beta_{12} \\ \beta_{12} & \beta_{22} \end{bmatrix}$ via (24/5/6)	$\begin{bmatrix} 1.238\times10^{11} & 1.958\times10^9 \\ 1.958\times10^9 & 8.873\times10^7 \end{bmatrix}$	$\begin{bmatrix} 4.592\times10^8 & 5.114\times10^6 \\ 5.114\times10^6 & 1.08\times10^5 \end{bmatrix}$
Stationary Point $X_s = [J_s, B_s]^T$ via (38)	$[2.968\times10^{-4}, 2.138\times10^{-3}]^T$	$[3.005\times10^{-4}, 2.216\times10^{-3}]^T$
Slope at X_s via (37)	$[9.313\times10^{-10}, 1.455\times10^{-11}]^T$	$[0, 1.421\times10^{-14}]^T$
Model Cost $\hat{R}_s^f$ at X_s	2.086	-7.654×10^{-3}
Quadratic Form $Q(X_s-X_m)$ via (39)	34.934	0.211
Normal Form of $\hat{G}$ $A = \begin{bmatrix} \lambda_1 & 0 \\ 0 & \lambda_2 \end{bmatrix}$	Eigenvalues $\begin{bmatrix} 1.238\times10^{11} & 0 \\ 0 & 5.774\times10^7 \end{bmatrix}$	Eigenvalues $\begin{bmatrix} 4.593\times10^8 & 0 \\ 0 & 5.109\times10^4 \end{bmatrix}$
Transformation/Modal Matrix T with $T^{-1}\hat{G}T = A$	Normalized Eigenvectors $\begin{bmatrix} 999.875 & -15.825 \\ 15.825 & 999.875 \end{bmatrix} \cdot 10^{-3}$	Normalized Eigenvectors $\begin{bmatrix} 999.938 & -11.137 \\ 11.137 & 999.938 \end{bmatrix} \cdot 10^{-3}$
Co-ordinate Rotation θ	-1.813°	-1.276°

TABLE 3: Details of Cost Surface Quadratic Model Fit at X_m based on actual BLMD Experimental Test Data shown in [1] for Zero Shaft Inertial Load Conditions

Target Data Record Length N_d with Time Step 20µs	4095 Points $\hat{t}$=0.082sec ~62.6%τ_m	4095 Points $\hat{t}$=0.082sec ~56.5%τ_m
Target Data Parameters $\tilde{X}_{opt}=[J_{opt}, \hat{B}_{opt}]^T$ "To be identified"	Shaft Velocity Reference Data for zero shaft Inertial load (NSL) Fig. 32; Ref [1] below	Current Feedback Reference Data for zero shaft Inertial load (NSL) Fig. 29; Ref [1] below
Quadratic Model Fulcrum $X_m=[J_m, B_m]^T$	Model Surface $\hat{R}^{\omega r}$: $[2.8 \times 10^{-4}, 2.14 \times 10^{-3}]^T$	Model Surface $\hat{R}^{Ifa}$: $[3.1 \times 10^{-4}, 2.14 \times 10^{-3}]^T$
Model Cost $\hat{R}^f_m$ at X_m	66.543	0.081
Constant β_0 via (21)	66.543	0.081
Gradient Vector B $[\beta_1, \beta_2]^T$ via (22/3)	$[-6.842 \times 10^6, -2.422 \times 10^5]^T$	$[1.865 \times 10^4, 449]^T$
Hessian Matrix $\hat{G}$	$\begin{bmatrix} 4.02 \times 10^{11} & 1.376 \times 10^{10} \\ 1.376 \times 10^{10} & 8.801 \times 10^8 \end{bmatrix}$	$\begin{bmatrix} 1.809 \times 10^9 & 4.061 \times 10^7 \\ 4.061 \times 10^7 & 2.843 \times 10^6 \end{bmatrix}$
Stationary Point $X_s=[J_s, B_s]^T$ via (38)	$[2.964 \times 10^{-4}, 2.159 \times 10^{-3}]^T$	$[3.00 \times 10^{-4}, 2.124 \times 10^{-3}]^T$
Slope at X_s via (37)	$[4.657 \times 10^{-9}, 1.746 \times 10^{-10}]^T$	$[4.002 \times 10^{-11}, 5.116 \times 10^{-13}]^T$
Model Cost $\hat{R}^f_s$ at X_s	8.232	-0.015
Quadratic Form $Q(X_s-X_m)$ via (39)	116.624	0.193
Normal Form of $\hat{G}$ $\Lambda=\begin{bmatrix} \lambda_1 & 0 \\ 0 & \lambda_2 \end{bmatrix}$	Eigenvalues $\begin{bmatrix} 4.025 \times 10^{11} & 0 \\ 0 & 4.086 \times 10^8 \end{bmatrix}$	Eigenvalues $\begin{bmatrix} 1.81 \times 10^9 & 0 \\ 0 & 1.931 \times 10^6 \end{bmatrix}$
Spectral Condition No. η	0.985×10^3	0.937×10^3
Contour sign check (51/2)	-1.645×10^{20}	-3.495×10^{15}
Contour Eccentricity e	999.999×10^{-3}	999.999×10^{-3}
Modal Matrix T	$\begin{bmatrix} 999.413 & -34.246 \\ 34.246 & 999.413 \end{bmatrix} \cdot 10^{-3}$	$\begin{bmatrix} 999.748 & -22.461 \\ 22.461 & 999.748 \end{bmatrix} \cdot 10^{-3}$
Co-ordinate Rotation θ	-1.9624°	-1.2874°

The corresponding set of partial derivatives with FC I_{fa} are obtained via (11) as

$$\frac{\partial I_{fa}}{\partial J} = \sin p\left(Kt - \tau\omega_r\right) \cdot p\left(\frac{\omega_r}{B} + \tau\frac{\partial\omega_r}{\partial J}\right)$$

(32)

$$\frac{\partial I_{fa}}{\partial B} = \sin p\left(Kt - \tau\omega_r\right) \cdot p\left(\frac{t}{B}\left[2K - \omega_r\right] - 2\left(\frac{\tau}{B}\right)\omega_r\right)$$

(33)

$$\frac{\partial^2 I_{fa}}{\partial J^2} = -I_{fa} \cdot p^2\left(\frac{\omega_r}{B} + \tau\frac{\partial\omega_r}{\partial J}\right)^2 + \sin p\left(Kt - \tau\omega_r\right) \cdot p\left(\frac{t}{B\tau}\right)\frac{\partial\omega_r}{\partial J}$$

(34)

$$\frac{\partial^2 I_{fa}}{\partial B^2} = -I_{fa} \cdot p^2 \left(\frac{t}{B}\left[2K - \omega_r\right] - 2\left(\frac{r}{B}\right)\omega_r\right)^2$$
$$- \sin p\left(Kt - \tau\omega_r\right) \cdot p\left(2\frac{t}{B}\left[\frac{3K - 2\omega_r}{B}\right] - 6\left(\frac{r}{B^2}\right)\omega_r - \tau\left(\frac{t}{B}\right)\frac{\partial\omega_r}{\partial J}\right)$$

(35)

$$\frac{\partial^2 I_{fa}}{\partial J \partial B} = I_{fa} \cdot p^2 \left(\frac{t}{B}\left[2K - \omega_r\right] - 2\left(\frac{r}{B}\right)\omega_r\right) \cdot \left(-\frac{\omega_r}{B} - \tau\frac{\partial\omega_r}{\partial J}\right)$$
$$- \sin p\left(Kt - \tau\omega_r\right) \cdot p\left(2\frac{\omega_r}{B^2} + \left(\frac{t}{B}\right)\frac{\partial\omega_r}{\partial J} + 2\left(\frac{t}{B}\right)\frac{\partial\omega_r}{\partial J}\right)$$

(36)

The variation of the directed contour gradient over the fitted cost surface model, given by

$$\nabla\hat{\Re}^f = \mathbf{B} + \hat{\mathbf{G}}(\mathbf{X} - \mathbf{X}_m)$$

(37)

is used to locate the global optimum $\mathbf{X}_0$ in the parameter hyperspace region of interest. The condition necessary [22] for the presence of a stationary point $\mathbf{X}_s$ is the existence of a vanishing gradient in the neighborhood of $\mathbf{X}_m$ located within the parameter tolerance band $\Delta\mathbf{X}$ with

$$\mathbf{X}_s = \mathbf{X}_m - \hat{\mathbf{G}}^{-1}\mathbf{B}$$

(38)

from (37) and the nature of which is determined by the local curvature from the sign of the quadratic form [28]

$$Q(\mathbf{X} - \mathbf{X}_m) = (\mathbf{X} - \mathbf{X}_m)^T \hat{\mathbf{G}}(\mathbf{X} - \mathbf{X}_m)$$

(39)

The parametric details, which include estimates of the gradient vectors and Hessian matrices at $\mathbf{X}_m$ for the indicated data record lengths, of the fitted models to the cost surfaces illustrated in Figs.13 and 17are summarized in Table 2. Similar parametric quantities, employing BLMD experimental test data, are given in Table 3 for cost surface models shown in Figs.1 and 2.

Novel Analysis Of Global Minimum Estimation And Response Surface Selectivity [18]

An estimate of the cost surface global minimum $\hat{X}_{opt}$ is provided in each case by inference from the vanishing gradient in (37) with location of the fitted model stationary point $\mathbf{X}_s$ in (38). A sufficient condition for the existence of a global minimizer at $\mathbf{X}_s$ is that $Q(\mathbf{X}_s - \mathbf{X}_m)$ must be positive-definite [28] in which $Q(\mathbf{X}_s - \mathbf{X}_m) > 0$ for $\mathbf{X}_s \neq \mathbf{X}_m$. This is verified by the sign of the eigenvalues

λ_i of $\hat{G}$ in Table 2which are determined from the characteristic equation

$$det\left[\hat{G} - \lambda I\right] = 0$$

(40)

The accuracy of global estimates returned in each case for the inertial parameter J in Table 2 admit to the quality and goodness of fit of the models employed for cost surface approximation in the vicinity of the global extremum. The contributory effect of parameter interaction in model approximation in both cases is not insignificant with coefficients β_{ij} comparable in magnitude to the geometric mean of the eigenvalues of Hessian $\hat{G}$ in Table 4 defined by

$$\hat{\lambda} = \sqrt[n]{\lambda_1 \cdot \lambda_2 \ldots \lambda_n}$$

(41)

The relative magnitudes of the Hessian curvature components provide information about the uniqueness of the solution X_s in (38), via the matrix condition number in Table 3, based on the infinity norm defined as

$$cond_\infty(\hat{G}) = \left\|\hat{G}\right\|_\infty \left\|\hat{G}^{-1}\right\|_\infty \quad where \quad \left\|\hat{G}\right\|_\infty = \max_{1 \le i \le n}\left(\sum_{j=1}^{n}\left|\beta_{ij}\right|\right)$$

(42)

Table 4: Results Derived From Cost Surface Quadratic Fit in Table 2

	Uniqueness of Global Minimum Estimate	
	Shaft Velocity Reference Data	Current Feedback Reference Data
Cond $_{\infty\hat{\phi}(x_m)}$	2.2126x10³	9.1878x10³
Spectral Condition No. η	2.144x10³	8.99x10³
Geometric Mean $\hat{\lambda}$	2.674x10⁹	4.844x10⁶
Cost Surface Selectivity and Fitted Model Re-evaluation at Global Minimum Estimate X_s		
Fitted Model Fulcrum X_s	[2.968x10⁻⁴, 2.138x10⁻³]ᵀ	[3.005x10⁻⁴, 2.216x10⁻³]ᵀ
Model Constant β_0 at X_s	0.376	4.275x10⁻⁴
Gradient Vector at X_s	[3.562x10⁴, 579.78]ᵀ	[11.535, 0.081]ᵀ
Re-evaluation of $\hat{G}$ at X_s	$\begin{bmatrix} 9.142x10^{10} & 1.437x10^9 \\ 1.437x10^9 & 3.145x10^7 \end{bmatrix}$	$\begin{bmatrix} 4.295x10^8 & 4.99x10^6 \\ 4.99x10^6 & 6.08x10^4 \end{bmatrix}$
Global Estimate Update X_{s1}	[2.996x10⁻⁴, 2.138x10⁻³]ᵀ	[2.998x10⁻⁴, 2.159x10⁻³]ᵀ
Slope at X_{s1} via (37)	[1.717x10⁻⁹, 2.547x10⁻¹¹]ᵀ	[8.413x10⁻¹², 9.948x10⁻¹⁴]ᵀ
Eigenvalues of $\hat{G}$ at X_s	$\begin{bmatrix} 9.144x10^{10} & 0 \\ 0 & 8.846x10^6 \end{bmatrix}$	$\begin{bmatrix} 4.295x10^8 & 0 \\ 0 & 2.819x10^3 \end{bmatrix}$
Modal Matrix T	$\begin{bmatrix} 999.876 & -15.723 \\ 15.723 & 999.876 \end{bmatrix} \cdot 10^{-3}$	$\begin{bmatrix} 999.933 & -11.618 \\ 11.618 & 999.933 \end{bmatrix} \cdot 10^{-3}$
Residual Cost $\hat{R}_s^r$ at X_{s1}	6.439x10⁻⁴	-3.996x10⁻⁶
Quadratic Form $Q(X_{s1}-X_s)$	0.738	8.629x10⁻⁴

The matrix condition number is much greater than unity in both cases, with the highest value associated with the FC response surface, which indicates a sizeable measure of ill conditioning in the extraction of the global estimate in (38). The curvature component β_{11} associated with the J-parameter is much greater than that associated with damping B by about three orders of magnitude which indicates greater selectivity of the solution J_s along the J axis. This suggests the presence of many potential solutions to (38) along the B-parameter co-ordinate direction due to poorer selectivity or smaller curvature component β_{22}. A more complete interpretation of the nature of the cost surface syncline containing the stationary point region is obtained from the spectral condition number η of $\hat{G}$ [22] as

$$\eta = \lambda_{max} / \lambda_{min}$$

(43)

The relative magnitude η of the eigenvalues indicate that a 'line minimum' of potential solutions, which explains the degree of ill conditioning in the global solution estimate, is feasible due to the 'long' elliptical shape of the contour map associated with the stationary point zone of convergence in both cases as shown in Figs 9 and 10. The elliptical character of the response surface model in the vicinity of the global minimum estimate can be visualized by a coordinate translation of the parameter axes to X_s as pivot with

$$V = X - X_S$$

(44)

resulting in the modified representation from (16) as

$$\hat{R}^f = \beta_0 + B^T(V + X_s - X_m) + \tfrac{1}{2}(V + X_s - X_m)^T \hat{G}(V + X_s - X_m)$$
$$= \hat{R}_s^f + \tfrac{1}{2} V^T \hat{G} V \qquad \text{(a)}$$
$$\text{where} \quad \hat{R}_s^f = \beta_0 + B^T(X_s - X_m) + \tfrac{1}{2}(X_s - X_m)^T \hat{G}(X_s - X_m) \qquad \text{(b)}$$

(45)

The normalized eigenvectors T_i, associated with the distinct eigenvalues of the symmetric Hessian $\hat{G}$ as

$$T^{-1}\hat{G}T = \Lambda$$

(46)

in Table 2, can be used as an orthonormal basis to transform the parameter axes along the principal directions of the elliptical shaped contour system.

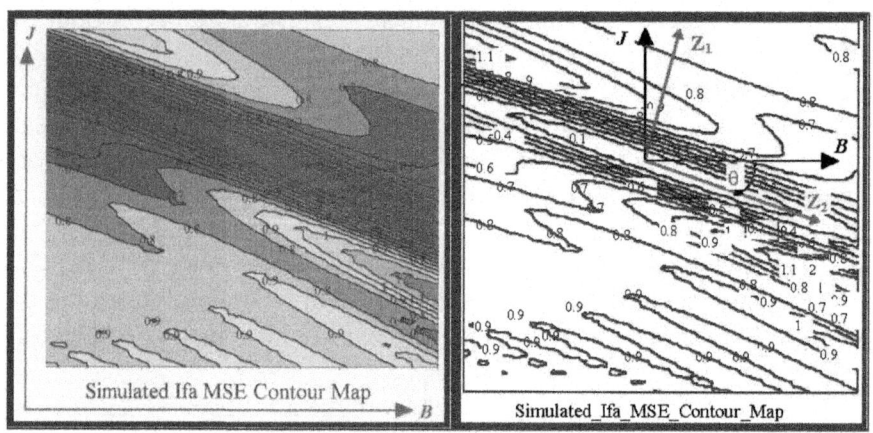

Simulated Ifa MSE Contour Map

Simulated_Ifa_MSE_Contour_Map

Figure 20: A: Simulated MSE-I_{fa} Contour Map B: I_{fa} Contour Map with Canonical Variables

This rotation of co-ordinates, with origin anchored to X_s, is displayed in Figs 20 and 10 for both simulated cost surfaces to eliminate the interactive terms β_{12} in $\hat{G}$. The X co-ordinate angular displacement θ in Figure 20B can be evaluated from the conical expression [29] as

$$\cot 2\theta = \frac{\beta_{22}-\beta_{11}}{2} \Big/ \beta_{12} \tag{47}$$

using (15) with values listed in Tables 2 and 3 which are very small. The rotation can also be deduced from the directional cosines of the unit column vectors $[\hat{a}_{z1} \quad \hat{a}_{z2}]$ constituting the modal matrix T via

$$\mathbf{T}=\begin{bmatrix}\hat{\mathbf{a}}_{z1} & \hat{\mathbf{a}}_{z2}\end{bmatrix}=\begin{bmatrix}\hat{\mathbf{a}}_{x1} & \hat{\mathbf{a}}_{x2}\end{bmatrix}\begin{bmatrix}\cos\theta & -\sin\theta \\ \sin\theta & \cos\theta\end{bmatrix} \tag{48}$$

The normal form of the response surface model can be expressed with substitution of the canonical variable

$$Z = T^T V \tag{49}$$

As $\quad \hat{R}^f = \hat{R}_s^f + \frac{1}{2}Z^T \Lambda Z$

$$=\hat{R}_s^f +\frac{1}{2}\sum_{k=1}^{2}\lambda_k z_k^2 =\hat{R}_s^f +\frac{\lambda_1}{2}z_1^2 +\frac{\lambda_2}{2}z_2^2 \tag{50}$$

with model cost $\hat{R}_s^f$ at the global estimate given in Table 4. The nature of the fitted quadratic model can be deduced from the shape of the embedded contours, in the vicinity of the global minimum estimate X_s, by inference from the sign of the scalar discriminant invariant which pertains to elliptical conic sections [29] as

$$\beta_{12}^2 - \beta_{11}\beta_{22} < 0 \tag{51}$$

in the $x_1 x_2$ frame or

$$-\lambda_1 \lambda_2 < 0 \qquad \text{with} \qquad \lambda_{12} = 0 \tag{52}$$

in $z_1 z_2$ normal co-ordinates. These contours are elliptical for both choices of observed BLMD target data with a negative discriminant in Table 3 based on the model cost at nominal parameter value X_m. Consequently the stationary region enveloping the global minimum is encircled and thus bounded by elliptical contours rather than contained within an open wedge shaped response surfaces with no define convergence zone. The degree of elliptical eccentricity e of the trapped stationary zone quantifies the extent of the 'line minimum' of global minimum convergence, congruent with the major axis, as notionally illustrated in Fig. 21.

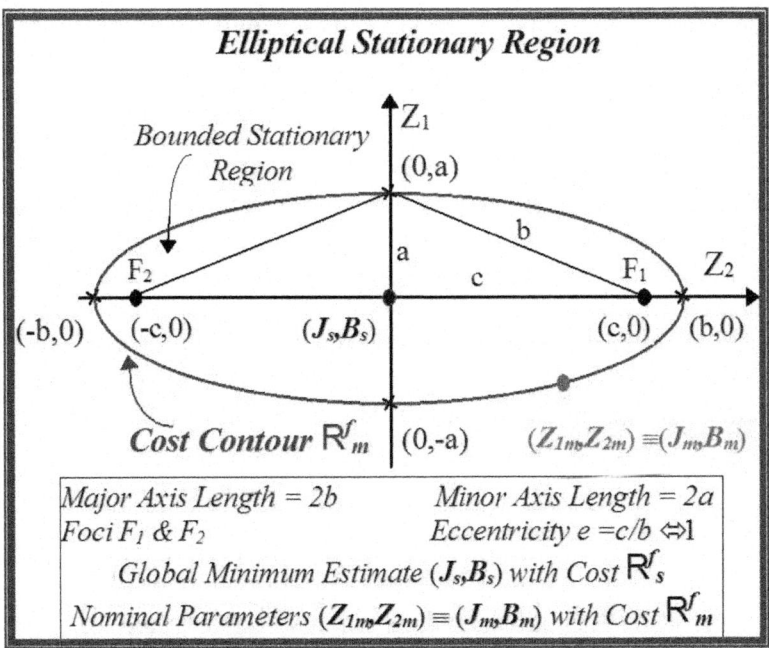

Figure 21: Elliptical Contour Bounded Stationary Zone

This can be determined from consideration of Fig. 21 by recasting the expression for the normal form of the cost surface model in (50) into that for an elliptical contour, evaluated at the nominal parameter value X_m with origin at X_s, as

$$\frac{z_1^2}{a^2} + \frac{z_2^2}{b^2} = 1$$

(53)

for model cost differential

$$\Delta\hat{R}_{m,s}^f = \left(\hat{R}_m^f - \hat{R}_s^f\right)$$

(54)

with intrinsic parameters

$$a^2 = 2\Delta\hat{R}_{m,s}^f / \lambda_1 \text{ and } b^2 = 2\Delta\hat{R}_{m,s}^f / \lambda_2$$

(55)

The eccentricity e, which measures the degree of 'flatness' of the oblate model contour $\hat{R}_m^f$ specified at X_m and thus the 'linear extension' of the global minimum X_0, is given in terms of the lateral displacement c of the elliptical foci from the global estimate X_s relative to the length 2b of the major axis as

$$e = \frac{c}{b} = \frac{\sqrt{b^2 - a^2}}{b} = \sqrt{1 - \left(\frac{a}{b}\right)^2} = \sqrt{1 - \left(\frac{\lambda_2}{\lambda_1}\right)^2}$$

(56)

The eccentricity of the contours for the model target data in Table 3 is almost unity, as a consequence of the large spectral condition number η in each case, indicating a very flat distended ellipse. The elliptical contours approximates an extended pencil-like global minimum predominantly in the B parameter co-ordinate direction because the inclination angle θ in (47) is less than $2°$. This is qualified by the magnitude of the axial ratio (AR) which defines the extent 2b of the 'line minimum valley' along the principal direction of the ellipse in relation to its girth 2a, given by the minor axis length, as

$$AR = \frac{b}{a} = \frac{\lambda_1}{\lambda_2} = \eta$$

(57)

This contrast of stationary region 'feature sizes' in parameter X-space is readily identified as the spectral condition number η in Table 3 with a 'line minimum' extension ratio of about three orders of magnitude for each target data training record used in the MSE cost surface description.

The slope and curvature matrix $\hat{G}$ of the fitted cost model including its associated eigenvalues are re-evaluated at the acquired global estimate X_s as summarized in Table 4 to gauge the response surface selectivity either along the parameter co-ordinate directions or the principal axes of the normal form. The second iterative estimate X_{s1}, along with the residual costs given in Table 4, is very close to the global minimum target X_0 listed in Table 2 for both cost surface models despite the large condition number in each case. A quantifiable measure of the fitted model selectivity in an ill conditioned stationary region, tagged by a large spectral number η, at discerning the global minimum can be obtained from the surface curvature κ_j along a particular parameter co-ordinate direction x_j as

$$\kappa_j = |\beta_{jj}| \Big/ \sqrt{1 + \frac{\partial \hat{R}^f}{\partial x_j}\Big|_{X_s}} \approx |\beta_{jj}|$$

(58)

where

$$\frac{\partial \hat{R}^f}{\partial x_j} << 1 \text{ at } X = X_s$$

(59)

or alternatively in normal form as

$$\kappa_j = \lambda_j$$

(60)

The degree of model selectivity is three orders of magnitude greater in the case of the motor inertia for actual measured target data employed in both response surface approximations as evidenced from the spectral condition number in Table 3 and in Table 4 for simulated target data trials. Consequently this selectivity margin renders a more accurate estimate in the extracted J-parameter which is mirrored by the arguments leading to the feature size ratio in (57). The cost surface selectivity improves along the principal axes of the normal form when the target data length N_d is extended as indicated by the increased magnitudes of the eigenvalues in Tables 4 and 3.

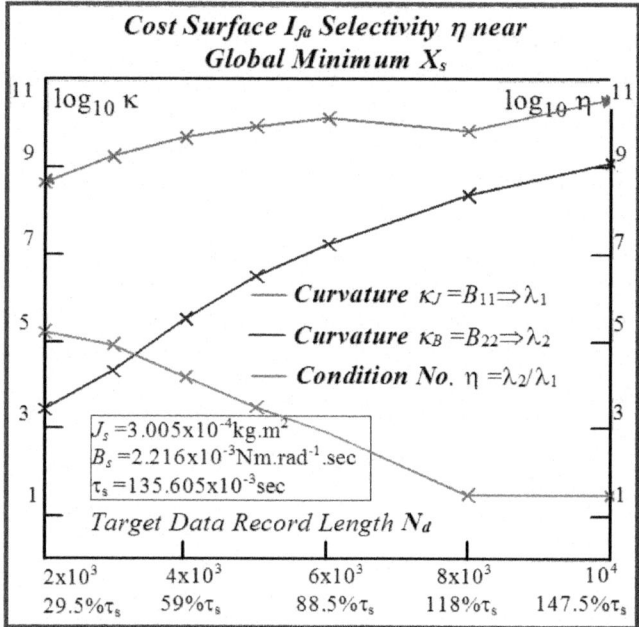

Figure 22: I_{fa} Cost Surface Selectivity

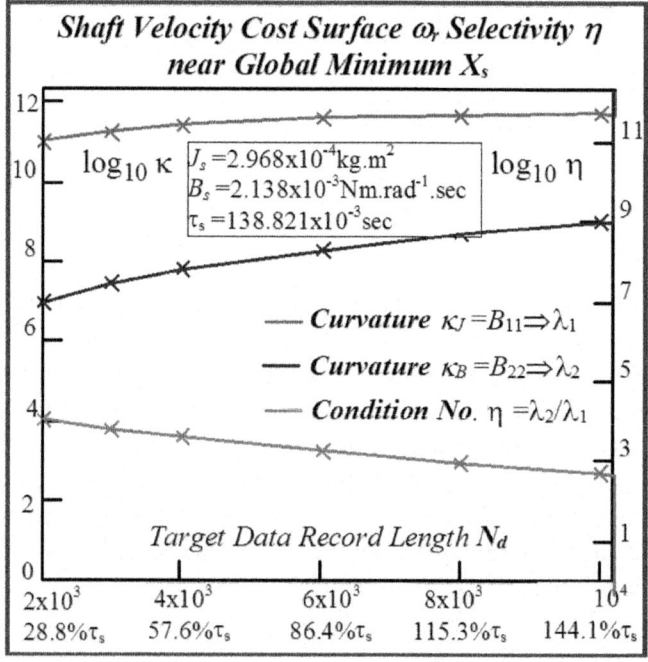

Figure 23: ω_r Cost Surface Selectivity

This trend in enhanced J parameter selectivity, which is a measure of the accompanying increase in curvature at the global extremum, is displayed in Figs. 22 and 23 for increasing data record lengths and is a manifestation of the narrowing of the cost surface fold containing the directed 'line minimum' principally in the B-parameter direction. The selectivity improvements are greater for increased FC step response data record lengths in Fig. 22than those for shaft velocity target data in Fig.23. This due to the appearance of more FC cycles with reduced periodicity as motor speed increases demanding a greater degree of fitted model accuracy, with smaller margins of error in terms of frequency and phase coherence at the global minimum value, in the extraction of the optimum parameter vector X_O during system identification. The shaft velocity step response by contrast losses its excitation persistence with transient speed decay as it evolves towards steady state conditions with increased data capture time. After a sufficient time elapse the target data transient information, responsible for velocity cost surface folding, is submerged by the steady state onset of maximum motor speed conditions. This irretrievable loss of target velocity signal amplitude variation with time results in a reduction of surface selectivity with parameter variation near the global minimum. These considerations admit to a better choice in the current feedback as a suitable candidate for MSE objective function formulation where accurate parameter extraction is essential during the identification phase of optimal controller design in high performance adaptive BMLD systems for electric vehicle mobility. Furthermore the increasing trend towards motor sensorless control [30] obviates the need for separate rotor position sensors with essential information obtained from the motor signature current via FC sensing at the inverter controller o/p. This adoption of sensorless operation in motor drive systems lends added importance to observed FC data as a suitable target function during parameter identification.

RESPONSE SURFACE NOISE AND PARAMETER QUANTI-ZATION

The computation 'noise' inherent in the MSE penalty function construct, based on simulated target data at nominal machine parameter values, is manifested as response surface roughness in parameter space. This is due to model nonlinearities and coarseness of evaluation of the PWM switching instants and results in 'false' local minima proliferation in the neighborhood of the global minimizer.

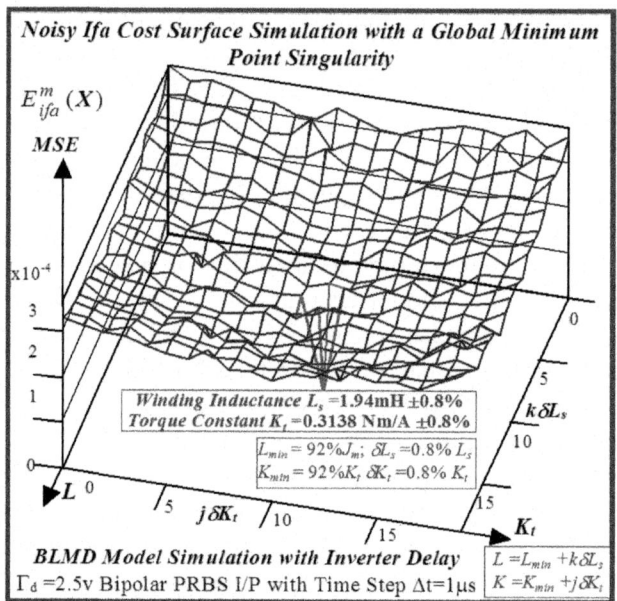

Figure 24: Noisy I_{fa} Cost Surface with PRBS I/P

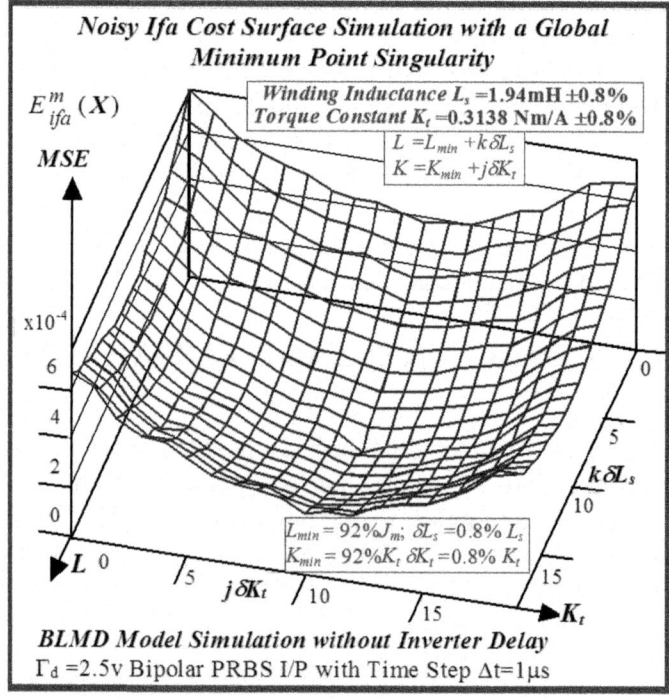

Figure 25: Noisy I_{fa} Cost Surface with PRBS I/P

A typical example of this is illustrated in Figs 24 and 25 for BLMD model simulations, with and without inverter turn-on delay δ considered, for small step change variations in the stator winding inductance L_s and torque constant K_t parameters. These response surfaces were obtained from BLMD simulation, using FC target data for nominal parameter values as in [1], with a 4095 bit maximal length 2.5 volt bipolar pseudorandom binary sequence (PRBS) input stimulus. The response surface in Fig.24has a very shallow paraboloidal shape for the small parameter tolerance ranges chosen with a rough noisy texture peppered with local minima in the vicinity of the point-like global minimum. The response surface for simulated FC target data is relatively smooth in the absence of inverter delay turn-on with a point-like singularity at the global minimum as shown in Fig.25. The cost functions pertaining to simulated step response FC I_{fa} and shaft velocity ω_r target data, displayed in Figs.26 and 27 for the dynamic parameters {J,B}, are also noisy with point-like multiminima scattered around the 'pinhole' stationary point as in the former case. These surfaces are parabolic for very small tolerance ranges selected near the global minimum as in the main lobe of Fig.1 for the FC corrugated surface.

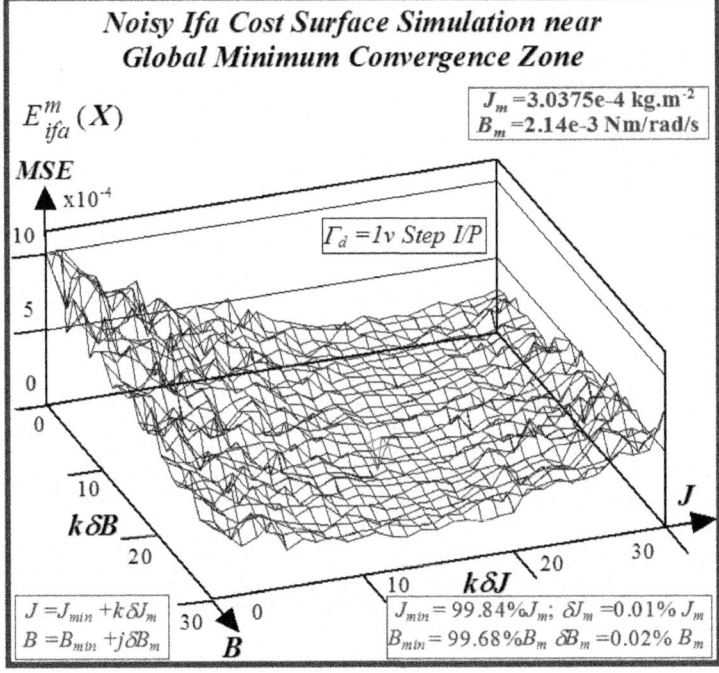

Figure 26: Noisy I_{fa} Cost Surface with Step I/P

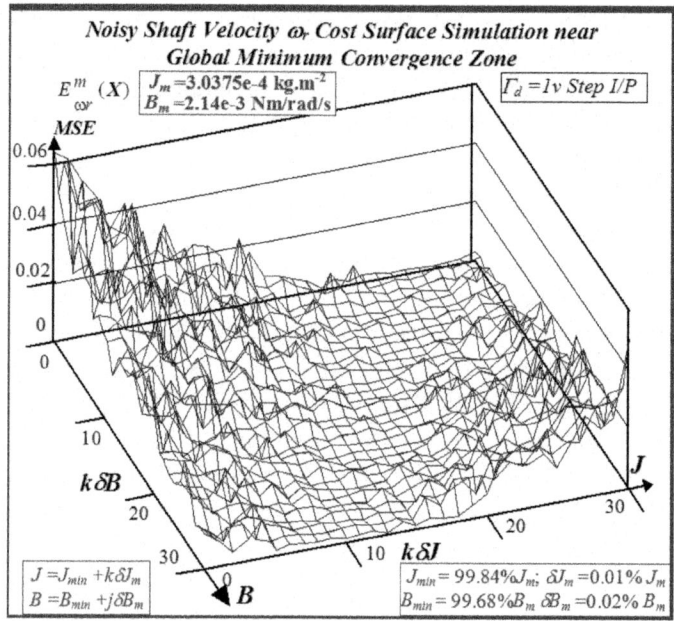

Figure 27: Noisy Velocity Cost Surface with Step I/P

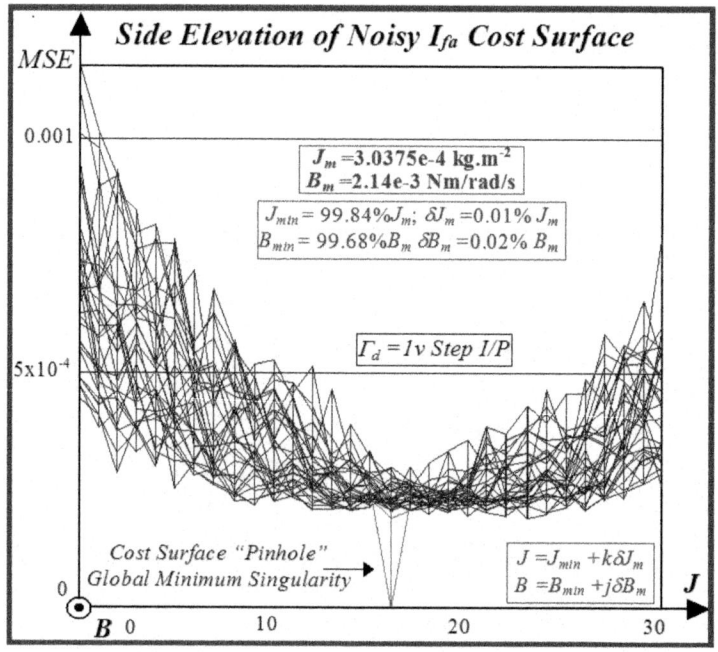

Figure 28: Noisy I_{fa} Cost Surface Side Elevation

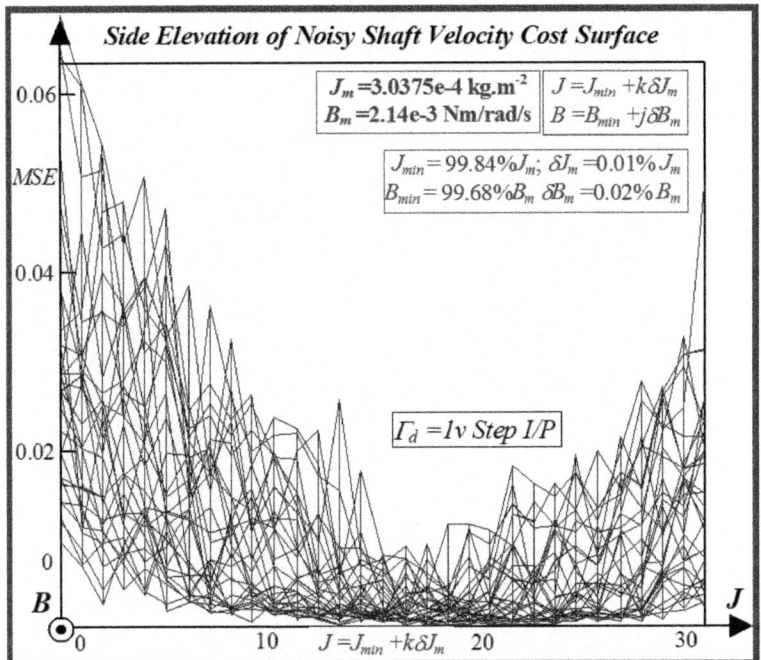

Figure 29: Noisy Velocity Cost Surface Side Elevation

The side elevations of the MSE cost functions in Figs.28 and 29 demonstrate very effectively the fractal landscape with multiminima plurality disposed about the global extremum in the FC case.

In the simulated velocity response surface shown in Fig.29 a stationary region exists at zero floor cost with no definite observable global minimum point. An alternative perspective of the minimum stationary regions is provided by the contour maps shown in Figs.30 and 31 for FC and shaft velocity target data respectively.

The existence of the point-like global minimum singularity with surface noisiness is clearly evident from the level contours in the FC surface relief map. In the case of the shaft velocity response surface the presence of 'noisy' local minima strewn over the 'river bed' syncline of the global minimum stationary region is clearly defined by the contour map in Fig.31. The occurrence of 'noisy' local minima in the above error surfaces presents a difficulty to any classical optimization method in acquiring the global minimizer where fine parameter resolution is concerned.

A more detailed examination of the effect of inverter delay, achieved through BLMD model simulation without current controller o/p saturation using a 1 volt torque demand step i/p, on the one dimensionalMSE response surface in Fig. 32 for very small inductance variation reveal a granulated profile which is less pronounced than that in Fig. 33 with the absence of delay.

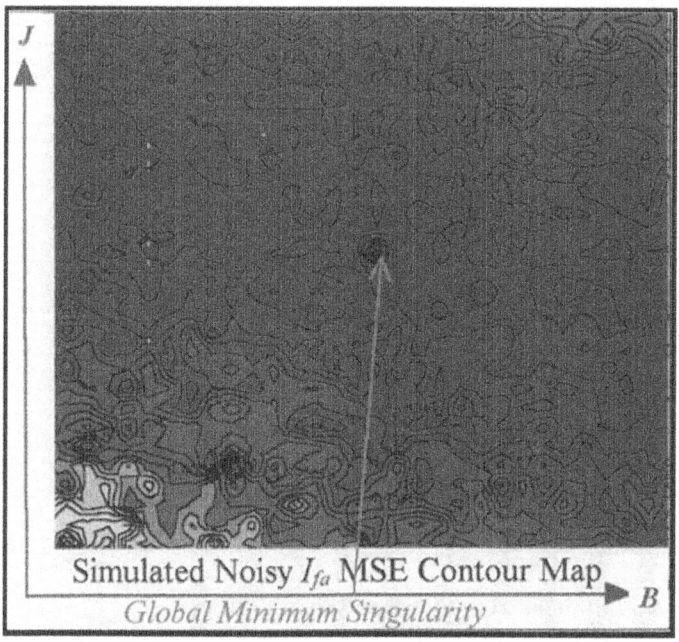

Figure 30: Noisy Local Minima I_{fa} Contour Map

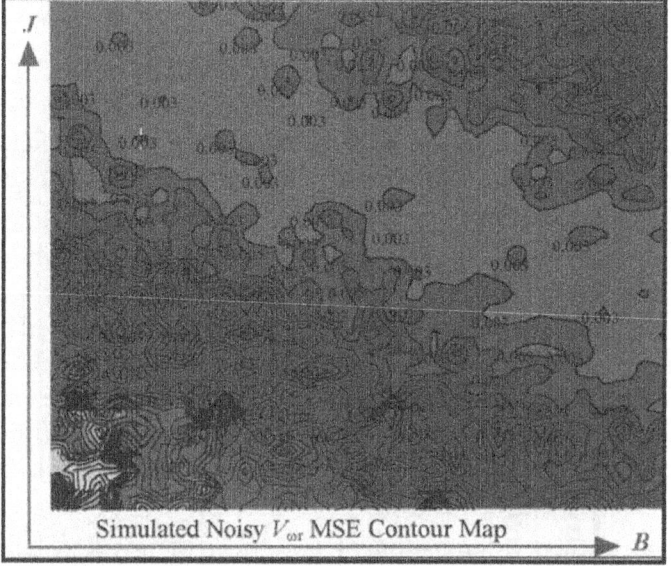

Figure 31: Noisy Local Minima ω_r Contour Map

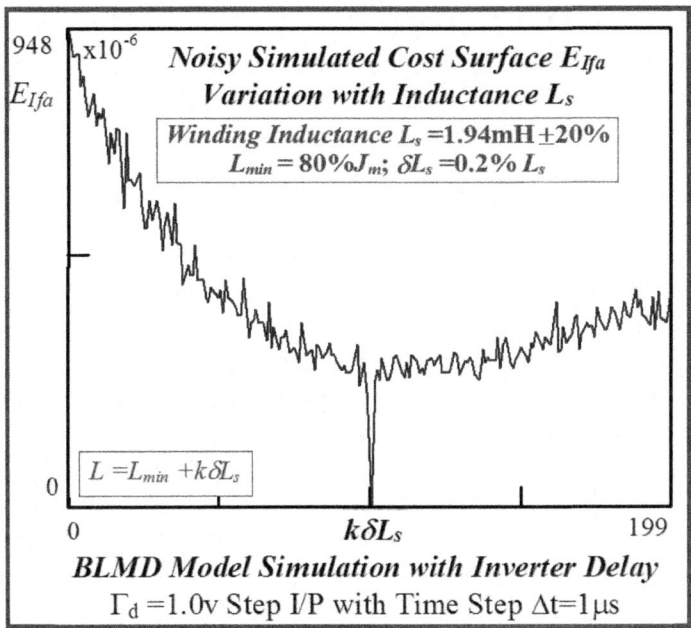

Figure 32: Cost Simulation with Delay δ

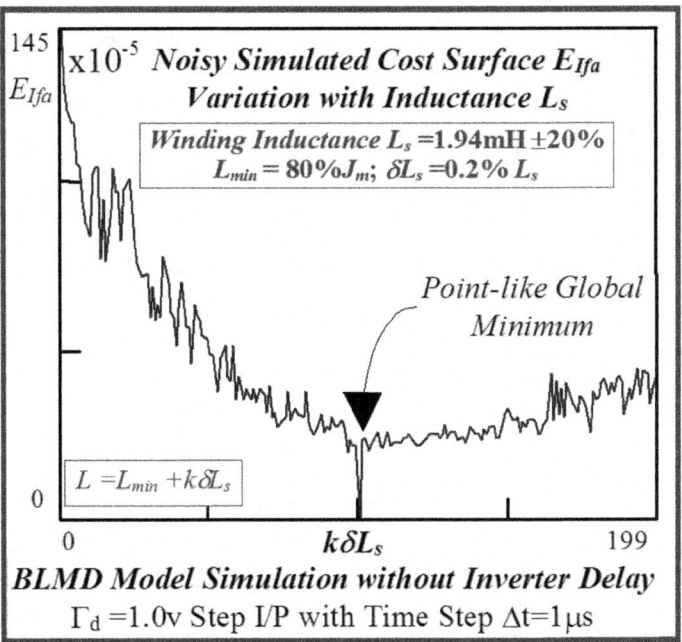

Figure 33: Cost Simulation without Delay

The use of a PWM switch transition time search, based on a single iteration of the regula-falsi method to keep simulation time overhead low, marginally reduces the response error as in Figs.34 and 35. The sensitivity [31] of the error response E with inductance L_s

$$S^E_{L_s} = \left(\frac{L_s}{E}\right)\left(\frac{\partial E}{\partial L_s}\right)$$

(61)

is very low in all cases and for a ±12% inductance variation gives a change of $\Delta E = 1.5 \times 10^{-4}$ in 1.75×10^{-4} for E. Poor cost surface selectivity will ensue in such cases of low sensitivity over a large parameter tolerance range with possible local minimum convergence if the search process is initiated far from the global minimizer with a noisy cost function.

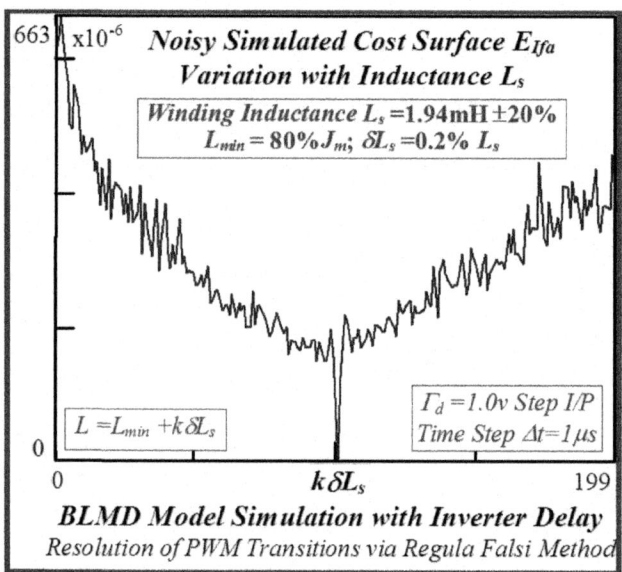

Figure 34: BLMD Delay & Reg.-Fal. Method

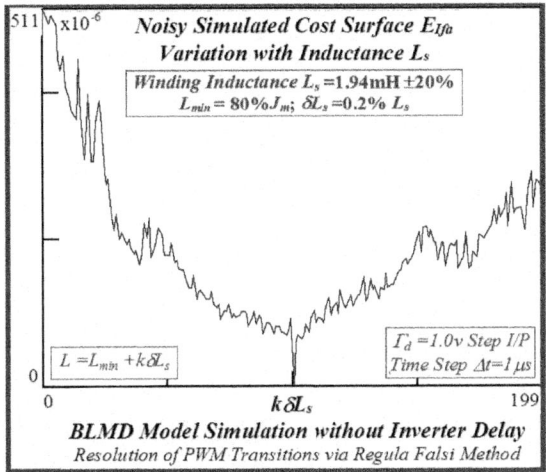

Figure 35: Zero Delay & Reg.-Fal. Method

Novel Theoretical Estimation of Pwm Edge Transition Computation Noise [18]

A measure of the computation 'noise', induced through inaccurate resolution of the PWM edge transition within a simulation time step Δt, can be ascertained from the associated error in random pulsed energy delivery by the inverter to the stator winding within Δt. Since there is one PWM edge transition every half-switching interval $T_S/2$ of the inverter the expectation in the power delivery error to the stator can be obtained [18], from the error in pulse energy dispatch during the time step interval Δt, as

$$E(\boldsymbol{P}) = E(\boldsymbol{E}) \Big/ \frac{T_s}{2} = \frac{E_{max}}{2} \Big/ \frac{T_s}{2} = U_d^2 \Delta t / T_S$$

(62)

The expected random voltage v_n error associated with inaccurate resolution in PWM inverter switching during BLMD simulation is thus given by

$$v_n = \sqrt{E(\boldsymbol{P})} = U_d \sqrt{\Delta t / T_S} \qquad \text{(a)}$$
$$v_n = 310\sqrt{1/200} = 21.92 \text{ volts} \qquad \text{(b)}$$

(63)

If the chosen simulation time step Δt is 1μs and the inverter switching parameters in [1] are substituted into (63)-(a) the expected uncertainty v_n in the inverter output voltage V_{jg} per phase j can be obtained as (63)-(b)

This value of voltage uncertainty in the inverter output is not insignificant as its magnitude is 7.1% of the inverter HT voltage U_d for a simulation time step size of $1\mu s$. The error can be reduced by decreasing the simulation step size Δt for a more accurate resolution of the pulse edge transition time, once its occurrence has been flagged, or alternatively by means of an accurate search using the regula-falsi method as described in [1].

The statistical considerations of pulsed energy delivery by the PWM inverter in [18], arising from BLMD simulation with a fixed time step size, illuminates the origin of computation 'noisiness' and its subsequent manifestation as cost surface roughness as shown in Figs.36 and 37. For fixed shaft load inertia changes, encountered in motive power applications and electric vehicles, the use of a coarse quantization step size δJ in the inertial parameter variable about the nominal value J_m results in smooth generated and noise-free response surfaces as shown in Figs.1 and 2 for actual FC and shaft velocity target data. However for a sufficiently small step size variation in the inertia J_m and damping B_m a 'noisy' cost surface with a proliferation of local minima results in both cases as shown in Figs 36 and 37 for corresponding target test data. The degree of resolution of the parameter step size, that can be obtained and then used in an identification search strategy, depends upon the onset of cost surface irregularity.

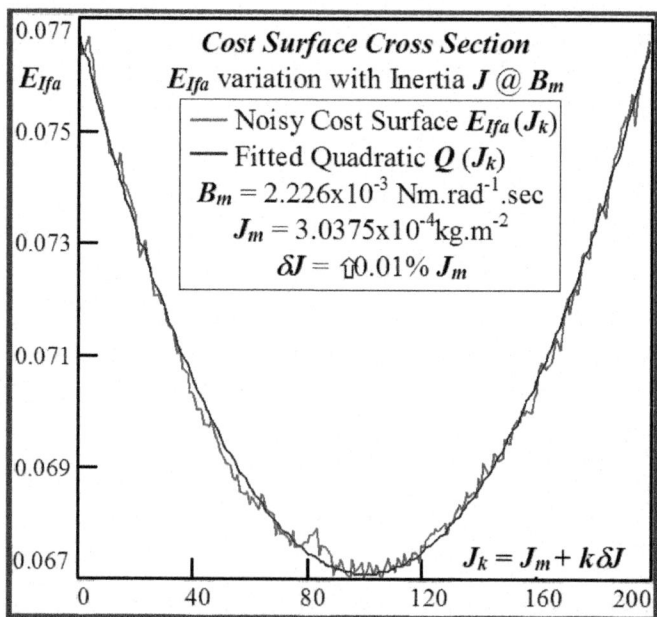

Figure 36: I_{fa} Surface Noise with Inertia Variation

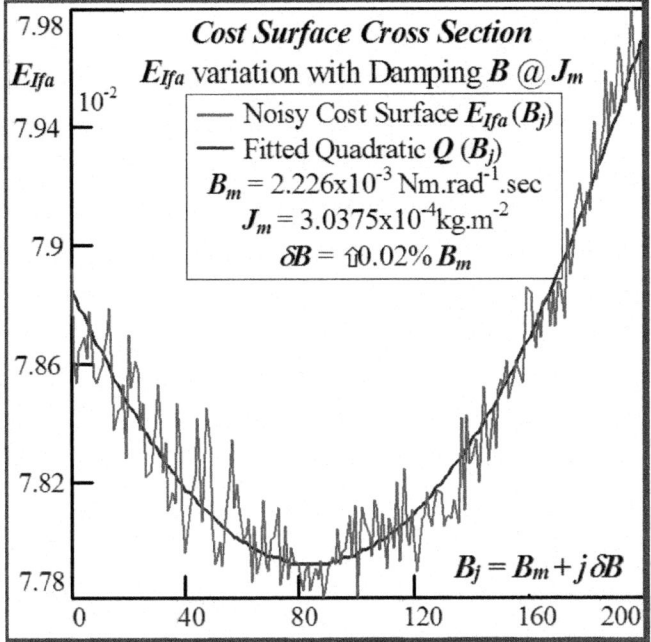

Figure 37: Ifa Surface Noise Variation with B

Novel Mathematical Analysis Of Quadratic Curve Fitting To Noisy Mse Cost Surface [18,21]

The accuracy of any classical identification scheme used in terms of parameter resolving capabilities can be gauged by fitting a quadratic [26] to the response surfaces in each test case and determining rms deviation of the PWM computation noise related residuals. The quadratic fit employed

$$Q(x_k) = b_0 + b_1 x_k + b_2 x_k^2; \qquad 1 \le k \le N$$

$$(64)$$

for N steps in the indexed parameter x_k as shown in Fig.38 with

$$x_k = x_m + (k - m)\delta x; \qquad x_m \in \{J_m, B_m\}$$

$$(65)$$

is based on an infinitesimal step size δx, in model simulation to reflect response surface roughness, and centred in a tolerance band $\pm \Delta x_m$ within indexed range m of the nominal value x_m as

$$m = \Delta x_m / \delta x$$

$$(66)$$

for

$$\Delta x_m = \left(x_m - x_{min} \right) = \left(x_{max} - x_m \right)$$

$$(67)$$

The nature of the residual error

$$W_f(x_k) = E_f(x_k) - Q_f(x_k)$$

$$(68)$$

associated with the least squares quadratic fit to the various cost functions

$$f \in \{ \omega_r, I_{fa} \}$$

$$(69)$$

for example in Figs.39 and 40 for the FC cost surface, is demonstrated by the autocorrelation (ACR) functions

$$\Re_W^f(j) = \frac{1}{N} \sum_{k=1}^{N} W_f(x_k) \cdot W_f(x_{k+j})$$

$$(70)$$

shown in Figs.41 and 42, which are mainly of the impulse type at zero offset, indicating a white noise-like characteristic.

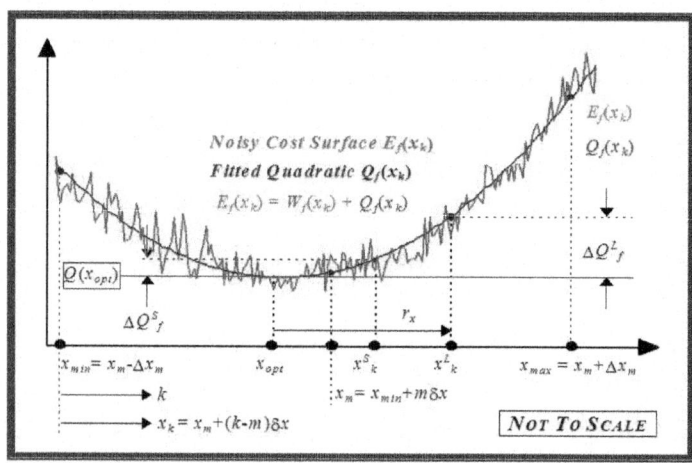

Figure 38: Parameter Step Size Resolution

Table 5: FC Cost Function $E_{Ifa}(x)$ and Quadratic Fit $Q_{Ifa}(x)$

FC Target Data with reference to details in Figs. 36 and 37		
Parameter varied x	Jm	Bm
Nominal value x_m	$3.0375 \times 10^{-4} \text{kg.m}^2$	$2.226 \times 10^{-3} \text{Nm.rad}^{-1}.\text{sec}$
Parameter Step Size δx	$0.01\% \ Jm$	$0.02\% \ Bm$
No. of Steps N	200	200
Tolerance Index m	100	100
b_0	96.864	3.286
b_1	-6.373×10^5	-2.891×10^3
b_2	1.049×10^9	6.516×10^5
Residual Error $W_{Ifa}(x) = E_{Ifa}(x) - Q_{Ifa}(x)$ illustrated in Figs. 39 and 40		
Standard Deviation $\hat{\sigma}$	1.916×10^{-4}	1.127×10^{-4}
Mean	-3.309×10^{-8}	-3.306×10^{-9}
Peak Absolute Deviation	5.203×10^{-4}	3.355×10^{-4}

The various cost function details with accompanying quadratic fits and corresponding residuals are summarized in Tables 5 and 6, based on FC and shaft velocity test data respectively, for independent parameter variation in the BLMD shaft inertia and damping factor. The quadratic polynomials fitted to the noisy shaft velocity cost surface sections are displayed in Figs. 43 and 44 with coefficients given inTable 6. The corresponding cost residuals associated with the fitted velocity profiles, which appear to be random, are shown for each of the dynamical parameter variables in Figs. 45 and 46.

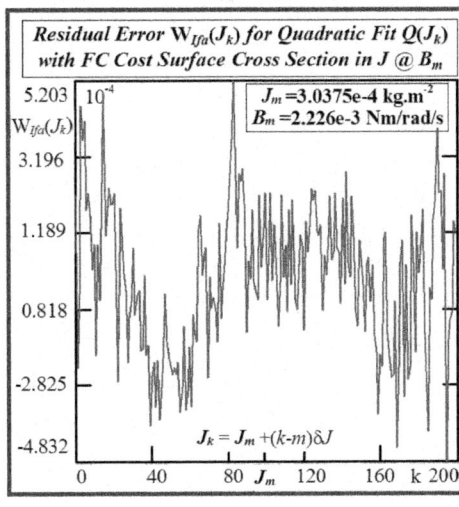

Figure 39: Quadratic Error in J @ B_m

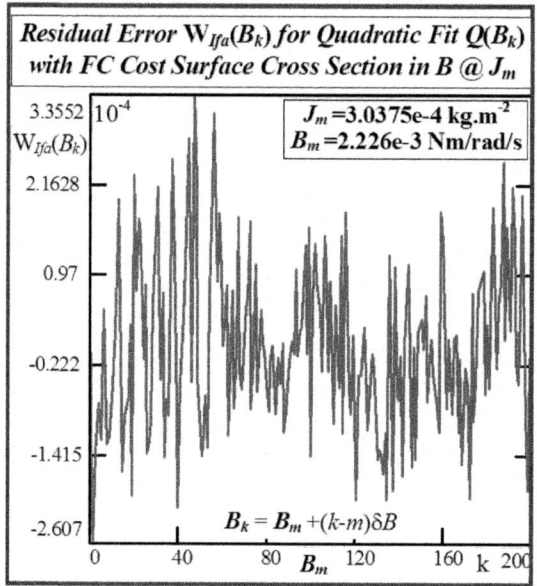

Figure 40: Quadratic Error in B @ J_m

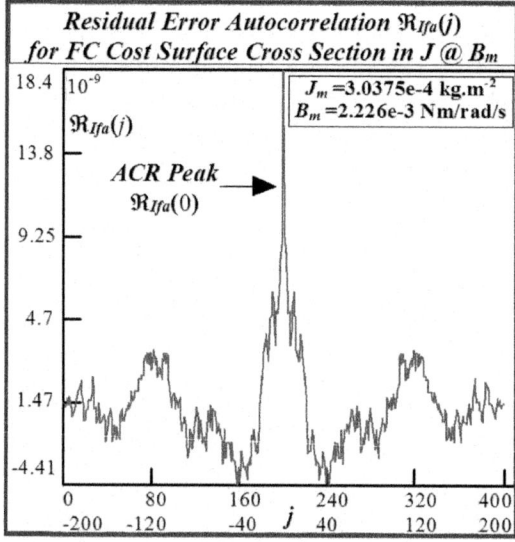

Figure 41: Error Autocorrelation in J @ B_m

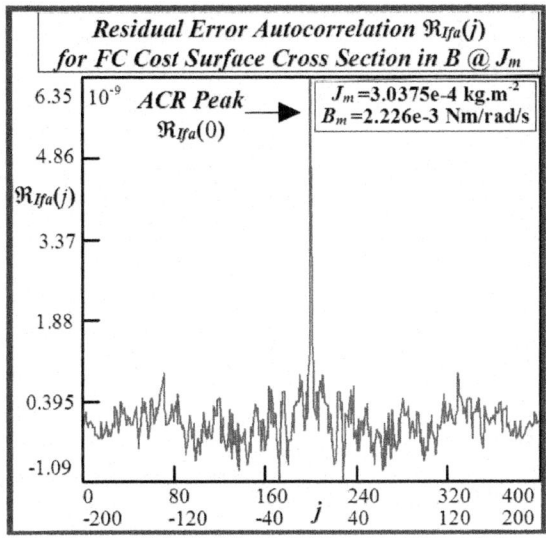

Figure 42: Error Autocorrelation in B @ J_m

The white noise-like nature of the error-of-fit in the case of the shaft velocity cost surface sections is demonstrated by the impulse characteristic of the ACR spike functions in Figs.47 and 48. The errors-of-fit can thus be considered as a random entity, with an ACR related noise signature, associated with the BLMD simulation model at very high parameter resolution for each of the observed target data records used in the MSE cost formulation. This manifestation is attributed to some residual uncertainty in the BLMD model simulation of the PWM edge transitions at the comparator o/p with dead time, despite the single iteration cycle of the regula-falsi search, which are magnified in the three phase inverter o/p before being fed to the stator winding.

TABLE 6: Shaft Velocity Cost Function $E_\omega(x)$ and Quadratic Fit $Q_\omega(x)$

Shaft Velocity Target Data with reference to details in Figs.43 and 44		
Parameter varied x	Jm	Bm
Nominal value x_m	$3.0375 \times 10^{-4} \text{kg.m}^2$	$2.14 \times 10^{-3} \text{Nm.rad}^{-1}.\text{sec}$
Parmeter Step Size δx	0.01% Jm	0.02% B_m
No. of Steps N	370	600
Tolerance Index m	100	100
b_0	1.698	0.157
b_1	-1.079×10^4	-1.098×10^2
b_2	1.747×10^7	2.465×10^4
Residual Error $W_\omega(x) = E_\omega(x) - Q_\omega(x)$ illustrated in Figs 45 and 46		
Standard Deviation $\hat{\sigma}$	1.028×10^{-5}	8.974×10^{-4}
Mean	6.982×10^{-10}	-1.415×10^{-11}
Peak Absolute Deviation	3.476×10^{-5}	2.871×10^{-5}

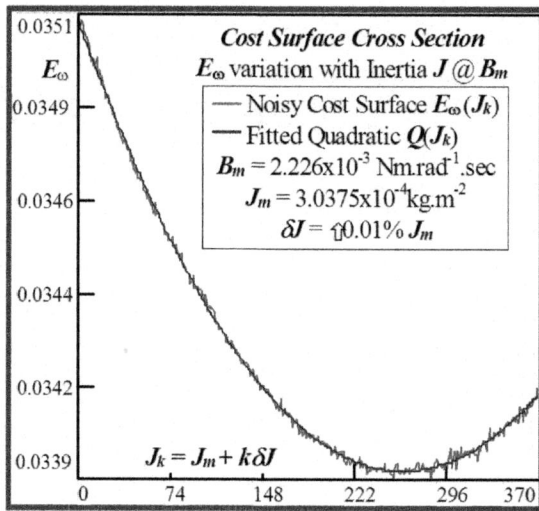

Figure 43: Velocity Cost Noise Variation with J

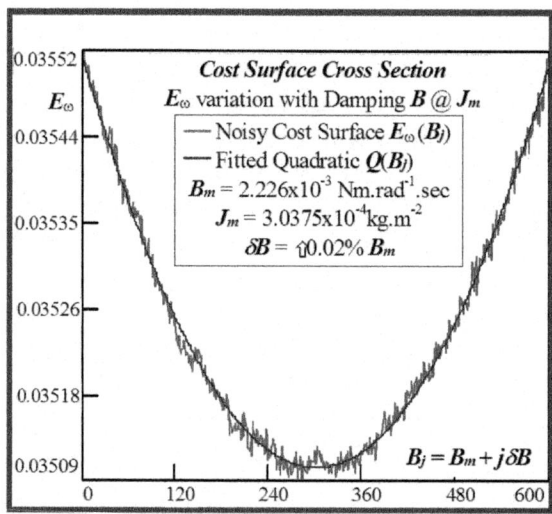

Figure 44: Shaft Vel. Surface Noise with B Variation

The effect of lowering the drive model simulation time step Δt, as shown in Figs. 49 and 50 for very small parameter variation in the vicinity of the global singularity, translates into a reduction of the MSE as well as gradual removal of response surface roughness. This tangible decrease in surface roughness with time step size, evident form Fig.51, is measured in terms of the standard deviation of the residual errors associated with various quadratic polynomials fitted to each of the FC cost sections. However the computational effort in terms of CPU time increases in proportion with the decrease in time step size for a given simulation trace length. The requirement for surface noise reduction with the elimination of false local minima plurality has to be balanced with a tradeoff in simulation run time in an attempt to reduce computation costs where BLMD model tractability is an issue in parameter identification and as a simulator in practical applications for performance related prediction of proposed embedded drive systems. A Taylor series expansion of the quadratic fit about the parabolic vertex x_{opt} as

$$Q_f(x) = Q_f(x_{opt}) + b_2(x - x_{opt})^2$$

(71)

with gradient

$$\left. \frac{\partial Q_f}{\partial x} \right|_{x_{opt}} = b_1 + 2b_2 x_{opt} = 0$$

(72)

can now be used to check the limit of parameter resolution and the "radius" of convergence for worst case conditions [19].

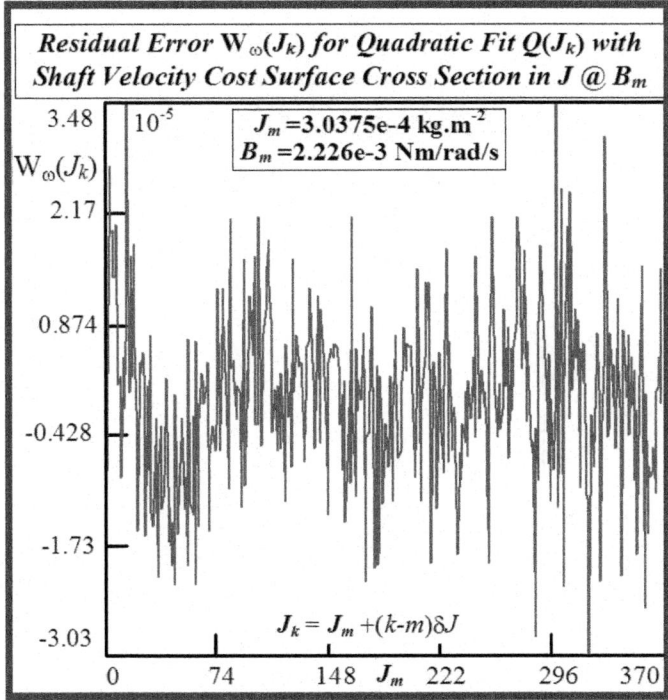

Figure 45: Quadratic Error in J @ B$_m$

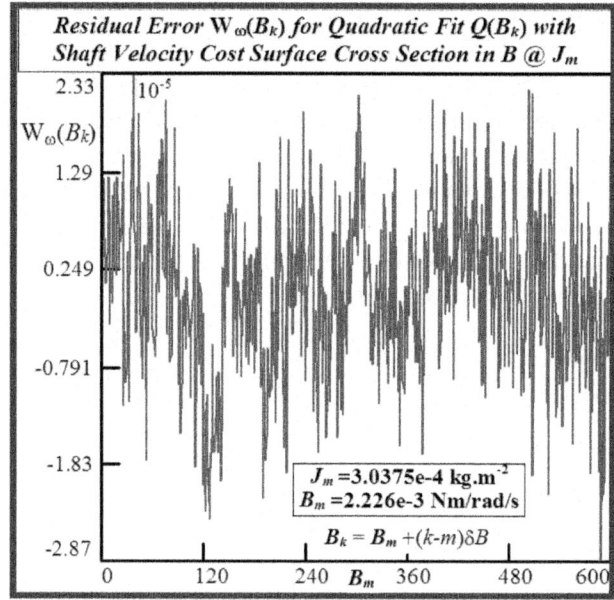

Figure 46: Quadratic Error in B @ J_m

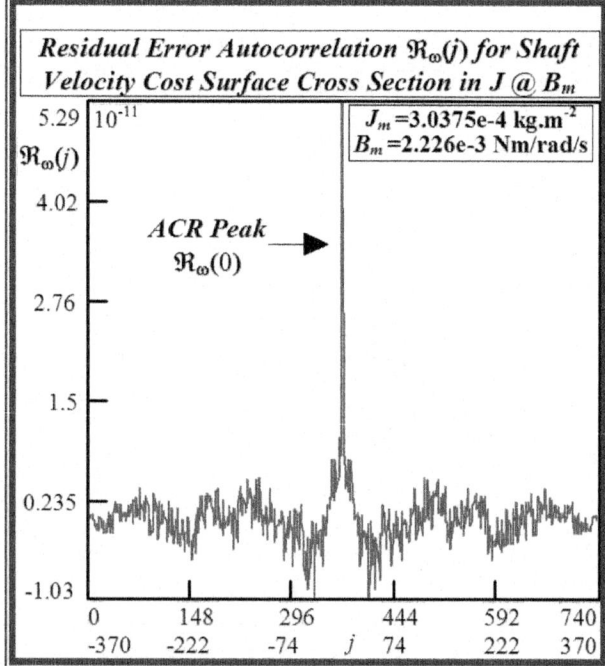

Figure 47: Error Autocorrelation in J @ B_m

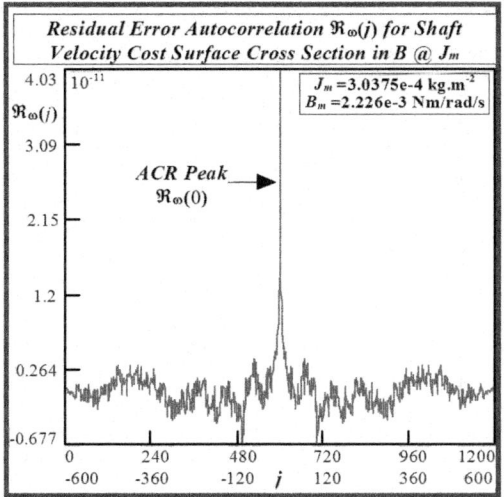

Figure 48: Error Autocorrelation in B @ J$_m$

At best the smallest parameter threshold step size required in simulation to overcome response surface noise, with rms sample estimate $\hat{\sigma}$, , is determined from that value x_k^s near the global minimum as inFig.38 such that

$$\Delta Q_f^S = \Delta Q_f(x_k^S) = Q(x_k^s) - Q(x_{opt}) = b_2(x_k^s - x_{opt})^2 \geq \hat{\sigma} \qquad (73)$$

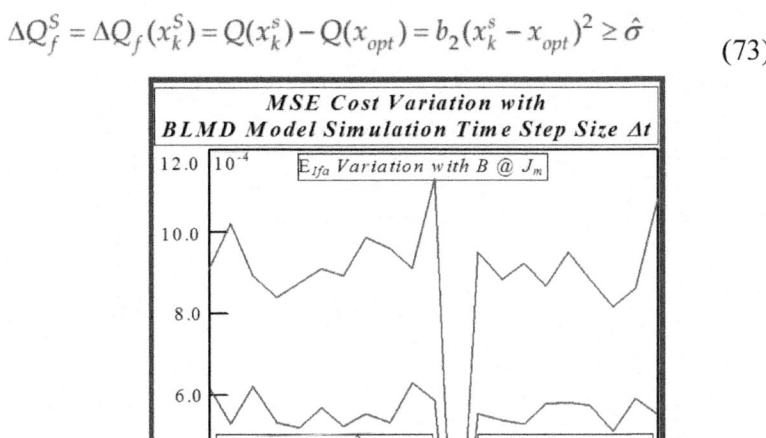

Figure 49: Error of Fit Vs Time Step

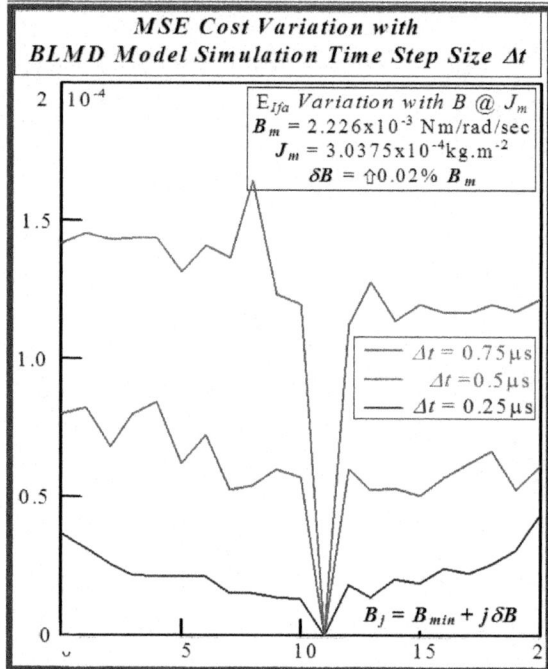

Figure 50: Error of Fit Vs Time Step

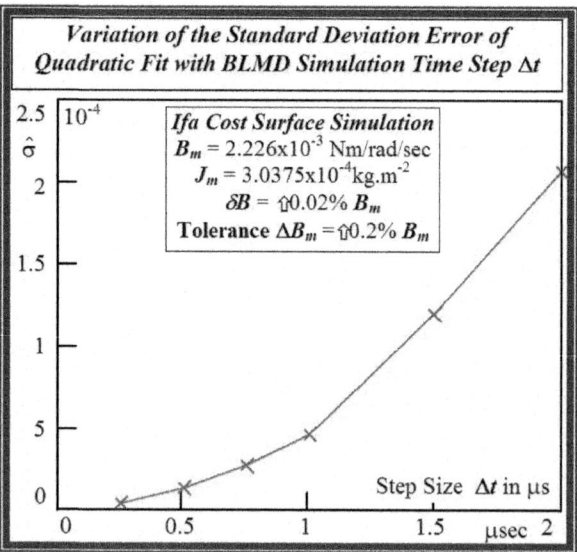

Figure 51: Error Variation with Time Step

Parameter Quantization and Radius of Convergence Estimation for System Identification

The largest threshold step size estimate can be determined by applying Chebyshev's theorem for statistical measurements [32] to the response surface noise sample [19, 26]. This theorem indicates that at least the fraction $1 - (1/h^2)$ of all the residuals W_{f_k} in any sample lie within h standard deviations of the mean $\hat{\mu}$ with probability

$$Prob\left\{(\hat{\mu} - h\hat{\sigma}) \le W_f(x_k) \le (\hat{\mu} + h\hat{\sigma})\right\} = 1 - (1/h^2)$$

(74)

and for h = 4, which exceeds the tabulated peak absolute deviation in all cases in Tables 5 and 6, is 94%. Thus a measure of the worst case parameter resolution is provided by the inequality

$$\Delta Q_f^L = Q_f(x_k^L) - Q_f(x_{opt}) = b_2(x_k^L - x_{opt})^2 \ge 4\hat{\sigma}$$

(75)

for some large x_k^L via the quadratic minimiser

$$x_{opt} = -b_1 / 2b_2$$

(76)

in (72) as

$$x_k^L \ge x_{opt} \pm \sqrt{\frac{4\hat{\sigma}}{b_2}}$$

(77)

The parameter resolution limit in terms of quantization step size δx necessary to overcome cost surface noisiness and local minimum trapping in BLMD parameter identification, which is also a measure of the convergence radius r_x about the global minimizer in Fig.38, is given by

$$\delta x^L = r_x = \sqrt{\frac{4\hat{\sigma}}{b_2}}$$

(78)

Table 7: Quantized Step Sizes for FC Cost Function in Figs. 36 & 37

Parameter varied x	Jm	Bm
Minimizer x_{opt}	3.038×10^{-4}kg.m^2	2.219×10^{-3}Nm.rad^{-1}.sec
Minimizer Offset $(m - k_{opt})$	−0.243	16.199
Threshold Locations k^L	72 & 129	24 & 143
Worst Relative Step Size $\frac{\delta x^L}{x_m}$	0.281%	1.182%

If measurements are referenced to the nominal value x_m at the centre of the parameter tolerance range the relative step sizes

$$\delta x_m^L = (k^L - m)\delta x$$

(79)

of which there are two pending the sign of the quadratic surd in (77), must be corrected by allowance for the global minimum offset

$$x_m - x_{opt} = (m - k_{opt})\delta x$$

(80)

Table 8: Quantized Step Sizes for Shaft Velocity Cost Function in Figs.43 & 44

Parameter varied x	Jm	Bm
Minimizer x_{opt}	$3.007 \times 10^{-4} \text{kg.m}^2$	$2.097 \times 10^{-3} \text{Nm.rad}^{-1}.\text{sec}$
Minimizer Offset $(m - k_{opt})$	-161.146	-201.448
Threshold Locations k^L	210 & 312	212 & 391
Worst Relative Step Size $\frac{\delta x^L}{x_m}$	0.505%	1.783%

The dynamical parameter threshold step sizes δx^L, which are by default the convergence radii measures for reliable global parameter estimation, are tabulated for the response surface cross sections in Tables 7 and 8. The resolution of the motor shaft inertia from the tabulated step sizes, which is the most likely to vary and more essential to identify in high performance applications, is higher when the FC cost function is used instead of the shaft velocity equivalent. Convergence of the inertia parameter estimates to the global minimum is enhanced in the former case with a lower uncertainty due to the smaller step size. The degree of selectivity of the fitted response surfaces with respect to the parameter variability [31] given by

$$V_x = \frac{\Delta x}{x}$$

(81)

can be determined through the sensitivity coefficient

$$S_x^{Q_f} = \left(\frac{x_{opt}}{Q_{opt}}\right)\left(\frac{\partial Q_f}{\partial x}\right)$$

(82)

in the vicinity of the global minimum x_{opt}. This measure can then be usefully employed as a performance index to decide on the best target test data available to use in a motor parameter identification strategy. The sensitivities for 2% parameter variability, greater than the largest threshold step size encountered, of the various fitted surfaces are summarized in Table 9. These sensitivity considerations indicate the suitability of FC test data in the objective function formulation for accuracy in parameter identification.

Table 9: Response Surface Sensitivity

Parameter varied	Jm	Bm
XX		
FC Response Surface Sensitivities Figs.36 and 37		
Sensitivity	28.88	0.82
Shaft Velocity Response Surface Sensitivities Figs.43 and 44		
Sensitivity	0.98	0.07

The above method of parameter quantization, employed to surmount cost surface noise and resultant avoidance of local minimum capture during system identification, reduces the search time in parameter space to global optimality. This is due to the reduction of N-Dimensional parameter space into a finite sized hypercube of countable lattice points N_C to be searched, within the imposed parameter tolerance bounds $^{\pm\Delta x_{m'}}$, using an interstitial 'distance' equivalent to the step size variability in Tables 7 and 8as

$$N_C = 2^N \prod_{j=1}^{N} \frac{\Delta x_m^j}{\delta x_j^L} = 2^2 \left(\frac{\Delta J_m}{\delta J^L}\right)\left(\frac{\Delta B_m}{\delta B^L}\right)$$

(83)

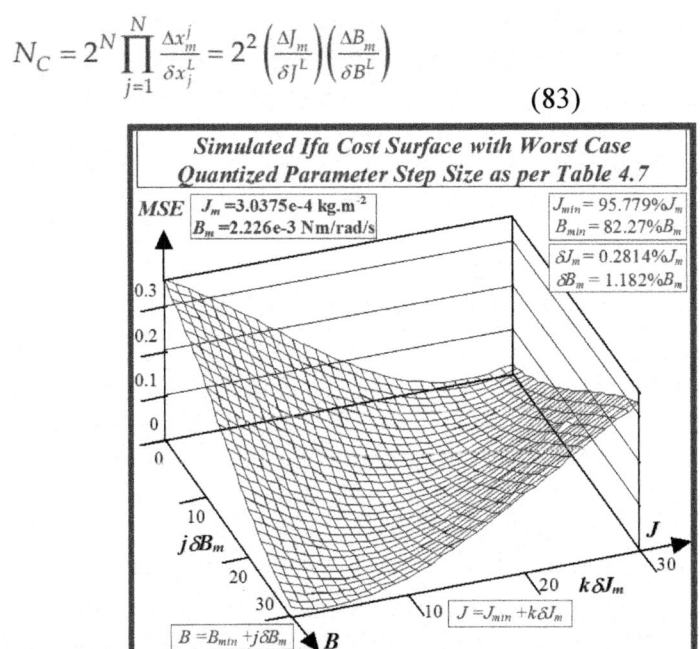

Figure 52: Simulated Ifa Cost Surface

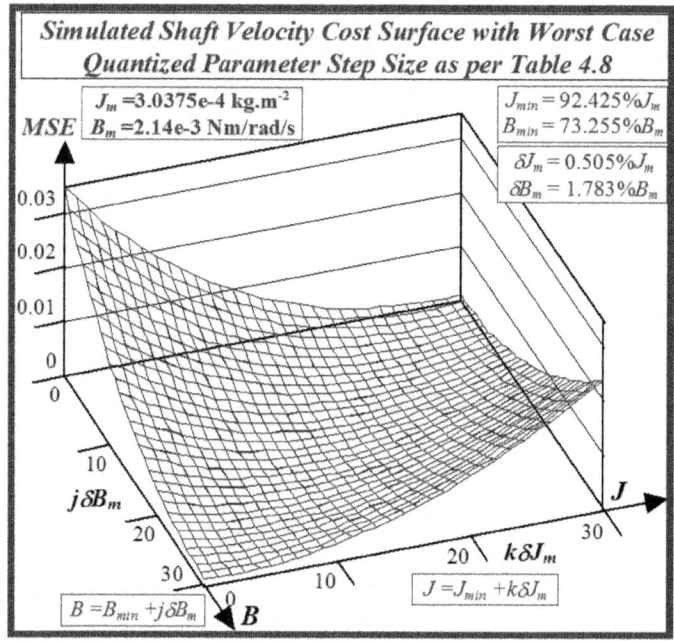

Figure 53: Simulated Shaft Velocity Cost Surface

The application of the tabulated parameter threshold sizes δx^L in the objective function simulation results in smooth noise-free response surfaces in the stationary region enclosing the global minimum as displayed in Figs.52 and 53. The degree of accuracy achieved by parameter quantization, with restricted step size during dynamical system identification, in acquiring the global extremum X_{opt} is determined from the critical values in Tables 7 and 8 as the estimate

$$\bar{X}_{opt} = X_{opt} \pm \delta X^L$$

(84)

The accuracy of the estimate in (84) can be improved with the selection of FC target data because of its greater cost surface sensitivity and smaller relative step size. If the length of the test data record is extended with more data values collected, accompanied by a corresponding transient response decay in the observed variables, the selectivity of the FC response surface improves with genuine local minima proliferation while the parabolic V_{cor} surface concavity decreases. Thus a more accurate global estimate $\bar{X}_{opt}$ is obtained for reference purposes with increased data record length and improved FC surface selectivity. A suitable identification method can then be applied in conjunction with the

BLMD model to the response surfaces corresponding to either motor shaft velocity $V_{\omega r}$ or winding FC I_{fa} target data, obtained in torque mode control for different shaft load inertia, in the parameter search process of the optimal estimate

$\hat{X}_{opt} = [\hat{J}_{opt}, \hat{B}_{opt}]^T$. The Powell conjugate direction method [22,23] and FSD [19] parameter extraction techniques can be applied, for example, to the respective $V_{\omega r}$ and I_{fa} cost surfaces to obtain $\hat{X}_{opt}$ [18].

CONCLUSIONS

Response surface simulation has been theoretically investigated and shown to be a useful graphical tool in motor parameter identification with a multiminima objective function and BLMD model validation for electric vehicle systems. This visual concept provides an intuitive insight into the topographical structure of the cost function to be minimized, the location of the global minimum, and the relevant identification search strategy to be adopted in parameter space to obtain an accurate estimate. It also provides an alternative parameter measurement strategy against which the accuracy of other parameter identification search techniques can be judged. A novel mathematical analysis of the competing shaft velocity and current feedback response surfaces, for identification purposes, has revealed the existence of a 'line' minimum of possible solutions principally in the B-parameter direction via a comparison of the eigenvalues derived from the quadratic model fit of the global stationary region. This analysis also shows that the global stationary region is closed and bounded by elliptical shaped MSE contours, which guarantees the existence of an optimal parameter vector solution. Furthermore a comparison of the quadratic model eigenvalues, for the competing cost surfaces, illustrates the dominance of the current feedback response selectivity and its acceptance as the most suitable objective function during SI for accurate parameter extraction.

The quantization of parameter space to remove 'false' local minima proliferation has been examined and demonstrated to be effective in surmounting cost surface 'noisiness' engendered during BLMD simulation, with a finite step size, of the PWM natural sampling process. A probability analysis has shown that the error incurred in resolution of PWM edge transition times during BLMD simulation, which is responsible for cost surface granularity, is dependent on the step size and is manifested as a random error voltage at the PWM inverter output. The effect of cost surface selectivity with choice of target data in MSE penalty cost function formulation, for usage in BLMD parameter identification, has been examined with motor current feedback being the preferred option.

ACKNOWLEDGEMENTS

The author wishes to acknowledge

- Eolas – The Irish Science and Technology Agency – for research funding.
- Moog Ireland Ltd for brushless motor drive equipment for research purposes.

REFERENCES

1.   R.A Guinee, Extended Simulation of an Embedded Brushless Motor Drive (BLMD) System for Adjustable Speed Control Inclusive of a Novel Impedance Angle Compensation Technique for Improved Torque Control in Electric Vehicle Propulsion Systems in Electric Vehicles - Modelling and Simulations, ISBN: 978-953-307-477-1, InTech ; 2011.

2.   R.A Guinee, Mathematical Modelling and Simulation of a PWM Inverter Controlled Brushless Motor Drive System from Physical Principles for Electric Vehicle Propulsion Applications in Electric Vehicles - Modelling and Simulations, ISBN: 978-953-307-477-1, InTech ; 2011.

3.   Miller, J.; (2010). Propulsion Systems for Hybrid Vehicles, IET, Renewable Energy, 2nd Edition.

4.   R.M. Crowder, Electric Drives and their Controls, 1995, Clarendon Press, Oxford.

5.   A.J. Critchlow, Introduction to Robotics, Macmillan Pub. Co. NY, 1985.

6.   H. Gross, Electrical Feed Drives for Machine Tools, 1983 by Siemens, J. Wiley & Sons.

7.   Moog Brushless Technology User Manual:D31X-XXX Motors,T158-01X Controllers,T157-001 Power Supply, Moog GmbH, D-7030 Böblingen, Feb 1989.

8.   H. Asada and K. Youcef-Toumi, Direct-Drive Robots Theory and Practice, 1987, MIT Press.

9.   R.P. Paul, Robot Manipulators: Mathematics, Programming and Control, The MIT Press, Camb, Mass, USA, 1986.

10.   W.E. Snyder, Industrial Robots: Computer Interfacing and Control, PHI, 1985

11.   M.A. El-Sharkawi and S. Weerasooriya, "Development and Implementation on Self-Tuning Tracking Controllers for DC Motors", IEEE Trans. on Energy Conv., Vol. 5, No. 1, Mar 1990.

12.   N.A. Demerdash, T.W. Nehl and E. Maslowski, "Dynamic modelling

of brushless dc motors in electric propulsion and electromechanical actuation by digital techniques", IEEE/IAS Conf. Rec. CH1575-0/80/0000-0570, pp. 570-579, 1980.

13. H. Dohmeki and M. Nasu, " Development of a Brushless DC Motor for Incremental Motion Systems", Proc. 14th IMCSD annual Symp., pp.63-71, 1985

14. J.Y.S. Luh, "Conventional Controller Design for Industrial Robots – A Tutorial", IEEE Trans. On Systems, Man, and Cybernetics, Vol. SMC-13, No. 3, May/June 1983.

15. K.J. Astrom and T. Hagglund, Automatic Tuning of PID Controllers, Instr. Soc. Amer, 1988, ISBN 1-55617-081-5.

16. Moog Brushless Technology:Brushless Servodrives User Manual D310.01.03 En/De/It 01.88, Moog GmbH, D-7030 Böblingen, Germany, 1988.

17. R.A. Guinee and C. Lyden, "Motor Parameter Identification using Response Surface Simulation and Analysis", Proc. of American Control Conference, ACC-2001,June 25-27, 2001, Arlington, VA, USA.

18. Guinee, R.A., "Response Surface Methodology", Modelling, Simulation, and Parameter Identification of a Permanent Magnet Brushless Motor Drive System, Chapter 3, pages 125 – 206, Ph. D. Thesis, 2003, National University of Ireland – University College Cork.

19. R.A. Guinee and C. Lyden, "A Novel Application of the Fast Simulated Diffusion Algorithm for Dynamical Parameter Identification of Brushless Motor Drive Systems". IEEE-ISCAS 2000, The 2000 IEEE International Symposium on Circuits and Systems, May 28-31, Geneva, Switzerland.

20. R.A. Guinee and C. Lyden, "A Novel Application of the Fast Simulated Diffusion Optimization Technique for Brushless Motor Parameter Extraction" UKACC International Conference on Control, Cambridge Univ., Sep 2000.

21. R.A. Guinee and C. Lyden, "Parameter Identification of a Brushless Motor Drive System using a Modified Version of the Fast Simulated Diffusion Algorithm", Proc. of American Control Conference – IEEE ACC-1999, San Diego, June 1999, pp.3467-3471

22. R. Fletcher, Practical Methods of Optimization, 2nd edition,1993, J.Wiley & Sons.

23. W.H. Press, B.F. Flannery, S.A. Teukolsky and W.T. Vetterling, Numerical Recipes in C, 1990, CUP.

24. Guinee and C. Lyden, "Accurate Modelling And Simulation Of A DC Brushless Motor Drive System For High Performance Industrial Applications", IEEE ISCAS'99 - IEEE International Symposium on Circuits and Systems, May/June 1999, Orlando, Florida

25. R.H. Myers and D.C. Montgomery, Response Surface Methodology - Process and Product Optimization Using Designed Experiments, 1995, J.Wiley & Sons, NY.

26. R.A. Guinee and C. Lyden, "A Novel Application of the Fast Simulated Diffusion Algorithm in Brushless Motor Parameter Identification", The 3rd IEEE European Workshop on Computer-Intensive Methods in Control and Data Processing, Sep 7-9, Prague, Czech Republic.

27. R.A. Guinee and C. Lyden, "Parameter Identification of a Motor Drive using a Modified Fast Simulated Diffusion Algorithm", Proc. of the IASTED Intern. Conf. on Modelling and Simulation, pp 224-228, May. 1998, Pittsburgh, Pa., USA.

28. H.A. Taha, Operations Research, Macmillan Publishing Co., NY., 1971.

29. Protter and Murray, Calculus and Analytic Geometry

30. J. Holtz, "Sensorless Position Control of Induction Motors - an Emerging Technology", invited paper, IEEE-IECON'98, Proc. of the 24th Annual Conf. of the IEEE Indus. Electronics Society, Aug 31 - Sep 4, 1998, Aachen, Germany.

31. G. Daryanani, Principles of Active Networks Synthesis and Design, 1976, J. Wiley & Sons.

32. F. Mosteller, R.E.K. Rourke and G.B. Thomas, Probability with Statistical Applications, 1961, Addison-Wesley Publ. Co.

Chapter 8

HIGH PERFORMANCE NOVEL SQUARE ROOT ARCHITECTURE USING ANCIENT INDIAN MATHEMATICS FOR HIGH SPEED SIGNAL PROCESSING

Arindam Banerjee, Aniruddha Ghosh, Mainuck Das

Department of ECE, JIS College of Engineering, Kalyani, India

ABSTRACT

Novel high speed energy efficient square root architecture has been reported in this paper. In this architecture, we have blended ancient Indian Vedic mathematics and Bakhshali mathematics to achieve a significant amount of accuracy in performing the square root operation. Basically, Vedic Duplex method and iterative division method reported in Bakhshali Manuscript have been utilized for that computation. The proposed technique has been compared with the well-known Newton- Raphson's (N-R) technique for square root computation. The algorithm has been implemented and tested using Modelsim simulator, and performance parameters such as the number of lookup tables, propagation delay and power consumption have been estimated using Xilinx ISE simulator. The functionality of the circuitry has been checked using Xilinx Virtex-5 FPGA board.

INTRODUCTION

Arithmetic circuits are indispensable in DSP and image processing applications. Many researchers have already designed different arithmetic circuits like adder, subtractor, multiplier, divider, squarer, square root architecture etc. [1] -[9] , but those techniques exhibit more propagation delay and power consumption. Square root design is a very challenging work in modern research area. Newton-Raphson's (N-R) method of square root computation has been the most efficient approach so far. After N-R method, there has been no remarkable approach for square root computation so far. We are showing an efficient technique using ancient Indian mathematics. In ancient times, mathematicians made calculations based on 16 Sutras (Formulae). The idea for designing the square root processor has been adopted from ancient Indian mathematics "Vedas" [10] [11]. The well-known duplex method is an ancient Indian method of extracting the square root. The algorithm incorporates digit by digit method for calculating the square root of a whole or decimal number one digit at a time. As per the proposed algorithm, the square root of any number is obtained in one step which reduces the iterations. Bakhshali method is used for finding a proximal square root described in ancient Indian mathematical manuscript called the Bakhshali manuscript. A prototype of 16 bit square root processors has been implemented and their functionality has been experimented on a Virtex-5 FPGA board. For calculating integer part of a number, we have used the Vedic method and for calculating the floating point part, we have adopted Bakhshali method. Many researchers have used the most popular Newton-Raphson's method for computing square root of a number. In our methodology, we are combining both Vedic duplex and Bakhshali approximation to generate more accurate result with comparatively less time and power consumption.

The paper has been organized as follows. Section 2 describes the back ground mathematics for calculating the square root. Section 3 gives the architectural description of the proposed technique. Section 4 describes the error calculation and accuracy analysis of the proposed technique. Result analysis has been described in Section 5 and Section 6 is the conclusion.

SQUARE ROOT ALGORITHM

Square root technique using Division method was introduced by ancient Indian Mathematicians and the procedure has been lucidly discussed in "Vedas". The technique stands upon the execution by mere observation. Square root of a perfect square can be easily determined by the Division method. But for those numbers, which are not perfect square, Division method is not enough to compute with high precision. In this paper an iterative method described in Bakhshali Mathematics of Quran has been proposed to achieve high speed and high precision square root calculation. The mathematical formulation of the iterative approach is shown below.

Consider a number X whose square root is to be calculated. The Bakhshali manuscript approach computes the square root of the perfect square which is nearest but less than the number X by the method of observational inspection. Consider A is the above mentioned perfect square whose square root is R.

So it can be expressed as

$$X = A \pm Y = R^2 \pm Y \tag{1}$$

where Y is the residue.

Equation (1) can be reformulated as

$$X = R^2 \left(1 \pm \frac{Y}{R^2} \right) \tag{2}$$

So the square root of X can be written as

$$X^{1/2} = R \left(1 \pm \frac{Y}{R^2} \right)^{1/2} \tag{3}$$

Using binomial expansion and the mathematical formulae of Bakhshali manuscript, Equation (3) can be expressed as

$$X^{1/2} \cong R \left[1 \pm \frac{Y}{2R^2} - \frac{1}{8}\left(\frac{Y}{R^2}\right)^2 \pm \frac{1}{16}\left(\frac{Y}{R^2}\right)^3 - \frac{1}{32}\left(\frac{Y}{R^2}\right)^4 \pm \cdots \right] \tag{4}$$

$$= R \left[1 \pm \frac{Y}{2R^2} - \frac{1}{2R^2}\left(\frac{Y}{2R}\right)^2 \pm \frac{1}{2R^2} \times \frac{1}{R}\left(\frac{Y}{2R}\right)^3 - \frac{1}{2R^2} \times \frac{1}{R^2}\left(\frac{Y}{2R}\right)^4 \pm \cdots \right] \tag{5}$$

$$= R \left[1 \pm \frac{Y}{2R^2} - \frac{1}{2R^2}\left(\frac{Y}{2R}\right)^2 \left\{ 1 \mp \frac{Y}{2R^2} + \left(\frac{Y}{2R^2}\right)^2 \mp \cdots \right\} \right] \tag{6}$$

$$= R \left[1 \pm \frac{Y}{2R^2} - \frac{1}{2R^2}\left(\frac{Y}{2R}\right)^2 \left(1 \pm \frac{Y}{2R^2} \right)^{-1} \right] \tag{7}$$

$$= R \left[1 \pm \frac{Y}{2R^2} - \frac{\left(\dfrac{Y}{2R}\right)^2}{2R^2\left(1 \pm \dfrac{Y}{2R^2}\right)} \right] \tag{8}$$

$$= R \pm \frac{Y}{2R} - \frac{\left(\dfrac{Y}{2R}\right)^2}{2R\left(1 \pm \dfrac{Y}{2R^2}\right)} \tag{9}$$

$$= R \pm \frac{Y}{2R} - \frac{\left(\dfrac{Y}{2R}\right)^2}{2\left(R \pm \dfrac{Y}{2R}\right)}$$

(10)

Equation (10) is the modified Bakshali expression for computing the square root of non-squared numbers.

In modified Bakshali approach the accuracy is more as comparable to in Bakshali approach which is shown in the table. So in respect of accuracy modified Bakshali approach is beneficial. Also Bakshali methodology contains some drawbacks mentioned below:

1) Bakshali approach does not gives the idea for computing the nearest squared number and its squareroot. To obtain thenearest square number and its square root we have to go for Vedic approach.

2) The division method for calculating the square root using Vedic mathematics is based on "Vilokanam" (Inspection) method which incorporates an extra squaring and extra subtraction operations.

The modified Bakshali approach eliminates the extra squaring and subtraction operations. thus the technique reduces the stage delay.

The square root determination technique combining the Vedic Duplex and Bakhshali approach is shown in the flow chart of Figure 1.

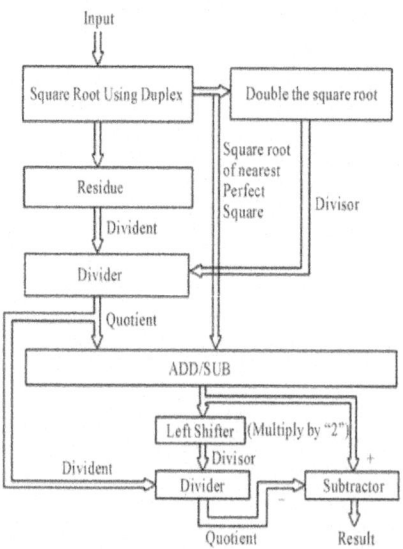

Figure 1: Architecture for square root determination technique.

Integral Square Root Determinant Using Vedic Methodology

Mathematical Modeling of the Algorithm to Calculate Perfect Square Root

Consider X is an N bit number whose square root is to be determined. X can be expressed as

$$X = \sum_{N-1}^{i=0} x_i 2^i = 2^{N/2} \sum_{i=0}^{\frac{N}{2}-1} x_{i+N/2} 2^i + \sum_{i=0}^{\frac{N}{2}-1} x_i 2^i \tag{11}$$

$$= 2^{N/2} \sum_{i=0}^{\frac{N}{2}-1} y_{i+N/2} 2^i + 2^{N/2} \sum_{i=0}^{\frac{N}{2}-1} \left(x_{i+\frac{N}{2}} - y_{i+\frac{N}{2}} \right) 2^i + \sum_{i=0}^{\frac{N}{2}-1} x_i 2^i, \tag{12}$$

Considering $\sum_{i=0}^{\frac{N}{2}-1} y_{i+N/2} 2^i$ to be the nearest perfect square number of $\sum_{i=0}^{\frac{N}{2}-1} x_{i+N/2} 2^i$. Now assume that the square root of $\sum_{i=0}^{\frac{N}{2}-1} y_{i+N/2} 2^i$ is $Z = 2^{N/4} \sum_{i=0}^{\frac{N}{4}-1} z_{i+N/4} 2^i$. Then, Equation (12) can be rewritten as

$$X = Z^2 + 2Z \left(\frac{2^{\frac{N}{2}} \sum_{i=0}^{\frac{N}{2}-1} \left(x_{i+\frac{N}{2}} - y_{i+\frac{N}{2}} \right) 2^i + \sum_{i=0}^{\frac{N}{4}-1} x_{i+n/4} 2^i}{2Z} \right) + \sum_{i=0}^{\frac{N}{4}-1} x_i 2^i \tag{13}$$

Now consider

$$\left(\frac{2^{\frac{N}{2}} \sum_{i=0}^{\frac{N}{2}-1} \left(x_{i+\frac{N}{2}} - y_{i+\frac{N}{2}} \right) 2^i + \sum_{i=0}^{\frac{N}{4}-1} x_{i+n/4} 2^i}{2Z} \right) = Q + \frac{R}{2Z} \tag{14}$$

where Q is the quotient and R is the remainder. Inserting Equation (14), Equation (13) can be reformulated as

$$X = Z^2 + 2ZQ + \left(R + \sum_{i=0}^{\frac{N}{4}-1} x_i 2^i \right) \tag{15}$$

Equation (15) can be rewritten as

$$X = Z^2 + 2ZQ + Q^2 + \left(R + \sum_{i=0}^{\frac{N}{4}-1} x_i 2^i - Q^2 \right)$$

$$= (Z+Q)^2 + \left(R + \sum_{i=0}^{\frac{N}{4}-1} x_i 2^i - Q^2 \right) \tag{16}$$

Assume $R + \sum_{i=0}^{\frac{N}{4}-1} x_i 2^i - Q^2 = P$. Then X can be represented as,

$$X = (Z+Q)^2 + p \tag{17}$$

$$\sqrt{X} = \sqrt{(Z+Q)^2 + p} = \sqrt{(Z+Q)^2 \left(1 + \frac{p}{(Z+Q)^2} \right)} \tag{18}$$

$$= (Z+Q)\sqrt{\left(1 + \frac{p}{(Z+Q)^2} \right)} \tag{19}$$

$$= (Z+Q) \pm \frac{p}{2(Z+Q)} - \frac{\left(\frac{p}{2(Z+Q)} \right)^2}{2\left((Z+Q) \pm \frac{p}{2(Z+Q)} \right)} \tag{20}$$

The procedure to calculate the square root by division method can be described in the following steps:

Step 1: Obtain the nearest square root of the N/2 Most Significant Bits (MSB). Assume that the output is Z.

Step 2: Determine the square of Z by combining Yavadunam and Duplex methodology.

Step 3: Subtract the squared output from N/2 MSB.

Step 4: Obtain the double of Z.

Step 5: Combine the output of the subtractor and the next N/4 bits. Divide the combination by 2Z. Assume the quotient as Q and the remainder as R.

Step 6: Determine the square of Q and subtract Q2 from $R + \sum_{i=0}^{\frac{N}{4}-1} x_i 2^i$. If the residue (p) is zero then X is the perfect square number whose square root is (Z + Q) otherwise (Z + Q) is the square root of the perfect square nearest to X. Again if "p" is positive then the nearest perfect square number is less than the given input else if "p" is negative then the nearest perfect square number is greater than the input.

Step 7: Compute $\left|\dfrac{p}{2(Z+Q)}\right|$ by Straight Division ("Dhvajanka") methodology.

Step 8: Determine the square of $\dfrac{p}{2(Z+Q)}$.

Step 9: Compute $\left((Z+Q)\pm\dfrac{p}{2(Z+Q)}\right)$ by add/sub unit.

Step 10: Finally subtract $\dfrac{\left(\dfrac{p}{2(Z+Q)}\right)^2}{2\left((Z+Q)\pm\dfrac{p}{2(Z+Q)}\right)}$ from $(Z+Q)\pm\dfrac{p}{2(Z+Q)}$ to obtain the exact square root.

Step 11: Compute the maximum power of first left most "1" of $(Z+Q)\pm\dfrac{p}{2(Z+Q)}-\dfrac{\left(\dfrac{p}{2(Z+Q)}\right)^2}{2\left((Z+Q)\pm\dfrac{p}{2(Z+Q)}\right)}$ by left shifting operation.

PROPOSED ARCHITECTURE

Square Root architecture following the methodology given in Step 1 to Step 6, consists of four sub sections: 1) Nearest Square Root Generation Unit (NSRGU) for first N/2 numbers; 2) Squaring Unit; 3) Sub-tractor; 4) Divider. Figure 2 exhibits a fully optimized system level architecture for square root generation using Duplex method adopted by ancient Indian mathematicians. In this architecture all the left shifters are N/4 bit left shifters. Quotient register is used to store the fixed point or integral result of the square root. The most significant two bits fed to the doubler whose output is fed to the subtraction unit.

Nearest Square Root Generation Unit

In ancient Indian Mathematics, the nearest square number for a given number and its square root was calculated by the method of inspection. In this paper an easy and efficient architecture has been provided to compute the nearest square root generation. It is obvious that $N = N/2$ in "Equation (12)", the number bits in square root must be of N/4 bits.

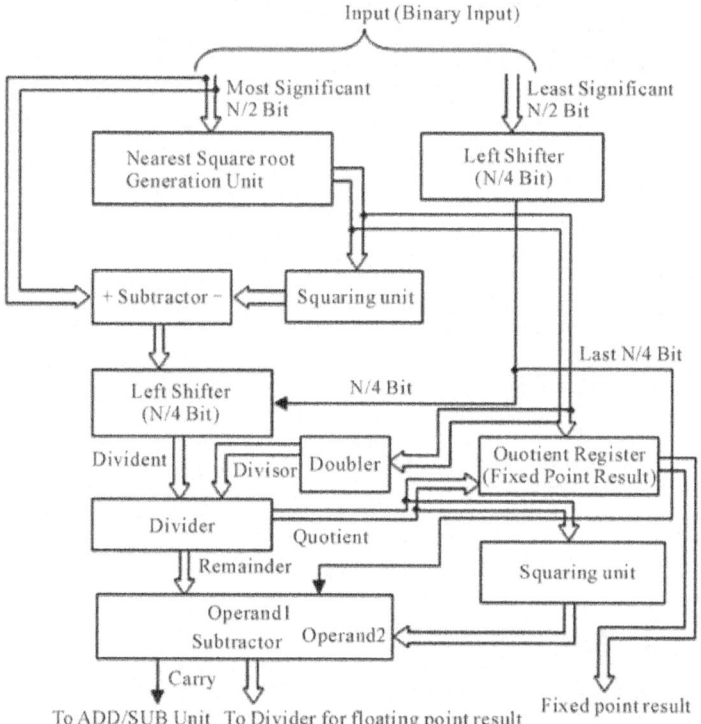

Figure 2: System level architecture for Integral part of Square root generator, N indicating number of input bits.

The architecture of the nearest square root generation unit is same as the figure depicted in Figure 2. The above mentioned Step 1 to Step 6 are needed to obtain the nearest square root for N/2 bit number.

Squaring Unit

Squaring algorithm and the corresponding architecture was implemented with the aid of "Yavadunam Sutra".

Mathematical Formulation of Yavadunam Sutra

If a number X is between 2n − 1 and 2n then the average of 2n − 1 and 2n is $A = \frac{2^{n-1} + 2^n}{2} = 3 \times 2^{n-2}$. If X > A

then 2n is chosen as radix and if $X \leq A$ then 2n − 1 is the selected radix.

In Binary mathematics, the number $X = \sum_{i=0}^{n-1} x_i 2^i$ can be reformulated as,

$$X = x_{n-1} 2^{n-1} + \sum_{i=0}^{n-2} x_i 2^i, \text{ when Radix} = 2n - 1 \tag{21}$$

$$X = 2^n - (2\text{'s complement of } X), \text{ when Radix} = 2n \tag{22}$$

From Equation (21), the square of the number X can be expressed as,

$$X^2 = \left(x_{n-1} 2^{n-1}\right)^2 + 2\left(x_{n-1} 2^{n-1}\right)\sum_{i=0}^{n-2} x_i 2^i + \left(\sum_{i=0}^{n-2} x_i 2^i\right)^2 \tag{23}$$

$$= 2^{n-1}\left(X + \sum_{i=0}^{n-2} x_i 2^i\right) \tag{24}$$

Similarly, the expression of X^2 that can be obtained from Equation (22) is given as,

$$X^2 = 2^n \left(X - 2\text{'s complement of } X\right) + \left(2\text{'s complement of } X\right)^2 \tag{25}$$

Equation (24) and Equation (25) are the mathematical formulations of Yavadunam Sutra in Binary mathematics.

The architecture of squaring algorithm using "Yavadunam Sutra" is shown in Figure 3. The basic building blocks of the architecture are 1) RSU, 2) Subtractor, 3) Add-Sub unit and 4) Duplex squaring architecture. The test bench waveform of RSU is shown in Figure 4. The Subtractor architecture using "Nikhilam" sutra has been elucidated in sub-section-(C).

The input bits X (of length n) is fed to the RSU. The RSU produces proper radix and exponent for the input. The radix is subtracted from the input and the result is added to or subtracted from the input again depending upon the control signal generated from the borrow output of the sub-tractor. The output of add/sub unit is the residue which is again squared by duplex operation discussed in [1] and it is also fed to the left shifter which shift the data depending upon the exponent value. Finally the outputs of duplex squarer and left shifter are added to obtain the squared result.

Subtractor

Mathematical Modeling of "Nikhilam" Sutra for Binary Subtraction

The Subtraction method has been implemented using normal binary arithmetic. It is a bit wise subtraction method. The rule is mathematically expressed in

Equation (26). Consider the number, $Y = \sum_{n-1}^{n-1} y_i 2^i$ is to be subtracted from the number $X = \sum_{i=0}^{n-1} x_i 2^i$. So, it can be written as

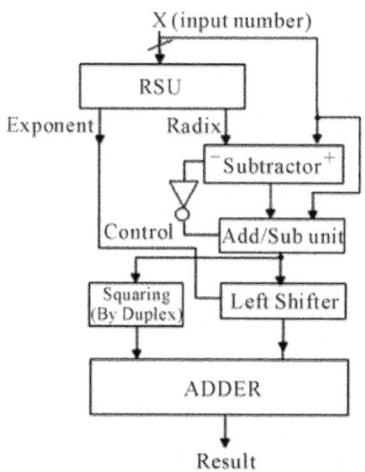

Figure 3: Architecture of squaring algorithm using "Yavadunam" sutra.

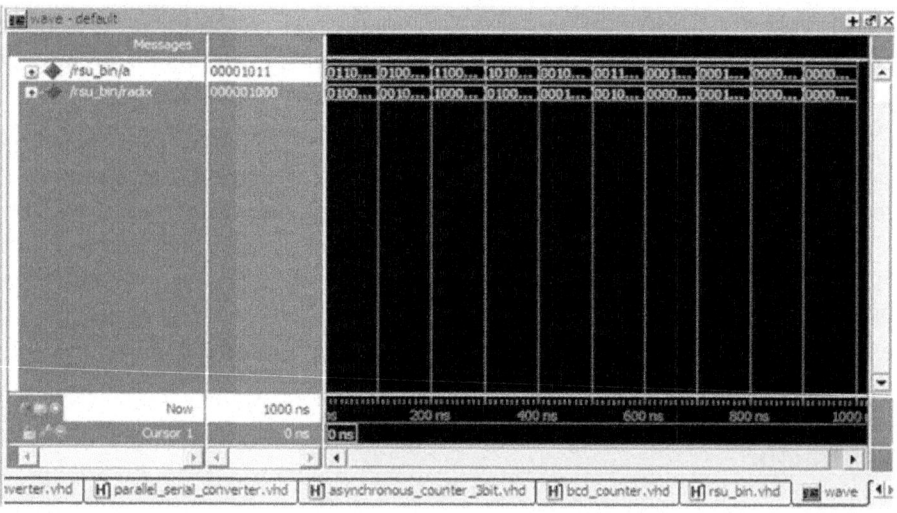

Figure 4: Test bench window for VHDL implementation of RSU.

$$X - Y = \sum_{i=0}^{n-1} \left(x_i 2^i - y_i 2^i \right)$$

$$(26)$$

There is a borrow bit generated at the end of operation which signifies that the result is positive or negative.

Divider Using "Dhvajanka" Sutra ("On Top of the Flag")

Mathematical Modeling of Division Operation

Consider the number $A = \sum_{i=0}^{n/2-1} a_i x^i$ to be divided by $B = \sum_{i=0}^{n/2-1} b_i x^i$ assuming that both the numbers are of same length (here the length is $(n/2) - 1$) and x is the radix. To execute the division operation easily and efficiently by "Dhvajanka" sutra ("On top of the flag") methodology described in the ancient Indian Vedic Mathematics, Dividend must have greater length than Divisor. Consider again a number A' which can be represented

$$A' = x^{\frac{n}{2}} \times A = x^{\frac{n}{2}} \times \sum_{i=0}^{n/2-1} a_i x^i = \sum_{i=0}^{n-1} a_i x^i, \, a_i \left(\forall i \in \left\{ 0,1,2,\cdots,\left(\frac{n}{2}\right)-1 \right\} \right) = 0$$

a s
.

$B = \sum_{i=0}^{n/2-1} b_i x^i$ can be mapped with $B' = \sum_{i=0}^{\frac{n}{2}-1} b_i' x^i$ where the condition of mapping is $b_i = b_i' \forall i \in \left\{ 0,1,2,\cdots,\left(\frac{n}{2}\right)-1 \right\}$. The quotient Q can be determined as

$$Q = \frac{A}{B} = \frac{A' \times x^{\left(-\frac{n}{2}\right)}}{B'} = \left(\frac{A'}{B'}\right) \times x^{\left(-\frac{n}{2}\right)} \tag{27}$$

The term $x^{\left(-\frac{n}{2}\right)}$, where x is the radix is to be extracted by Exponent Extraction Unit (EEU). In Binary number system, the radix "x" is equal to 2. Then from Equation (27), it is obvious that the result of $\frac{A}{B}$ is needed to be shifted right by n/2 terms which has been shown in Figure 5, to get the actual quotient. The mathematical modeling of division operation by "Vertically and Crosswise" methodology has been described in the following subsection.

The architecture shown in Figure 6 consists of some elementary building blocks like left shifter, right shifter, incrementer and demultiplexer. Incrementer calculates the exponent of the number which is controlled by the shifted bit. The clock applied to the shifter is also controlled by the shifted bit. If the shifted bit is "1" then the shifters stops further shifting.

Mathematical Modeling of "On Top of the Flag" Sutra

Consider the number $A' = \sum_{i=0}^{n-1} a_i' x^i$ is divided by $B' = \sum_{i=0}^{\frac{n}{2}-1} b_i' x^i$. So A' can be expressed in terms of B' as

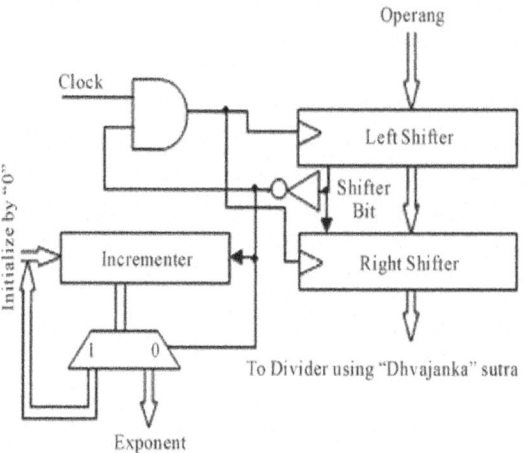

Figure 5: RTL representation of exponent extraction unit.

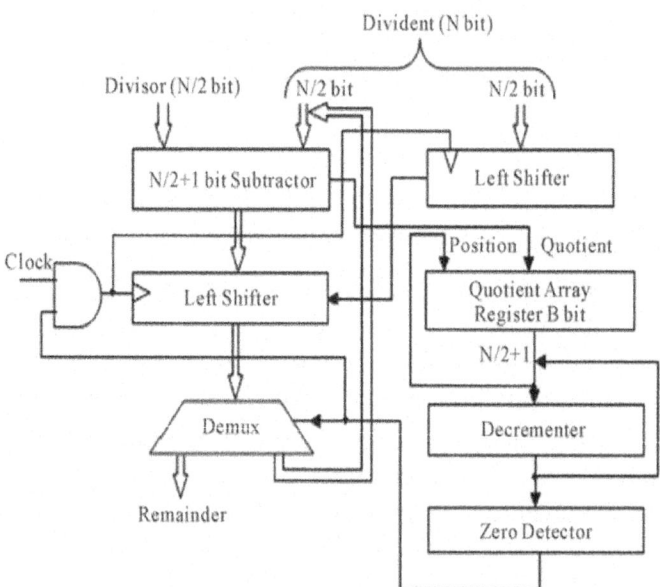

Figure 6: Divider architecture using "Dhvajanka" sutra.

$$A' = \sum_{i=0}^{n-1} a'_i x^i = a'_{n-1}x^{n-1} + a'_{n-2}x^{n-2} + a'_{n-3}x^{n-3} + a'_{n-4}x^{n-4} + \cdots + a'_3 x^3 + a'_2 x^2 + a''_1 x^1 + a'_0 x^0 \tag{28}$$

$$= \left(b'_{\frac{n}{2}-1} x^{\frac{n}{2}-1} + b'_{\frac{n}{2}-2} x^{\frac{n}{2}-2} + \cdots + b'_0 x^0 \right)$$

$$\times \left(\frac{a'_{n-1}}{b'_{\frac{n}{2}-1}} x^{\frac{n}{2}} + \frac{\left(a'_{n-2} - \left(\frac{a'_{n-1}}{b'_{\frac{n}{2}-1}}\right)b'_{\frac{n}{2}-2}\right)}{b'_{\frac{n}{2}-1}} x^{\frac{n}{2}-1} + \frac{\left(a'_{n-3} - \left(\frac{a'_{n-1}}{b'_{\frac{n}{2}-1}}\right)b'_{\frac{n}{2}-3}\right) - \frac{\left(a'_{n-2} - \left(\frac{a'_{n-1}}{b'_{\frac{n}{2}-1}}\right)b'_{\frac{n}{2}-2}\right) \times b'_{\frac{n}{2}-2}}{b'_{\frac{n}{2}-1}}}{b'_{\frac{n}{2}-1}} x^{\frac{n}{2}-2} + \cdots \right)$$

$$\tag{29}$$

Here "x" signifies the radix.

Figure 6 shows the architectural description of division operation using "On top of the flag" Sutra. The Dividend of N bits is divided into two equal N/2 bits. The most significant part is fed to the subtractor which is of N/2 + 1 bits size. Single bit zero padding is done in subtractor module to the left. The division procedure incorporated here is non-restoring type of division. The carry output after the subtraction is fed to the quotient array register as quotient.

The quotient array register stores the quotients based on the iterated value which is used as position signal. The value N/2 + 1 is stored in the decrementer initially to count and check the specified iteration. After each iteration, both the left shifters are updated by shifting until the decremented value reaches zero.

The Divider architecture combining the exponent extraction and the "Dhvajanka" Sutra is shown inFigure 7. The basic building blocks of the architecture are 1) Exponent Extraction Unit, 2) Divider using "Dhvajanka" sutra, 3) Subtractor, 4) Quotient Array Register and 5) Bidirectional Shift Register.

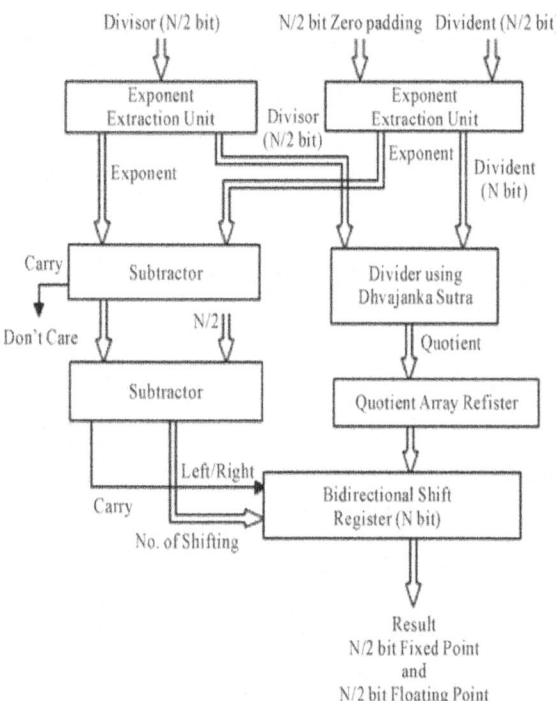

Figure 7: Divider architecture combining exponent extraction and "Dhvajanka" sutra.

Zero padding is needed to represent a number to higher number of bits. For example if a four bit number "0111" is represented in eight bit format then the representation becomes "00000111". Here, zero padding has been executed based on the requirement. Bidirectional shift register performs the operation of shifting both to the left or right depending upon the control signal "carry" from the second subtractor. If carry = 0 then the input bits are shifted to the left and if carry = 1 then input bits are shifted to the right. The result of the second subtractor is fed to the shifter to control number shifting. The contents of the Quotient Array Register are the input to the Bidirectional Shift register.

Illustration: Consider a fourth order function $f(x) = a_3 x^3 + a_2 x^2 + a_1 x^1 + a_0 x^0$ and a second order function $g(x) = b_1 x^1 + b_0 x^0$. We have to compute $\dfrac{f(x)}{g(x)}$ with the help of "On top of the flag" sutra. Now,

$$\frac{f(x)}{g(x)} = \frac{a_3 x^3 + a_2 x^2 + a_1 x^1 + a_0 x^0}{b_1 x^1 + b_0 x^0} = Q(x) + \frac{R}{g(x)} \Rightarrow f(x) = Q(x)g(x) + R$$

$$Q(x) = \left(\frac{a_1}{b_1}x^2 + \frac{\left(a_2 - \frac{a_3}{b_1}b_0\right)}{b_1}x^1 + \frac{\left(a_1 - \frac{\left(a_2 - \frac{a_3}{b_1}b_0\right)}{b_1}b_0\right)}{b_1}x^0 \right)$$

where,   and $R = a_0 - \dfrac{\left(a_1 - \dfrac{\left(a_2 - \dfrac{a_3}{b_1}b_0\right)}{b_1}b_0\right)}{b_1}b_0$

Adder-Subtractor Unit Using "Nikhilam" Sutra for Subtraction

From Equation (5), it is obvious that to achieve the final result an Adder-Subtractor unit is required which has been designed with the help of well-known binary addition/subtraction methodology. The architecture of Adder-Subtractor unit is shown in Figure 8. The carry signal from the square root determinant shown in Figure 2 is fed to the adder-subtractor unit at the control pin to assign the type of operation (addition/subtraction). If the carry signal is low then addition operation is executed else subtraction operation is executed.

ERROR CALCULATION AND ACCURACY ANALYSIS

The exact expression for square root can be recapitulated from Equation (3) as,

$$X^{1/2} = R\left(1 \pm \frac{Y}{R^2}\right)^{1/2} = R\left[1 \pm \frac{Y}{2R^2} - \frac{1}{8}\left(\frac{Y}{R^2}\right)^2 \pm \frac{1}{16}\left(\frac{Y}{R^2}\right)^3 - \frac{5}{4 \times 32}\left(\frac{Y}{R^2}\right)^4 \pm \frac{7}{4 \times 64}\left(\frac{Y}{R^2}\right)^5 - \cdots\right] \quad (30)$$

The approximated expression for the square root can be expressed as,

$$X^{1/2} \cong R\left[1 \pm \frac{Y}{2R^2} - \frac{1}{8}\left(\frac{Y}{R^2}\right)^2 \pm \frac{1}{16}\left(\frac{Y}{R^2}\right)^3 - \frac{1}{32}\left(\frac{Y}{R^2}\right)^4 \pm \frac{1}{64}\left(\frac{Y}{R^2}\right)^5 - \cdots\right] \quad (31)$$

So the computational error (ec) in determining the square root can be expressed as,

$$e_c = \left(X^{\frac{1}{2}}_{exact} - X^{\frac{1}{2}}_{approximated} \right) \cong R \times \frac{1}{32}\left(\frac{Y}{R^2}\right)^4 \left[1 \pm \frac{1}{2}\left(\frac{Y}{R^2}\right)\right]^{-1} \quad (32)$$

The percentage of error in calculating the square root is,

$$e_c(\%) \cong \left(\frac{R \times \frac{1}{32}\left(\frac{Y}{R^2}\right)^4 \left\{1 \pm \frac{1}{2}\left(\frac{Y}{R^2}\right)\right\}^{-1}}{R\left(1 \pm \frac{Y}{R^2}\right)^{\frac{1}{2}}} \right) \times 100$$

$$= \left(\frac{\frac{1}{32}\left(\frac{Y}{R^2}\right)^4}{\left\{1 \pm \frac{1}{2}\left(\frac{Y}{R^2}\right)\right\} \times \left(1 \pm \frac{Y}{R^2}\right)^{\frac{1}{2}}} \right) \times 100 < \frac{\frac{1}{32}\left(\frac{Y}{R^2}\right)^4}{\left\{1 \pm \frac{1}{2}\left(\frac{Y}{R^2}\right)\right\}^2} \times 100 \quad (33)$$

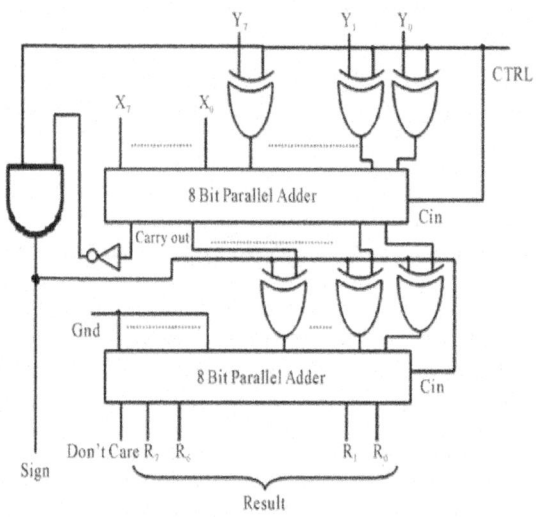

Figure 8: 8 bit binary adder-subtractor architecture.

If $x = \dfrac{Y}{R^2}$, then can easily be shown that $\dfrac{x^4}{32\left\{1-\dfrac{x}{2}\right\}^2} < \dfrac{1}{4\times32} = \dfrac{1}{128} < \dfrac{1}{100}$.

So, $e_e(\%) \cong \left| \dfrac{R\times\dfrac{1}{32}\left(\dfrac{Y}{R^2}\right)^4\left\{1\pm\dfrac{1}{2}\left(\dfrac{Y}{R^2}\right)\right\}^{-1}}{R\left(1\pm\dfrac{Y}{R^2}\right)^{\frac{1}{2}}}\right| \times100 < \dfrac{x^4}{32\left\{1-\dfrac{x}{2}\right\}^2}\times100 < \dfrac{1}{100}\times100 = 1\%$. It is

obvious that the error is minimum if and only if $X^{1/2} = R$ that is if Y = 0.

Putting Y = 0 in Equation (33), the expression becomes $e_e(\%) = \left(\dfrac{0}{1}\right)\times100 = 0\%$.

Table 1 exhibits the comparative study of the computational error in proposed algorithm with respect to "Bakshali" algorithm.

RESULT ANALYSIS

The square root architecture has been implemented using VHDL and verified using Modelsim simulator which is shown in Figure 9 and tested in Xilinx simulator with Xilinx Vertex-5 FPGA using XC5VLX30 device with a package FF676 at a speed grade (−3). Table 2 shows the comparative study of the N-R method and the proposed method with respect to LUT count, delay and power consumption. The power has been measured using Xilinx Xpower Analyzer tool.

Table 1: Coefficients of the quotient expression.

Digit	Exact square root	Square root by Bakhshali method	Square root by modified Bakhshali (MB) method
194	13.9283882771841193384677389285131	13.9284276329730875185420639966011	3.9283882783882783882783882783887
180	13.4164078649987381784550420123881	13.4164095217103813092351774300131	3.4164133738601823708206686930008
175	13.2287565553229529525080787681961	13.2287567084078711985688729874681	3.2287567084078711985688729874681
190	13.7840487520902217679559125529341	13.7840689950717805871009213627571	3.7840488527017024426350851221281
174	13.1909059582729191709368077327221	13.1909060327427674366449876653941	3.1909060327427674366449876653941
193	13.8924439894498045084325470410291	13.8924776880577985550361240968911	3.8924439955930958501652589056121
183	13.5277492584686828974258748173811	13.5277534965034965034965034965021	3.5277516019600452318130418394231
186	13.6381816969858558927585975519	3 13.6381906825568797399783315276211	3.6381824981301421091997008227
188	13.7113092008020882498717428981	7 13.7113229907347554406377935789611	3.7113095238095238095238095238041
192	13.8564064605510183482195707320471	3.8564351161304069891327509055991	3.8564064801178203240058910161971

Table 2: Comparison of LUT count, delay and power of two methods.

Square root technique	4 bit			8 bit			16 bit		
	LUT (no. of slices)	Delay (ns)	Power (mw)	LUT (no. of slices)	Delay (ns)	Power (mw)	LUT (no. of slices)	Delay (ns)	Power (mw)
N-R method	17	3.56	2.78	36	10.03	4.63	252	33.13	5.32
Proposed method	16	3.47	2.56	34	7.47	3.15	103	27.43	3.96

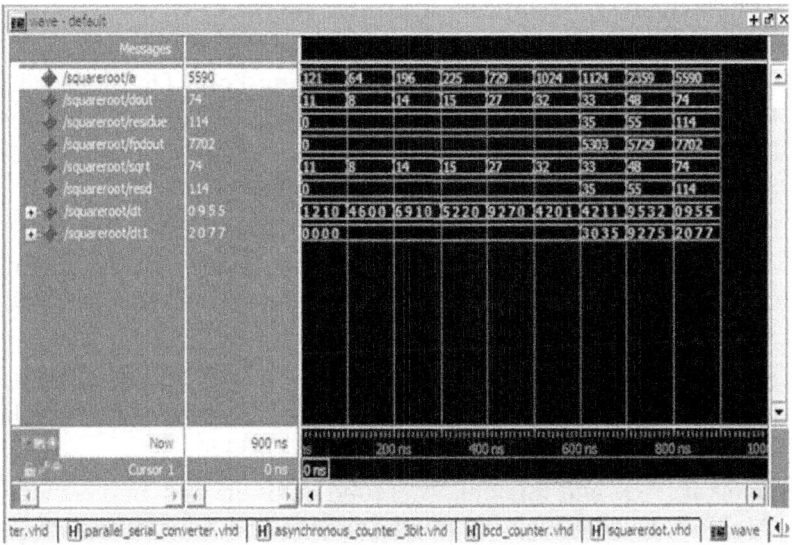

Figure 9: Test bench window of VHDL implementation of square root architecture.

CONCLUSION

From the result analysis, it is obvious that the architecture provides comparatively more accuracy than the well- known N-R approximation. Moreover, the proposed architecture produces less amount of propagation delay than N-R method and the power consumption is also very less as shown inTable 2. The area with respect to LUT count is also less than N-R method. It has also less circuit complexity than N-R technique which has been elucidated in the result analysis

ACKNOWLEDGEMENTS

We would like to thank our dear colleagues for their precious support to execute the research in the department.

REFERENCES

1.   Oberman, S.F. and Flynn, M.J. (1997) Design Issues in Division and Other Floating Point Operations. IEEE Transactions on Computers, 46, 154-161.

2.   Pineiro, J.A. and Bruguera, J.D. (2002) High-Speed Double-Precision Computation of Reciprocal, Division, Square Root, and Inverse Square Root. IEEE Transactions on Computers, 51, 1377-1388.

3.   Kwon, T.J. and Draper, J. (2008) Floating-Point Division and Square Root Implementation Using a Taylor-Series Expansion Algorithm with Reduced Look-Up Tables. 51st Midwest Symposium on Circuits and Systems, Knoxville, 10- 13 August 2008, 954-995.

4.   Park, I. and Kim, T. (2009) Multiplier-Less and Table-Less Linear Approximation for Square and Square-Root. IEEE International Conference on Computer Design, Lake Tahoe, 4-7 October 2009, 378-383.

5.   Ramamoorthy, C., Goodman, J. and Kim, K. (1972) Some Properties of Iterative Square-Rooting Methods, Using High-speed Multiplication. IEEE Transaction on Computers, C-21, 837-847.

6.   Thakkar, A.J. and Ejnioui, A. (2006) Design and Implementation of Double Precision Floating Point Division and Square Root on FPGAs. IEEE Aerospace Conference, Big Sky.

7.   Ye, M., Liu, T., Ye, Y., Xu, G. and Xu, T. (2010) FPGA Implementation of CORDIC-Based Square Root Operation for Parameter Extraction of Digital Pre-Distortion for Power Amplifiers. 6th International Conference on Wireless Communications Networking and Mobile Computing,

Chengdu, 23-25 September 2010, 1-4.

8. Ercegovac, M.D. and McIlhenny, R. (2009) Design and FPGA Implementation of Radix-10 Algorithm for Square Root with Limited Precision Primitives. Conference Record of the Forty-Third Asilomar Conference on Signals, Systems and Computers, Pacific Grove, 1-4 November 2009, 935-939.

9. Wang, X., Zhang, Y., Ye, Q. and Yang, S. (2009) A New Algorithm for Designing Square Root Calculators Based on FPGA with Pipeline Technology. 9th International Conference on Hybrid Intelligent Systems, 1, 99-102.

10. Maharaja, B.K.T. (1994) Vedic Mathematics. Motilal Banarasidass Publisher, Delhi.

11. Saha, P., Banerjee, A., Bhattacharyya, P. and Dandapat, A. (2011) High Speed ASIC Implementation of Complex Multiplier using VEDIC Mathematics. Proceeding of the 2011 IEEE Students' Technology Symposium, Kharagpur, 14-16 January 2011, 237-341.

Chapter 9

IMAGE MATHEMATICS—MATHEMATICAL INTERVENING PRINCIPLE BASED ON "YIN YANG WU XING" THEORY IN TRADITIONAL CHINESE MATHEMATICS (I)

Yingshan Zhang[1], Weilan Shao[2]

[1]School of Finance and Statistics, East China Normal University, Shanghai, China

[2]The College English Teaching and Researching Department, Xinxiang University, Xinxiang, China

ABSTRACT

By using mathematical reasoning, this paper demonstrates the mathematical intervening principle: "Virtual disease is to fill his mother but real disease is to rush down his son" and "Strong inhibition of the same time, support the weak" based on "Yin Yang Wu Xing" Theory in image mathematics of Traditional Chinese Mathematics (TCMath). We defined generalized relations and generalized reasoning, introduced the concept of steady multilateral systems with two non-compatibility relations, and discussed its energy properties. Later based on the intervention principle in image mathematics of TCMath and treated the research object of the image mathematics as a steady multilateral system, it has been proved that the mathematical intervening principle is true. The kernel of this paper is the existence and reasoning of the non-compatibility relations in steady multilateral systems, and it accords with the oriental thinking model.

MAIN DIFFERENCES BETWEEN TRADITIONAL CHINESE MATHEMATICS AND WESTERN MATHEMATICS

In Western Mathematics (mathematics; Greek: $\mu\,\alpha\,\theta\,\eta\,\mu\,\alpha\,\tau\,\iota\,\kappa$), the word comes from the ancient Greek in the west of the $\mu\,\theta\,\eta\,\mu\,\alpha$ (mathēma), its have learning, studying, science, and another relatively narrow meaning and technical sense-"mathematics study", even in its neck comes in. The adjective $\mu\,\alpha\,\theta\,\eta\,\mu\,\alpha\,\tau\,\iota\,\kappa$ (mathē- matikos), meaning and learning about or for the hard, also can be used to index of learning. In English on the function of the plural form of, and in French surface mathematiques plural form, can be back to Latin neutral plural mathematica, due to the Greek plural τ α μ α θ η μ α τ ι κ (tamathēmatika), the Greek by Aristotle brought refers to "all things several" concept. (Latin: Mathemetica) original intention is number and count of the technology. According to the understanding of the now in Western mathematics, mathematics is the real world number relationship and the form of the space science. Say simply, is the study of the form and number of science. The start of the study is always from the Axiomatic system. Any Axiomatic system of form and number comes from the observations. Because of the demand on life and labor, and even the most original nationality, also know simple count, and the fingers or physical count development to use digital count. Basic mathematics knowledge and use always individual and community essential to life. The basic concept of the refining is as early as in ancient Egypt, Mesopotamia, and in ancient India of ancient mathematical text and see in considerable. Since then, its development will continue to have a modest progress, and the 16th century until the Renaissance, because of the new scientific and found interactions and the generated mathematical innovation leads to the knowledge of the acceleration, until today.

Mathematics is used in different areas of the world, including science, engineering, medicine and economics, etc. The application of mathematics in these fields are usually called applied mathematics, sometimes also stir up new mathematical discovery, and led to the development of new subject. Mathematician also study didn't any application value of pure mathematics, even if the application is found in often after.

The Boolean school, founded in the 1930s in France by Boolean, thinks: mathematics, at least pure mathematics, is the study of the theory of abstract structure. Structure, it is the initial concept and the deduction system of Axiom. Boolean school also thinks that, there are three basic kinds of abstract structure: algebraic structure (group, ring, the domain...), sequence structure (partial order, all the sequence...), the topological structure (neighborhood, limit, connectivity, dimension...). Mathematics is a kind of transformation, an abstract model, and a sign system, the real world converted into mathematical model, using mathematical language describe them later, after operation, the results can back, explained in the real world specific scientific.

There is a basic logic that the human is everything, they can be observed to establish an Axiom system, make nature operations according to the Axiom system by human assumption. But in the traditional Chinese philosophy, always think that humans are the small, they can not establish a set of rules, let nature operation in accordance with the running rules of the human assumption.

In front to nature, human's only doing things is its behavior request, with its development with nature.

In other words, the western mathematics thinks the mathematical object of study is a simple system, which can be observed to establish an Axiom system, and then logical analysis. It is because a simple system can be assumed. But the image mathematics of the traditional Chinese mathematics thinks the mathematical object of study is a complex system, human can't do specific research object hypothesis, humans can only be clear, for general object of study (model-free), what kind of logic structure analysis can reach the humans to understand the research object of certain relations. It is because a complex system cannot be assumed.

Simple said: the western mathematics deals directly with the research object through the directly observed, but the image mathematics of the traditional Chinese mathematics researches object through the relationship between indirect processing analysis.

In fact, Western mathematics late nineteenth century was introduced into China, initially, "mathematics" to be directly translated as "arithmetic", and then said "arithmetic learn", and then they changed to "mathematics" words. But the ancient Chinese for this concept, in 3000 years ago, has been officially use the word "Gua" as the form, the "Xiang" as the number. In the Yi-Jing, this "mathematics" concept is defined as "Image Mathematics". It is part of the Traditional Chinese Mathematics (TCMath). This article mainly concerns image mathematical content in TCMath, so also said the image mathematics as TCMath. Image mathematics generally contains hexagrams ("Gua") and images ("Xiang") two content. The hexagrams ("Gua") is only the hexagrams mathematical symbols, which there is not the size since the size of number is about the definition of human beings. In general, the research object in the complex system is independent of human definition. The image ("Xiang ") is the study way or the calculation method for some mathematical indexes in order to know the objective existence of the fixed a state. The way or the calculation method for some mathematical indexes is independent of the complex system and only is Human's some methods of operation in order to study the relationship of the complex system.

The ancients speak of "mathematics" in Chinese is a word as the way for running (intervening and controlling) a complex system through the analysis of the use of hexagrams ("Gua") and images ("Xiang"). This and what we now understand the mathematical completely different things. In other words, in the image mathematics of Traditional Chinese Mathematics (TCMath), both intervening and controlling of an engineering are believed to as a complex system. It is because to run an engineering is difficult and complex in which there are the loving relation, the killing relation and the equivalent relation among many Axiom systems. The loving and killing relations are non-compatibility relations, which can compose the whole energy of the system greater than or less than the sum of each part energy of the system, respectively, rarely equal conditions. Mathematics means managing or controlling or intervening for the complex system through the analysis of the use of hexagrams ("Gua") and images ("Xiang"), and so on. Pursue the goal is the harmonious sustainable of the complex system in order to compose the complex system not outward expansion development. Generally speaking, the assumption involved the behavior of people is not needed since the system is complex.

But, in Western mathematics, mathematics means first through the observation to establish one Axiom system, then performing mathematical inference from the Axiom system. Both obtaining and inference of reasoning are believed to as a simple system, because all the conclusions and definition of are compatible with the Axiom system. Compared with Axiom system speaking, there are true and false. Major mathematical analysis method is to judge true or false from simple assumptions or simple model. It is because to obtain the true and false relationship of one Axiom system is easy and simple in which there is only a compatibility relation or a generalized equivalent relation under one Axiom system assumption. The compatibility relation or generalized equivalent relation can compose the whole energy of the system equal to the sum of each part energy of the system. Thus to obtain or to analyze under one Axiom system assumption can compose the simple system outward expansion development. Therefore, pursue the goal is for obtaining or analyzing in order to compose the simple system outward expansion development. Generally speaking, the various hypothetical models involves the behavior of people. This phenomenon in the image mathematics of TCMath is not allowed since a complex system cannot be supposed. Both true and false cannot be judged if the Axiom system has not been assumed.

Western mathematics using simple assumptions or simple models treats directly mathematical complex system from Microscopic point of view, always destroy the original mathematical complex system's balance, and has none beneficial to mathematical complex system's immunity. Western mathematical intervention method can produce imbalance of mathematical complex system, having strong side effects. Excessively using methods of mathematical intervention for a complex system can easily paralysis the mathematical complex system's immunity, which the debate of mathematical schools under different from Axiom systems is a product of Western mathematics since there are a number of Axiom systems in nature which are different to the people of faith. Using the method of mathematical intervention for a complex system too little can easily produce the mathematical intervention resistance problem.

The image mathematics of TCMath studies the world from the Macroscopic point of view, and its target is in order to maintain the original balance of mathematical complex system and in order to enhance the mathematiccal complex system's immunity. The image mathematics believes that each mathematical intervention has onethird of badness. She never encourage government to use mathematical intervention in long term. The ideal way is Wu Wei Er Wu Bu Wu—by doing nothing, everything is done. The image mathematics has over 5000-year history. It has almost none side effects or mathematical intervention resistance problem.

After long period of practicing, our ancient mathematical scientists use "Yin Yang Wu Xing" Theory extensively in the image mathematics to explain the origin of mathematical complex system, the law of mathematiccal complex system, mathematical changes, mathematiccal diagnosis, mathematical prevention, mathematical self-protection and mathematical intervening. It has become an important part of the image mathematics. "Yin Yang Wu Xing" Theory has a strong influence to the formation and development of traditional Chinese mathematical theory. As is known to all, China in recent decades, economy and related mathematical work have made great strides in development. Its reason is difficult to say the introduction of western mathematics, the fact that the Chinese traditional culture is in all kinds of mathematical decision plays a role. Her many mathematical intervening methods come from the traditional Chinese medicine since both human body and mathematical research objects of image mathematics are all complex systems. But, many Chinese and foreign schoolars still have some questions on the reasoning of image mathematics, such as Traditional Chinese Medicine which is due to image mathematics. In this article, we will start to the western world for presentation of image mathematics introduced some mathematical and logic analysis concept.

Zhang's theories, multilateral matrix theory [1] and multilateral system theory [2-19], have given a new and strong mathematical reasoning method from macro (Global) analysis to micro (Local) analysis. He and his colleagues have made some mathematical models and methods of reasoning [20-35], which make the mathematical reasoning of image mathematics possible based on "Yin Yang Wu Xing" Theory [36-38]. This paper will use steady multilateral systems to demonstrate the intervening principle of image mathematics: "Real (mathematical) disease is to rush down his son but virtual (mathematical) disease is to fill his mother" and "Strong inhibition of the same time, support the weak". The article proceeds as follows. Section 2 contains basic concepts and main theorems of steady multilateral systems while the intervening principle of image mathematics is demonstrated in Section 3. Some discussions in image mathematics are given in Section 4 and conclusions are drawn in Section 5.

BASIC CONCEPT OF STEADY MULTILATERAL SYSTEMS

In the real world, we are enlightened from some concepts and phenomena such as "biosphere", "food chain", "ecological balance" etc. With research and practice, by using the theory of multilateral matrices [1] and analyzing the conditions of symmetry [20-24] and orthogonality [25-35] what a stable complex system must satisfy, in particular, with analyzing the basic conditions what a stable working procedure of good product quality must satisfy [10,29], we are inspired and find some rules and methods, then present the logic model of analyzing stability of complex systems-steady multilateral systems [2-19]. There are a number of essential reasoning methods based on the stable logic analysis model, such as "transition reasoning", "atavism reasoning", "genetic reasoning" etc. We start and still use concepts and notations in papers [3-6].

Generalized Relations and Reasoning

Let V be a non-empty set and $V \times V = \{(x,y) : x \in V, y \in V\}$. The non-empty subset $R \subset V \times V$ is called a relation of V. Image mathematics mainly researches general relation rules for general V rather than for special V. So the general V cannot be supposed. We can only do matter is to research the structure of the set of relations $\Re = \{R_0, \cdots, R_{m-1}\}$.

For any relation set $\Re = \{R_0, \cdots, R_{m-1}\}$, we can define both an inverse relationship of $R_i \in \Re$ and a relation multiplication between $R_i \in \Re$ and $R_j \in \Re$ as follows:

$$R_i^{-1} = \{(x,y) : (y,x) \in R_i\}$$

and $R_i * R_j = \{(x,y) :$ there is at least a $u \in V$ such that $(x,u) \in R_i$ and $(u,y) \in R_j\}$.

The relation $R_i \in \Re$ is called reasonable if $R_i^{-1} \in \Re$. A generalized reasoning of general V is defined as for $R_i * R_j \neq \emptyset$ there is a relation $R_k \in \Re$ such that $R_i * R_j \subset R_k$.

The generalized reasoning satisfies the associative law of reasoning, i.e. $(R_i * R_j) * R_k = R_i * (R_j * R_k)$. This is the basic requirement of reasoning in TCMath. But there are a lot of reasoning forms which do not satisfy the associative law of reasoning in Western Science. For example, in true and false binary of proposition logic, the associative law does not hold on its reasoning because

$$(false * false) * false = true * false = false$$
$$\neq true = false * true = false * (false * false).$$

Equivalence Relations

Let V be a non-empty set and R_0 be its a relation. We call it an equivalence relation, denoted by ~, if the following three conditions are all true:

1)   Reflexive: $(x,x) \in R_0$ for all $x \in V$, i.e. $x \sim x$;

2)   Symmetric: if $(x,y) \in R_0$, then $(y,x) \in R_0$, i.e., if $x \sim y$, then $y \sim x$;

3)   Conveyable (Transitivity): if $(x,y) \in R_0$, $(y,z) \in R_0$, then $(x,z) \in R_0$, i.e., if $x \sim y$, $y \sim z$, then $x \sim z$.

Furthermore, the relation R is called a compatibility relation if there is a non-empty subset $R_1 \subset R$ such that R_1 satisfies at least one of the conditions above. And the relation R is called a non-compatibility relation if there doesn't exist any non-empty subset $R_1 \subset R$ such that R_1 satisfies any one of the conditions above. Any one of compatibility relations can be expanded into an equivalent relation to some extent [2].

Western Science only considers the reasoning under one Axiom system such that only compatibility relation reasoning is researched. However there are many Axiom systems in Nature. Traditional Chinese Science mainly researches the reasoning among many Axiom systems in Nature. Of course, she also considers the reasoning under one Axiom system but she only expands the reasoning as the equivalence relation reasoning.

Two Kinds of Opposite Non-Compatibility Relations

Equivalence relations, even compatibility relations, can not portray the structure of the complex systems clearly. In the following, we consider two non-compatibility relations.

In image mathematics, any Axiom system is not considered, but should first consider to use a logic system. Believe that the rules of Heaven and the behavior of Human can follow the same logic system. This logic system is equivalent to a group of computation. The method is to abide by the selected logic system to the research object classification, without considering the specific content of the research object, namely classification taking images.

Analysis of the relationship between research object, make relationships with computational reasoning comply with the selected logic system operation. And then in considering the research object of the specific content of the conditions, according to the logic of the selected system operation to solve specific problems. In mathematics, the method of classification taking images is explained in the following Definition 2.1.

Definition 2.1. Suppose that there exists a finite group $G^m = \{g_0, \cdots, g_{m-1}\}$ of order m where g_n is identity. Let V be a none empty set satisfying that $V = V_{g_0} + \cdots + V_{g_{m-1}}$ where the notation means that $V = V_{g_0} \cup \cdots \cup V_{g_{m-1}}$, $V_{g_i} \cap V_{g_j} = \varnothing, \forall i \neq j$ (the following the same).

In image mathematics, the V_{g_j} is first called a factor image of group element g_j for any j, and $V = V_{g_0} + \cdots + V_{g_{m-1}}$ is called a factor space (all "Gua (卦)"). We do not consider the factor size (class variables) and only consider it as mathematical symbols ("Gua"), such as, 0 or 1, because the size is defined by a human behavior for V, but we have no assumption of V.

A mathematical index of the unknown multivariate function $f\left(x_{g_0}, \cdots, x_{g_{m-1}}\right)$, $j = 0, \cdots, m-1$ $\forall x_{g_j} \in V_{g_j}$ is called a function image of V. All mathematical indexes of the unknown multivariate function f compose of the formation of a new set, namely image space

$$F(V) = F_{\omega_0}(V) + \cdots + F_{\omega_{g-1}}(V)$$

Where $G^g = \{\omega_0, \cdots, \omega_{g-1}\}$ is also a finite group of order g. The $F_{\omega_j}(V)$ is also called an Axiom system for any j if $F_{\omega_j}(V) \neq \varnothing$ because any Axiom system is the assumption of

$$F(V) = F_{\omega_0}(V) + \cdots + F_{\omega_{g-1}}(V)$$

(or, equivalent, V) in which there is only the compatibility relations, i.e., pursuing the same mathematics index $I(F_{\omega_j}(V))$. We do not consider the special multivariate function f (i.e., special function image) and only consider the calculation way of the general mathematical indexes of f from the factor space $V = V_{g_0} + \cdots + V_{g_{m-1}}$ in order to know some causal relations, because we have no assumption of f. But the size of the data image should be considered if we study specific issues by the general rules of data images.

Say simply, a study of the hexagrams ("Gua") in image mathematics is to learn the generalized properties of the inputs $x_{g_0}, \cdots, x_{g_{m-1}}$ of any multivariate function f for the given factor space $V = V_{g_0} + \cdots + V_{g_{m-1}}$, such as there are non-size, non-order relation, orthogonal relations and symmetrical relations, and so on. A study of the image ("Xiang") in image mathematics is to learn the generalized properties of all outputs f for the image space $F(V) = F_{\omega_0}(V) + \cdots + F_{\omega_{g-1}}(V)$, such as there are size specific meaning, a sequence of relationship, killing relations, loving relations, equivalent relations, and so on.

Without loss of generality, we put the function image space $F(V) = F_{\omega_0}(V) + \cdots + F_{\omega_{g-1}}(V)$ and the factor image space $V = V_{g_0} + \cdots + V_{g_{m-1}}$, still keep for V because of no assumption of V. In order to study the generalized relations and generalized reasoning, image mathematics researches the following relations.

Denoted $V_{g_i} \times V_{g_j} = \{(x,y) : x \in V_{g_i}, y \in V_{g_j}\}$, where the note $\times$ is the usual Cartesian product or cross join. Define relations

$$R_{g_r} = \sum_{g \in G^m} V_g \times V_{gg_r}, r = 0, \cdots, m-1,$$

where $R_{g_0} = R_{g_0}^{-1} = R_{g_0^{-1}}$ is called an equivalence relation of V if g_0 is identity; denoted by $\sim$; $R_{g_s} = R_{g_s}^{-1} = R_{g_s^{-1}}$ is called a symmetrical relation of V if $g_s = g_s^{-1}, s \neq 0$; denoted by $\longleftrightarrow$ or $\leftrightarrow$; $R_{g_n} = R_{g_n^{-1}}$ is called a neighboring relation of V if $g_1 \neq g_1^{-1}$; denoted by $\xrightarrow{R_1}$ or $\rightarrow$; $R_{g_a} = R_{g_a^{-1}}^{-1} \neq R_{g_{-1}}, R_{g_1}, R_{g_1^{-1}}$ is called an alternate (or atavism) relation of V if $g_a \neq g_a^{-1}$, g_1, $g_1^{-1}, a \geq 2$; denoted by $\xrightarrow{R_a}$ or $\Rightarrow$. #

In this case, the equivalence relations and symmetrical relations are compatibility relations but both neighboring relations and alternate relations are non-compatibility relations. For the given relation set $\Re = \{R_{g_0}, \cdots, R_{g_{m-1}}\}$, these relations R_{g_i} are all reasoning relations since the relation $R_{g_i}^{-1} = R_{g_i^{-1}} \in \Re$ if $R_{g_i} \in \Re$.

The equivalence relation R_{g_0}, symmetrical relations R_{g_s}, neighboring relation R_{g_n} and alternate relations R_{g_a} are all the possible relations for the method of classification taking images. In this paper, we mainly consider the equivalence relation R_{g_0}, neighboring relation R_{g_n} and alternate relations R_{g_a}.

There is an unique generalized reasoning between the two kinds of opposite non-compatibility relations for case m = 5. For example, let V be a none empty set, there are two kinds of opposite relations: the neighboring relation R_1, denoted ® and the alternate (or atavism) relation R_2, denoted $\Rightarrow$. The logic reasoning architecture [2-19] of "Yin Yang Wu Xing" Theory in Ancient China is equivalent to the following reasoning:

1) If $x \to y$, $y \to z$, then $x \Rightarrow z$; i.e., if $(x, y) \in R_1$, $(y, z) \in R_1$, then $(x, z) \in R_2$; or, $R_1 * R_1 \subset R_2$; $\Leftrightarrow$ if $x \to y$, $x \Rightarrow z$, then $y \to z$; i.e., if $(x, y) \in R_1$, $(x, z) \in R_2$, then $(y, z) \in R_1$; or, $R_1^{-1} * R_2 \subset R_1$; $\Leftrightarrow$ if $x \Rightarrow z$, $y \to z$, then $x \to y$; i.e., if $(x, z) \in R_2$, $(y, z) \in R_1$, then $(x, y) \in R_1$; or, $R_2 * R_1^{-1} \subset R_1$.

2) If $x \Rightarrow y$, $y \Rightarrow z$ then $z \to x$; i.e., if $(x, y) \in R_2$, $(y, z) \in R_2$, then $(z, x) \in R_1$; or, $R_2 * R_2 \subset R_1^{-1}$; $\Leftrightarrow$ if $z \to x$, $x \Rightarrow y$, then $y \Rightarrow z$; i.e., if $(z, x) \in R_1$, $(x, y) \in R_2$ then $(y, z) \in R_2$; or, $R_1 * R_2 \subset R_2^{-1}$; $\Leftrightarrow$ if $y \Rightarrow z$, $z \to x$ then $y \Rightarrow z$; i.e., if $(y, z) \in R_2$, $(z, x) \in R_1$, then $(y, z) \in R_2$; or, $R_2 * R_1 \subset R_2^{-1}$.

Let $R_3 = R_2^{-1}$ and $R_4 = R_1^{-1}$. Then above reasoning is equivalent to the calculating as follows:

$$R_i * R_j \subset R_{\mathrm{mod}(i+j,5)}, \forall i, j \in \{1, 2, 3, 4\}$$

where the $\mathrm{mod}(i+j, 5)$ is the addition of module 5.

Two kinds of opposite relations can not be exist separately. Such reasoning can be expressed in Figure 1. The first triangle reasoning is known as a jumping-transition reasoning, while the second triangle reasoning is known as an atavism reasoning. Reasoning method is a triangle on both sides decided to any third side. Both neighboring relations and alternate relations are not compatibility relations, of course, none equivalence relations, called non-compatibility relations.

Genetic Reasoning

Let V be a none empty set with the equivalent relation R_0, the neighboring relation R_1 and the alternate relations $R_a \neq R_1^{-1}$, $a > 1$. Then a genetic reasoning is defined as follows:

1) If $x \sim y$, $y \to z$, then $x \to z$; i.e., if $(x, y) \in R_0$, $(y, z) \in R_1$, then $(x, z) \in R_1$; or, $R_0 * R_1 \subset R_1$;

2) If $x \sim y$, $y \Rightarrow z$, then $x \Rightarrow z$; i.e., if $(x, y) \in R_0$, $(y, z) \in R_2$, then $(x, z) \in R_2$; or, $R_0 * R_2 \subset R_2$;

3) If $x \to y$, $y \sim z$, then $x \to z$; i.e., if $(x, y) \in R_1$, $(y, z) \in R_0$, then $(x, z) \in R_1$; or, $R_1 * R_0 \subset R_1$;

4) If $x \Rightarrow y$, $y \sim z$, then $x \Rightarrow z$; i.e., if $(x, y) \in R_2$, $(y, z) \in R_0$, then $(x, z) \in R_2$; or, $R_2 * R_0 \subset R_2$.

The genetic reasoning is equivalent to that

$$
\begin{array}{ccccc}
x & & & x & \\
\Downarrow & \searrow & & \Downarrow & \nwarrow \\
z & \leftarrow & y & y & \Rightarrow & z
\end{array}
$$

Figure 1: Triangle reasoning.

$$R_0 * R_j = R_j * R_0 = R_j, \forall j \in \{0,1,\cdots,m-1\} = G_0^m$$

since $R_0 * R_j \supset R_j$, $R_j * R_0 \supset R_j$. The genetic reasoning is equivalent to that there is a group $G_0^m = \{0,\cdots,m-1\}$ with the operation * such that V can be cut into $V = V_0 + \cdots + V_{m-1}$ where V_1 may be an empty set and the corresponding relations of reasoning can be written as the forms as follows

$$R_r = \sum_{i=0}^{m-1} V_i \times V_{i*r}, r = 0,\cdots,m-1,$$

satisfying $R_i * R_j \subset R_{i*j}, \forall i,j \in G_0^m$.

Steady Multilateral Systems

For a none empty set V and its a relation set $\Re = \{R_0,\cdots,R_{m-1}\}$, the form $(V,\Re)$ (or simply, V) is called a multilateral system [2-19], if $(V,\Re)$ satisfies the following properties:

1) $R_0 + \cdots + R_{m-1} \subset V \times V$, i.e. $R_i \cap R_j = \varnothing, \forall i \neq j$.

2) $R_0 * R_j = R_j * R_0 = R_j, \forall j \in \{0,1,\cdots,m-1\} = G_0^m$

3)   The relation $R_i^{-1} \in \Re$ if $R_i \in \Re$.

4)   For $R_i * R_j \neq \varnothing$, there is a relation $R_k \in \Re$ such that $R_i * R_j \subset R_k$.

The 4) is called the reasoning, the 1) the uniqueness of reasoning, the 2) the hereditary of reasoning (or genetic reasoning) and the 3) the equivalent property of reasoning of both relations $R_i \in \Re$ and $R_i^{-1} \in \Re$, i.e., the reasoning of $R_i \in \Re$ is equivalent to the reasoning of $R_i^{-1} \in \Re$. In this case, the two-relation set $\{R_i, R_i^{-1}\}$ is a lateral relation of $(V,\Re)$. The R_0 is called an equivalence relation. The multilateral system $(V,\Re)$ can be written as. Furthermore, the V and $\Re$ are called the state space and relation set considered of $(V,\Re)$, respectively. For a multilateral system $(V,\Re)$, it is called complete (or, perfect) if " $\subset$ " changes into "=". And it is called complex if there exists at least a non-compatibility relation $R_i \in \Re$. In this case, the multilateral system is also called a logic analysis model of complex systems.

Let R_i be a non-compatibility relation. A complex multilateral system

$$(V, \Re) = \left(V_0 + \cdots + V_{n-1}, \{R_0, \cdots, R_{m-1}\}\right)$$

is said as a steady multilateral system (or, a stable multilateral system) if there exists a number n such that $R_i^{*n} = R_0$ where $R_i^{*n} = R_1 * * * R_1$. The condition is equivalent to there is a the chain $x_1, \cdots, x_n \in V$ such that $(x_1, x_2) \in R_1, \cdots, (x_n, x_1) \in R_1$, or $x_1 \to x_2 \to \cdots \to x_n \to x_1$. The steady multilateral system is equivalent to the complete multilateral system. The stability definition given above, for a relatively stable system, is most essential. If there is not the chain or circle, then there will be some elements without causes or some elements without results in a system. Thus, this system is to be in the state of finding its results or causes, i.e., this system will fall into an unstable state, and there is not any stability to say.

Theorem 2.1. The system $(V, \Re)$ is a multilateral system if and only if there exists a finite group $G^m = \{g_0, \cdots, g_{m-1}\}$ of order m where g_0 is identity such that the relation set $\Re = \{R_{g_0}, \cdots, R_{g_{m-1}}\}$ satisfying $R_{g_i} * R_{g_j} \subset R_{g_i g_j}, \forall i, j \in \{0, 1, \cdots, m-1\}$.

In this case, the multilateral system $(V, \Re)$ can be written as $\left(V_{g_0} + \cdots + V_{g_{m-1}}, \{R_{g_0}, \cdots, R_{g_{m-1}}\}\right)$ satisfying $R_{g_r} = \sum_{g \in G^m} V_g \times V_{g g_r}, r = 0, \cdots, m-1,$ where V_{g_i} may be an empty set.

Theorem 2.2. If the multilateral system

$$(V, \Re) = \left(V_0 + \cdots + V_{n-1}, \{R_0, \cdots, R_{m-1}\}\right)$$ is a steady multilateral system, then $n = m$ and $\Re = \{R_0, \cdots, R_{m-1}\}$ is a finite group of order m about the relation multiplication $R_i * R_j = R_k$ where V_i must be a non empty set.

Definition 2.2. Suppose that a multilateral system $(V, \Re)$ can be written as $\left(V_{g_0} + \cdots + V_{g_{m-1}}, \{R_{g_0}, \cdots, R_{g_{m-1}}\}\right)$ satisfying $R_{g_r} = \sum_{g \in G^m} V_g \times V_{g g_r}, r = 0, \cdots, m-1$ and $R_{g_i} * R_{g_j} \subset R_{g_i g_j}, \forall i, j \in \{0, 1, \cdots, m-1\}$.

The group $G^m = \{g_0, \cdots, g_{m-1}\}$ of order m where g_0 is identity is called the representation group of the multilateral system $(V, \Re)$. The representing function of R_{g_r} is defined as follows

$$I\left(R_{g_r}\right) = \left\{(x,y): x^{-1}y = g_r, x, y \in G^m\right\}, r = 0, \cdots, m-1$$

.

Let multilateral systems $\left(V^i, \Re^i\right), i=1,2$ be with two representation groups G_i, $i=1,2$, respectively. Both multilateral systems $\left(V^i, \Re^i\right), i=1,2$ are called isomorphic if the two representation groups G_i, $i=1,2$ are isomorphic.

Theorems 2.1 and 2.2 and Definitions 2.1 and 2.2 are key concepts in multilateral system theory because they show the classification taking images as the basic method. In the following, introduce two basic models to illustrate the method.

Theorem 2.3 Suppose that $G_0^2 = \{0,1\}$ with multiplication table

*	0	1
0	0	1
1	1	0

i.e., the multiplication of G_0^2 is the addition of module 2. In other words, $i * j = \mod(i+j, 2)$.

And assume that $(V, \Re) = \left(V_0 + V_1, \{R_0, R_1\}\right)$ satisfying

$$R_r = \sum_{i=0}^{1} V_i \times V_{\mod(i+r,2)}, r = 0, 1,$$

$$R_i * R_j \subset R_{\mod(i+j,2)}, \forall i, j \in G_0^2$$

.

Then $(V, \Re)$ is a steady multilateral system with one equivalent relation R_0 and one symmetrical relation R_1 which is a simple system since there is not any noncompatibility relation. In other words, the relations R_i's are the simple forms as follows:

$$I(R_0) = \{(0,0), (1,1)\}, \quad I(R_1) = \{(0,1), (1,0)\}$$

where (i, j) is corresponding to $V_i \times V_j$.

It will be proved that the steady multilateral system in Theorem 2.3 is the reasoning model of "Tao" in TCMath if there are two energy functions $\varphi(V_0)$ and

$\varphi(V_1)$ satisfying $\varphi(V_0) > \varphi(V_1)$, called Dao model, denoted by V^2.

Theorem 2.4. For each element x in a steady multilateral system V with two non-compatibility relations, there exist five equivalence classes below:

$$X = \{y \in V | y \sim x\}, \; X_S = \{y \in V | x \to y\},$$

$$X_K = \{y \in V | x \Rightarrow y\}, \; K_X = \{y \in V | y \Rightarrow x\},$$

$$S_X = \{y \in V | y \to x\}$$

which the five equivalence classes have relations in Figure 2.

It can be proved that the steady multilateral system in Theorem 2.4 is the reasoning model of "Yin Yang Wu Xing" in TCMath if there are five energy functions (Defined in Section 3) $\varphi(K_x)$, $\varphi(X_K)$, $\varphi(X_S)$, $\varphi(X)$, and $\varphi(S_x)$ satisfying $\varphi(X_K) > \varphi(X_S) > \varphi(X) > \varphi(K_x) > \varphi(S_x)$ called Wu-Xing model, denoted by V^5.

By Definition 2.2, the Wu-Xing model V^5 can be written as follows:

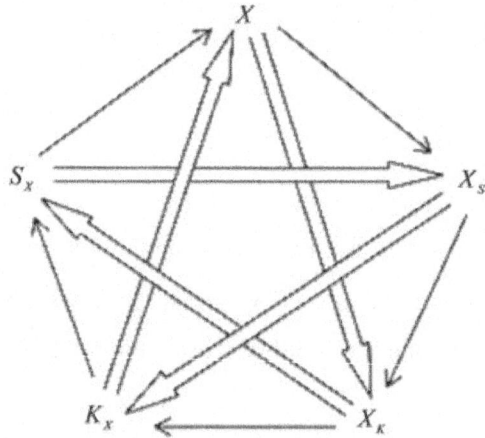

Figure 2: The method of finding Wu-Xing.

Define $V_0 = X$, $V_1 = X_S$, $V_2 = X_K$, $V_3 = K_X$, $V_4 = S_X$, corresponding to wood, fire, soil, metal, water, and assume $V = V_0 + V_1 + \cdots + V_4$ and $\Re = \{R_0, R_1, \cdots, R_4\}$, satisfying

$$R_r = \sum_{i=0}^{4} V_i \times V_{\mathrm{mod}(i+r,5)}, \, r \in G_0^5,$$

$$R_i * R_j \subset R_{\mathrm{mod}(i+j,5)}, \, \forall i, j \in G_0^5$$

i.e., the relation multiplication of V^5 is isomorphic to the addition of module 5. Then V^5 is a steady multilateral system with one equivalent relation R_0 and

two non-compatibility relations $R_1 = R_4^{-1}$ and $R_2 = R_3^{-1}$.

These Theorems can been found in [1-6,11-16]. Figure in Theorem 2.4 is the Figure of "Yin Yang Wu Xing" Theory in Ancient China. The steady multilateral system V with two non-compatibility relations is equivalent to the logic architecture of reasoning model of "Yin Yang Wu Xing" Theory in Ancient China. What describes the general method of complex systems can be used in mathematical complex system.

RELATIONSHIP ANALYSIS OF STEADY MULTILATERAL SYSTEMS

Energy of a Multilateral System

Energy concept is an important concept in Physics. Now, we introduce this concept to the multilateral systems (or image mathematics) and use these concepts to deal with the multilateral system diseases (mathematical index too bad or too good).In mathematics, a multilateral system is said to have Energy (or Dynamic) if there is a none negative function $\varphi(*)$ which makes every subsystem meaningful of the multilateral system.

For two subsystems V_i and V_j of multilateral system V, denote $V_i \to V_j$ (or $V_i \Rightarrow V_j$, or $V_i \sim V_j$, or $V_i \leftrightarrow V_j$) means $x_i \to x_j$, $\forall x_i \in V_i$, $x_j \in V_j$ (or $x_i \Rightarrow x_j$, $\forall x_i \in V_i$, $x_j \in V_j$, or $x_i \sim x_j$, $\forall x_i \in V_i$, $x_j \in V_j$, or $x_i \leftrightarrow x_j$, $\forall x_i \in V_i$, $x_j \in V_j$).

For subsystems V_i and V_j where $V_i \cap V_j = \varnothing$, $\forall i \neq j$. Let $\varphi(V_i)$, $\varphi(V_j)$ and $\varphi(V_i, V_j)$ be the energy function of V_i, the energy function of V_j and the total energy of both V_i and V_j, respectively.

For an equivalence relation $V_i \sim V_j$, if $\varphi(V_i, V_j) = \varphi(V_i) + \varphi(V_j)$ (the normal state of the energy of $V_i \sim V_j$), then the neighboring relation $V_i \sim V_j$ is called that V_i likes V_j which means that V_i is similar to V_j. In this case, the V_i is also called the brother of V_j while the V_j is also called the brother of V_i. In the causal model, the V_i is called the similar family member of V_j while the V_j is also called the similar family member of V_i. There are not any causal relation considered between V_i and V_j. For a symmetrical relation $V_i \leftrightarrow V_j$, if

$$\varphi(V_i, V_j) = \varphi(V_j, V_i) = \varphi(V_i) + \varphi(V_j) + \tau(V_i, V_j)$$

(the normal state of the energy of $V_i \leftrightarrow V_j$) where the $\tau(V_i, V_j)$ is an interaction of V_i and V_j satisfying $\tau(V_i, V_j) = \tau(V_j, V_i)$, then the symmetrical relation $V_i \leftrightarrow V_j$ is called that V_i is corresponding to V_j which means that V_i is positively (or non-negatively) corresponding to V_j if $\tau(V_i, V_j) > 0$ (or $\tau(V_i, V_j) \geq 0$) and that V_i is negatively corresponding to V_j if $\tau(V_i, V_j) < 0$. In this case, the V_i is also called the counterpart of V_j while the V_j is also called the counterpart of V_i. In the causal model, the V_i is called the reciprocal causation of V_j while the V_j is also called the reciprocal causation of V_i. There is a reciprocal causation relation considered between V_i and V_j.

For an neighboring relation $V_i \rightarrow V_j$, if $\varphi(V_i, V_j) > \varphi(V_j, V_i) > \varphi(V_i) + \varphi(V_j)$

(the normal state of the energy of $V_i \rightarrow V_j$), then the neighboring relation $V_i \rightarrow V_j$ is called that V_i bears (or loves) V_j [or that V_j is born by (or is loved by) V_i] which means that V_i is beneficial on V_j each other. In this case, the V_i is called the mother of V_j while the V_j is called the son of V_i. In the causal model, the V_i is called the beneficial cause of V_j while the V_j is called the beneficial effect of V_i.

For an alternate relation $V_i \Rightarrow V_j$, if $\varphi(V_i, V_j) < \varphi(V_j, V_i) < \varphi(V_i) + \varphi(V_j)$

(the normal state of the energy of $V_i \Rightarrow V_j$), then the alternate relation $V_i \Rightarrow V_j$ is called as that V_i kills (or hates) V_j [or that V_j is killed by (or is hated by) V_i] which means that V_i is harmful on V_j each other. In this case, the V_i is called the bane of V_j while the V_j is called the prisoner of V_i. In the causal model, the V_i is called the harmful cause of V_j while the V_j is called the harmful effect of V_i.

In the future, if not otherwise stated, any equivalence relation is the liking relation, any symmetrical relation is the reciprocal causation relation, any neighboring relation is the born relation (or the loving relation), and any alternate relation is the killing relation.

Suppose V is a steady multilateral system having energy, then V in the multilateral system during normal operation, its energy function for any subsystem of the multilateral system has an average (or expected value in Statistics), this state is called normal when the energy function is nearly to the average. Normal state is the better state.

That a subsystem of a multilateral system is not running properly (or, abnormal), is that the energy deviation from the average of the subsystems is too large (or too big), the high (mathematical real disease or economic overheating or real disease) or the low (mathematical virtual disease or economic downturn or virtual disease). Both mathematical real disease and mathematical virtual disease are all diseases of mathematical complex systems.

In a subsystem of a multilateral system being not running properly, if this sub-system through the energy of external forces increase or decrease, making them return to the average (or expected value), this method is called intervention (or making a mathematical treatment) to the multilateral system.

The purpose of intervention is to make the multilateral system return to normal state. The method of intervention is to increase or decrease the energy of a subsystem.

What kind of intervening should follow the principle to treat it? For example, Western economics emphasizes direct intervening, but the indirect intervening of oriental economics is required. In mathematics, which is more reasonable?

Based on this idea, many issues are worth further discussion. For example, if an economic intervening has been done to an economic society, what situation will happen?

Intervention Rule of a Multilateral System

For a steady multilateral system V with two non-compatibility relations, suppose that there is an external force (or an intervening force) on the subsystem X of V which makes it the energy $\varphi(X)$ change increment $\Delta\varphi(X)$, then the energies $\varphi(X_S)$, $\varphi(X_K)$, $\varphi(K_X)$, $\varphi(S_X)$ of other subsystems X_S, X_K, K_V, S_V (defined in Theorem 2.4) of V will be changed by the increments $\Delta\varphi(X_S)$, $\Delta\varphi(X_K)$, $\Delta\varphi(K_X)$ and $\Delta\varphi(S_X)$, respectively.

It is said that the multilateral system has the capability of intervention reaction if the multilateral system has capability to response the intervention force.

If a subsystem X of multilateral system is intervened, then the energies of the subsystems X_S and S_X which have neighboring relations to X will change in the same direction of the force outside on X. We call them beneficiaries. But the energies of the subsystems X_K and K_X which have alternate relations to X will change in the opposite direction of the force outside on X. We call them victims.

In general, there is an essential principle of intervenetion: any beneficial subsystem of X changes in the same direction of X, and any harmful subsystem of X changes in the opposite direction of X. The size of the energy changed is equal, but the direction opposite.

Intervention Rule: In the case of virtual disease or economic downturn, the intervening method of intervention is to increase the energy. If the intervening has been done on X, the energy increment (or, increase degree) $|\Delta\varphi(X_S)|$ of the son X_S of X is greater than the energy increment $|\Delta\varphi(S_X)|$ of the mother S_X of X, i.e., the best beneficiary is the son X_S of X. But the energy decrease degree $|\Delta\varphi(X_K)|$ of the prisoner X_K of X is greater than the energy decrease degree $|\Delta\varphi(K_X)|$ of the bane of X, i.e., the worst victim is the prisoner X_K of X.

In the case of real disease or economic overheating, the intervening method of intervention is to decrease the energy. If the intervening has been done on X, the energy decrease degree $|\Delta\varphi(S_X)|$ of the mother S_X of X is greater than the energy decrease degree $|\Delta\varphi(X_S)|$ of the son of X, i.e., the best beneficiary is the mother S_X of X. But the energy increment $|\Delta\varphi(K_X)|$ of the bane K_X of X is greater than the energy increment $|\Delta\varphi(X_K)|$ of the prisoner X_K of X, i.e., the worst victim is the bane K_X of X.

In mathematics, the changing laws are as follows.

1)   If $\Delta\varphi(X) = \Delta > 0$, then $\Delta\varphi(X_S) = \rho_1\Delta$, $\Delta\varphi(X_K) = -\rho_1\Delta$, $\Delta\varphi(K_X) = -\rho_2\Delta$, $\Delta\varphi(S_X) = \rho_2\Delta$.

2)   If $\Delta\varphi(X) = -\Delta < 0$, then $\Delta\varphi(X_S) = -\rho_2\Delta$, $\Delta\varphi(X_K) = \rho_2\Delta$, $\Delta\varphi(K_X) = \rho_1\Delta$, $\Delta\varphi(S_X) = -\rho_1\Delta$.

where $1 \geq \rho_1 \geq \rho_2 \geq 0$. Both ρ_1 and ρ_2 are called intervention reaction coefficients, which are used to represent the capability of intervention reaction. The larger ρ_1 and ρ_2, the better the capability of intervention reaction. The state $\rho_1 = \rho_2 = 1$ is the best state but the state $\rho_1 = \rho_2 = 0$ is the worst state.

This intervention rule is similar to force and reaction in Physics. In other words, if a subsystem of multilateral system V has been intervened, then the energy of subsystem which has neighboring relation changes in the same direction of the force, and the energy of subsystem which has alternate relation changes in the opposite direction of the force. The size of the energy changed is equal, but the direction opposite.

In general, ρ_1 and ρ_2 are decreasing functions of the intervention force Δ since the intervention force $\Delta\varphi(X)$ is easily to transfer all if Δ is small but the intervention force is not easily to transfer all if Δ is large. The energy function of complex system, the stronger the more you use. In order to magnify ρ_1 and ρ_2, should set up a mathematical complex system of the intervention reaction system, and often use it.

Mathematical intervening resistance problem is that such a question, beginning more appropriate mathematiccal intervening method, but is no longer valid after a period. It is because the capability of intervention reaction is bad, i.e., the intervention reaction coefficients ρ_1 and ρ_2 is too small. In the state $\rho_1 = \rho_2 = 1$, any mathematical intervening resistance problem is non-existence but in the state $\rho_1 = \rho_2 = 0$, mathematical intervening resistance problem is always existence. At this point, the paper advocates the essential principle of intervening to avoid mathematical intervening resistance problems.

Self-Protection Rule of a Multilateral System

If there is an intervening force on the subsystem X of a steady multilateral system V which makes the energy $\varphi(X)$ changed by increment $\Delta\varphi(X)$ such that the energies $\varphi(X_S), \varphi(X_K), \varphi(K_X), \varphi(S_X)$ of other subsystems X_S, X_K, K_V, S_V (defined in Theorem 2.4) of V will be changed by the increments $\Delta\varphi(X_S)$, $\Delta\varphi(X_K)$, $\Delta\varphi(K_X)$, $\Delta\varphi(S_X)$, respectively, then can the multilateral system V has capability to protect the worst victim to restore?

It is said that the steady multilateral system has the capability of self-protection if the multilateral system has capability to protect the worst victim to restore. The capability of self-protection of the steady multilateral system is said to be better if the multilateral system has capability to protect all victims to restore.

In general, there is an essential principle of self-protection: any harmful subsystem of X should be protected by using the same intervention force but any beneficial subsystem of X should not.

Self-protection Rule: In the case of virtual disease or economic downturn, the intervening method of intervention is to increase the energy. If the intervening has been done on X by the increment $\Delta\varphi(X) = \Delta > 0$, the worst victim is the prisoner X_K of X which has the increment $(-\rho_1\Delta)$. Thus the intervening principle of self-protection is to restore the prisoner X_V of X and the restoring method of self-protection is to increase the energy $\varphi(X_K)$ of the prisoner X_K of X by using the intervention force on X according to the intervention rule. In general, the increase degree is $(\rho_3\Delta)$ where $\rho_3 \le \rho_1$.

In the case of real disease or economic overheating, the intervening method of intervention is to decrease the energy. If the intervening has been done on X by the increment $\Delta\varphi(X) = -\Delta < 0$, the worst victim is the bane K_x of X which has the increment $(\rho_1\Delta)$. Thus the intervening principle of self-protection is to restore the bane K_x of X and the restoring method of self-protection is to decrease the energy $\varphi(K_x)$ of the bane K_x of X by using the same intervention force on X according to the intervention rule. In general, the decrease degree is $(-\rho_3\Delta)$ where $\rho_3 \le \rho_1$.

In mathematics, the following self-protection laws hold.

1) If $\Delta\varphi(X) = \Delta > 0$, then the energy of subsystem X_K will decrease the increment $(-\rho_1\Delta)$, which is the worst victim. So the capability of self-protection increases the energy of subsystem X_K by increment $(\rho_3\Delta)$ where $\rho_3 \le \rho_1$, in order to restore the worst victim by according to the intervention rule.

2) If $\Delta\varphi(X) = -\Delta < 0$, then the energy of subsystem K_X will increase the increment $(\rho_1\Delta)$, which is the worst victim. So the capability of self-protection decreases the energy of subsystem K_x by increment $(-\rho_3\Delta)$ where $\rho_3 \le \rho_1$, in order to restore the worst victim by according to the intervention rule.

In general, $0 \le \rho_3 \le \rho_1 \le 1$. The ρ_1 is the intervenetion reaction coefficient. The ρ_3 is called an self-protection coefficient, which is used to represent the capability of self-protection. The larger ρ_3, the better the capability of self-protection. The state $\rho_3 = \rho_1 = 1$ is the best state but the state $\rho_3 = 0$ is the worst state of self-protection. According to the general economy of the protection principle, ρ_3 should be not greater than ρ_1 since the purpose of protection is to restore the victims and not reward the victims.

The self-protection rule can be explained as: the general principle of self-protection subsystem is that the most affected is protected firstly, the protection method and intervention force are in the same way.

In general, ρ_3 is also a decreasing function of the intervention force Δ since the worst victim is easily to restore all if Δ is small but the worst victim is not easily to restore all if Δ is large. The energy function of complex system, the stronger the more you use. In order to magnify ρ_3, should set up an economic society of the selfprotection system, and often use it.

Theorem 3.1. Suppose that a steady multilateral system V which has energy function $\varphi(*)$ and capabilities of intervention reaction and self-protection is with intervention reaction coefficients $\rho_1 = \rho_1(\Delta)$ and $\rho_2 = \rho_2(\Delta)$, and with self-protection coefficient $\rho_3 = \rho_3(\Delta)$. If the capability of self-protection wants to restore both subsystems X_K and K_X, then the following statements are true.

1) In the case of virtual disease, the treatment method is to increase the energy. If an intervention force on the subsystem X of steady multilateral system V is implemented such that its energy $\varphi(X)$ has been changed by increment $\Delta\varphi(X) = \Delta > 0$, then all five subsystems will be changed finally by the increments as follows:

$$\Delta\varphi(X)_2 = \Delta\varphi(X) + \Delta\varphi(X)_1 = (1 - \rho_2\rho_3)\Delta > 0,$$
$$\Delta\varphi(X_S)_2 = \Delta\varphi(X_S) + \Delta\varphi(X_S)_1 = (\rho_1 + \rho_2\rho_3)\Delta > 0,$$
$$\Delta\varphi(X_K)_2 = \Delta\varphi(X_K) + \Delta\varphi(X_K)_1 = -(\rho_1 - \rho_3)\Delta \le 0,$$
$$\Delta\varphi(K_X)_2 = \Delta\varphi(K_X) + \Delta\varphi(K_X)_1 = -(\rho_2 - \rho_1\rho_3)\Delta \le 0,$$
$$\Delta\varphi(S_X)_2 = \Delta\varphi(S_X) + \Delta\varphi(S_X)_1 = (\rho_2 - \rho_1\rho_3)\Delta \ge 0$$
$$\forall\Delta\varphi(X) = \Delta > 0. \tag{1}$$

2) In the case of real disease, the treatment method is to increase the energy. If an intervention force on the subsystem X of steady multilateral system V is implemented such that its energy $\varphi(X)'$ has been changed by increment $\Delta\varphi(X)' = -\Delta < 0$, then all five subsystems will be changed by the increments as follows:

$$\Delta\varphi(X)'_2 = \Delta\varphi(X) + \Delta\varphi(X)'_1 = -(1 - \rho_2\rho_3)\Delta < 0,$$
$$\Delta\varphi(X_S)'_2 = \Delta\varphi(X_S) + \Delta\varphi(X_S)'_1 = -(\rho_2 - \rho_1\rho_3)\Delta \le 0,$$
$$\Delta\varphi(X_K)'_2 = \Delta\varphi(X_K) + \Delta\varphi(X_K)'_1 = (\rho_2 - \rho_1\rho_3)\Delta \ge 0,$$
$$\Delta\varphi(K_X)'_2 = \Delta\varphi(K_X) + \Delta\varphi(K_X)'_1 = (\rho_1 - \rho_3)\Delta \ge 0,$$
$$\Delta\varphi(S_X)'_2 = \Delta\varphi(S_X) + \Delta\varphi(S_X)'_1 = -(\rho_1 + \rho_2\rho_1)\Delta < 0$$
$$\forall\Delta\varphi(X)' = -\Delta < 0. \tag{2}$$

where the $\Delta\varphi(*)_1$'s and $\Delta\varphi(*)'_1$'s are the increments under the capability of self-protection.

Corollary 3.1. Suppose that a steady multilateral system V which has energy function $\varphi(*)$ and capabilities of intervention reaction and self-protection is with intervention reaction coefficients $\rho_1 = \rho_1(\Delta)$ and $\rho_2 = \rho_2(\Delta)$, and with self-protection coefficient $\rho_3 = \rho_3(\Delta)$. Then the capability of self-protection can make both subsystems X_K and K_X to be restored at the same time, i.e., the capability of self-protection is better, if and only if $P_2 = P_1 P_3$ and $P_3 = P_1$.

Side effects of mathematical intervening problems were the question: in the mathematical intervening process, destroyed the normal balance of non-fall ill subsystem or non-intervention subsystem. By Theorem 3.1 and Corollary 3.1, it can be seen that if the capability of self-protection of the steady multilateral system is better, i.e., the multilateral system has capability to protect all the victims to restore, then a necessary and sufficient condition is $P_2 = P_1 P_3$ and $P_3 = P_1$. General for a complex system of mathematical complex system, the condition $P_2 = P_1 P_3$ is easy to meet since it can restore two subsystems by Theorem 3.1, the condition $P_3 = P_1$ is difficult to meet it only can restore one subsystem by Theorem 3.1. At this point, the paper advocates the principle to avoid any side effects of intervening.

Mathematical Reasoning of Intervening Principle by Using the Neighboring Relations of Steady Multilateral Systems

Intervening principle by using the neighboring relations of steady multilateral systems is "Virtual disease or mathematical virtual disease is to fill his mother but mathematical real disease is to rush down his son". In order to show the rationality of the intervening principle, it is needed to prove the following theorems.

Theorem 3.2. Suppose that a steady multilateral system V which has energy function and capabilities of intervention reaction and self-protection is with intervention reaction coefficients $P_1 = P_1(\Delta)$ and $P_2 = P_2(\Delta)$, and with self-protection coefficient $P_3 = P_3(\Delta)$ satisfying $P_2 = P_1 P_3$ and $P_3 = P_1$. Then the following statements are true.

In the case of virtual disease, if an intervention force on the subsystem X of steady multilateral system V is implemented such that its energy $\varphi(X)$ increases the increment $\Delta\varphi(X) = \Delta > 0$, then the subsystems S_X, X_K and K_X can be restored at the same time, but the subsystems X and X_S will increase their energies by the increments

$$\Delta\varphi(X)_2 = (1 - \rho_2\rho_3)\Delta\varphi(X)$$
$$= (1 - \rho_2\rho_3)\Delta = (1 - \rho_1^3)\Delta > 0$$

and

$$\Delta\varphi(X_S)_2 = (\rho_1 + \rho_2\rho_3)\Delta\varphi(X)$$
$$= (\rho_1 + \rho_2\rho_3)\Delta = (\rho_1 + \rho_1^3)\Delta > 0,$$

respectively.

On the other hand, in the case of real disease, if an intervention force on the subsystem X of steady multilateral system V is implemented such that its energy $\varphi(X)'$ decreases, i.e., by the increment $\Delta\varphi(X)' = -\Delta < 0$, the subsystems X_S, X_K and K_X can also be restored at the same time, and the subsystems X and X_S will decrease their energies, i.e., by the increments

$$\Delta\varphi(X)'_2 = (1 - \rho_2\rho_3)\Delta\varphi(X)'$$
$$= -(1 - \rho_2\rho_3)\Delta = -(1 - \rho_1^3)\Delta < 0$$

and

$$\Delta\varphi(S_X)'_2 = (\rho_1 + \rho_2\rho_3)\Delta\varphi(X)'$$
$$= -(\rho_1 + \rho_2\rho_3)\Delta = -(\rho_1 + \rho_1^3)\Delta < 0,$$

respectively.

Theorem 3.3. For a steady multilateral system V which has energy function $\varphi(*)$ and capabilities of interveneing reaction and self-protection, assume intervention reaction coefficients are ρ_1 and ρ_2, and let the selfprotection coefficientbe ρ_3, which satisfy $\rho_2 = \rho_1\rho_3$, $\rho_3 = \rho_1$ and $\rho_1 \geq \rho_0$ where $\rho_0 \approx (<) 0.5897545123$ (the following the same) is the solution of $2\rho_1^3 + \rho_1 = 1$. Then the following statements are true.

1) If an intervention force on the subsystem X of steady multilateral system V is implemented such that its energy $\varphi(X)$ has been changed by increment $\Delta\varphi(X) = \Delta > 0$, then the final increment $(\rho_1 + \rho_2\rho_3)\Delta$ of the energy $\varphi(X_S)$ of the subsystem X_S changed is greater than or equal to the final increment $(1 - \rho_2\rho_3)\Delta$ of the energy $\varphi(X)$ of the subsystem X changed based on the capability of self-protection.

2) If an intervention force on the subsystem X of steady multilateral system V is implemented such that its energy $\varphi(X)$ has been changed by increment $\Delta\varphi(X) = -\Delta < 0$, then the final increment $-(\rho_1 + \rho_2\rho_3)\Delta$ of the energy $\varphi(S_X)$ of the subsystem S_X changed is less than or equal to the final increment $-(1 - \rho_2\rho_3)\Delta$ of the energy $\varphi(X)$ of the subsystem X changed based on the capability of self-protection.

Corollary 3.2. For a steady multilateral system V which has energy function $\varphi(*)$ and capabilities of intervening reaction and self-protection, assume intervention reaction coefficients are P_1 and P_2, and let the self-protection coefficient be P_3, which satisfy $P_2 = P_1 P_3$, $P_3 = P_1$ and $P_1 < P_0$. Then the following statements are true.

1)   In the case of virtual disease, if an intervention force on the subsystem X of steady multilateral system V is implemented such that its energy $\varphi(X)$ has been changed by increment $\Delta\varphi(X) = \Delta > 0$, then the final increment $(P_1 + P_2 P_3)\Delta$ of the energy $\varphi(X_S)$ of the subsystem X_S changed is less than the final increment $(1 - P_2 P_3)\Delta$ of the energy $\varphi(X)$ of the subsystem X changed based on the capability of self-protection.

2)   In the case of real disease, if an intervention force on the subsystem X of steady multilateral system V is implemented such that its energy $\varphi(X)$ has been changed by increment $\Delta\varphi(X) = -\Delta < 0$, then the final increment $-(P_1 + P_2 P_3)\Delta$ of the energy $\varphi(S_X)$ of the subsystem S_X changed is greater than the final increment $-(1 - P_2 P_3)\Delta$ of the energy $\varphi(X)$ of the subsystem X changed based on the capability of self-protection. #

By Theorems 3.2 and 3.3 and Corollary 3.2, the intervention method of "Virtual disease or economic downturn is to fill his mother but real disease or economic overheating is to rush down his son" should be often used in case: $P_2 = P_1 P_3$, $P_3 = P_1$ and $P_1 \geq P_0$ since in this time, $(P_1 + P_2 P_3)\Delta \geq (1 - P_2 P_3)\Delta$.

Mathematical Reasoning of Intervening Principle by Using the Alternate Relations of Steady Multilateral Systems

Intervening principle by using the alternate relations of steady multilateral systems is "Strong inhibition of the same time, support the weak". In order to show the rationality of the intervening Principle, it is needed to prove the following theorems.

Theorem 3.4. Suppose that a steady multilateral system V which has energy function $\varphi(*)$ and capabilities of intervention reaction and self-protection is with intervention reaction coefficients $P_1 = P_1(\Delta)$ and $P_2 = P_2(\Delta)$, and with self-protection coefficient $P_3 = P_3(\Delta)$. Then the following statements are true.

Assume there are two subsystems X and X_K of V with an alternate relation such that X encounters virtual disease, and at the same time, X_K befalls real disease. If an intervention force on the subsystem X of steady multilateral system V is implemented such that its energy $\varphi(X)$ has been changed by increment $\Delta\varphi(X) = \Delta > 0$, and at the same time, another intervention force on the subsystem X_K of steady multilateral system V is also implemented such that its energy $\varphi(X_{\kappa})$ has been changed by increment $\Delta\varphi(X_{\kappa})' = -\Delta < 0$, then all other subsystems: S_X, K_X and X_S can be restored at the same time, and the subsystems X and X_K will increase and decrease their energies by the same size but the direction opposite, i.e., by the increments

$$\Delta\varphi(X)_3 = (1 - \rho_2\rho_3)\Delta\varphi(X)$$
$$= (1 - \rho_2\rho_3)\Delta = (1 - \rho_1^3)\Delta > 0$$

and

$$\Delta\varphi(X_K)_3 = (1 - \rho_2\rho_3)\Delta\varphi(X_K)$$
$$= -(1 - \rho_2\rho_3)\Delta = -(1 - \rho_1^3)\Delta < 0$$

respectively.

Assume there are two subsystems X and K_X of V with an alternate relation such that X encounters real disease, and at the same time, K_X befalls virtual disease. If an intervention force on the subsystem X of steady multilateral system V is implemented such that its energy $\varphi(X)$ has been changed by increment $\Delta\varphi(X) = -\Delta < 0$, and at the same time, another intervention force on the subsystem K_X of steady multilateral system V is also implemented such that its energy $\varphi(K_X)$ has been changed by increment $\varphi(K_X) = \Delta > 0$, then all other subsystems: S_X, X_K and X_S can be restored at the same time, and the subsystems X and K_X will decrease and increase their energies by the same size but the direction opposite, i.e., by the increments

$$\Delta\varphi(X)_3 = (1 - \rho_2\rho_3)\Delta\varphi(X)$$
$$= -(1 - \rho_2\rho_3)\Delta = -(1 - \rho_1^3)\Delta < 0$$

and

$$\Delta\varphi(K_X)_3 = (1 - \rho_2\rho_3)\Delta\varphi(K_X)$$
$$= (1 - \rho_2\rho_3)\Delta = (1 - \rho_1^3)\Delta > 0$$

respectively.

By Theorems 3.3 and 3.4 and Corollary 3.2, the method of "Strong inhibition of the same time, support the weak" should be used in case: $\rho_2 = \rho_1\rho_3$, $\rho_3 = \rho_1$ and $\rho_1 < \rho_0$ since $(\rho_1 + \rho_2\rho_3)\Delta < (1 - \rho_2\rho_3)\Delta$.

RATIONALITY OF INTERVENING PRINCIPLE OF TRADITIONAL CHINESE MATHEMATICS AND "YIN YANG WU XING" THEORY

Chinese Traditional Mathematics and "Yin Yang Wu Xing" Theory

Ancient Chinese "Yin Yang Wu Xing" [38] Theory has been surviving for several thousands of years without dying out, proving it reasonable to some extent. If we regard ~ as the same category, the neighboring relation ® as beneficial, harmony, obedient, loving, etc. and the alternate relation $\Rightarrow$ as harmful, conflict, ruinous, killing, etc., then the above defined stable logic analysis model is similar to the logic architecture of reasoning of "Yin Yang Wu Xing". Both "Yin" and "Yang" mean that there are two opposite relations in the world: harmony or loving ® and conflict or killing $\Rightarrow$, as well as a general equivalent category ~. There is only one of three relations ~, ® and $\Rightarrow$ between every two objects. Everything X makes something ($X_s \neq \varnothing$), and is made by something ($S_x \neq \varnothing$); Everything restrains something ($X_\kappa \neq \varnothing$), and is restrained by something ($K_x \neq \varnothing$); i.e., one thing overcomes another thing and one thing is overcome by another thing. The ever changing world V, following the relations: ~, ® and $\Rightarrow$, must be divided into five categories by the equivalent relation ~, being called "Wu Xing": wood (X), fire (X_S), soil (X_K), metal (K_X), water (S_X). The "Wu Xing" is to be "neighbor is friend": wood (X) ® fire (X_S) ® soil (X_K) ® metal (K_X) ® water (S_V) ®wood (X), and "alternate is foe": wood (X) $\Rightarrow$soil (X_K) $\Rightarrow$water (S_X) $\Rightarrow$fire (X_S) $\Rightarrow$metal (K_X) $\Rightarrow$wood (X). In other words, the ever changing world must be divided into five categories:

$$V = X + X_S + X_K + K_X + S_X$$

satisfying

$$X \to X_S \to X_K \to K_X \to S_X \to X$$

and

$$X \Rightarrow X_K \Rightarrow S_X \Rightarrow X_S \Rightarrow K_X \Rightarrow X$$

where elements in the same category are equivalent to one another. We can see, from this, the ancient Chinese "Yin Yang Wu Xing" theory is a reasonable logic analysis model to identify the stability and relationship of complex mathematical systems.

Image mathematics firstly uses the verifying relationship method of "Yin Yang Wu Xing" Theory to explain the relationship between mathematical complex system and environment. Secondly, based on "Yin Yang Wu Xing" Theory, the relations of development processes of mathematical complex system can be shown by the neighboring relation and alternate relation of five subsets. Then a normal mathematical complex system can be shown as a steady multilateral system in which there are the loving relation and the killing relation and the liking relation. The loving relation in image mathematics can be explained as the neighboring relation, called "Sheng". The killing relation in image mathematics can be explained as the alternate relation, called "Ke". The liking relation can be explained as the equivalent relation, called "Tong-Lei". Constraints and conversion between five subsets are equivalent to the two kinds of triangle reasoning. So a normal mathematical complex system can be classified into five equivalence classes corresponding five mathematical indexes, respectively.

For example, in image mathematics, a mathematical complex system is similar to a human body. A mathematical index system of normal complex system following the "Yin Yang Wu Xing" Theory was classified into five equivalence classes as follows [29]:

Consider a complex system, its input $x_1, \cdots, x_m, \omega$ and output y can be written as

$$y = f(x_1, \cdots, x_m, \omega) = g(x_1, \cdots, x_m) + \varepsilon_g$$
$$g(x_1, \cdots, x_m) = E(f(x_1, \cdots, x_m, \omega) | x_1, \cdots, x_m),$$
$$\varepsilon_g = f(x_1, \cdots, x_m, \omega) - E(f(x_1, \cdots, x_m, \omega) | x_1, \cdots, x_m), \qquad (2)$$

where not only all output functions

$y = f = f(x_1, \cdots, x_m, \omega)$, $g(x_1, \cdots, x_m)$ and ε_g are not known, but also the input variables $x_1, \cdots, x_m, \omega$ are not known. The problem is called model-free.

The inputs $x_1, \cdots, x_m$ are called controllable if they are observed and controlled by human. So

$$g(x_1, \cdots, x_m) = E(f(x_1, \cdots, x_m, \omega) | x_1, \cdots, x_m)$$

can be observed if the controllable inputs $x_1, \cdots, x_m$ can be choose.

The input ω is called uncontrolled if it is not observed or controlled by human. So the freedom model error

$$\varepsilon_g = f(x_1, \cdots, x_m, \omega) - E(f(x_1, \cdots, x_m, \omega) | x_1, \cdots, x_m)$$

can not be assumed. But we can show the following properties:

$$E\left(\varepsilon_g \middle| x_1,\cdots,x_m\right) = E\left(f\left(x_1,\cdots,x_m,\omega\right)\middle| x_1,\cdots,x_m\right)$$

$$- E\left[E\left(f\left(x_1,\cdots,x_m,\omega\right)\middle| x_1,\cdots,x_m\right)\middle| x_1,\cdots,x_m\right] = 0,$$

$$Var\left(\varepsilon_g^2 \middle| x_1,\cdots,x_m\right) =: \sigma_{x_1,\cdots,x_m}^2 \geq 0,$$

$$Cov\left(\left(g\left(x_1,\cdots,x_m\right),\varepsilon_{g(x_1,\cdots,x_m)}\right)\middle| x_1,\cdots,x_m\right)$$

$$= E\left(\left(g - Eg\right)\left(\varepsilon_g - E\varepsilon_g\right)\middle| x_1,\cdots,x_m\right) = 0.$$

The condition is not hypothesis since they can be obtained if f makes the calculation meaningful, such as, if f is continuous. In general, we can consider the inputs $x_1,\cdots,x_m,\omega$ as independent random variables with continuous distributions $F_1(x_1),\cdots,F_m(x_m),F_\omega(\omega)$, respectively, since the inputs $x_1,\cdots,x_m$ are controllable which can be selected independently by human under some similar conditions. If not independent, by factor analysis can select orthogonal factor. This operation of experiment under some similar conditions is equivalent to the uncontrolled input ω is independent of the controllable input $x_1,\cdots,x_m$ since $Cov(g(x_1,\cdots,x_m),\varepsilon_g) = 0.$ This operation of experiment that the inputs $x_1,\cdots,x_m$ can be selected independently by human is equivalent to that the inputs $x_1,\cdots,x_m$ are independent random variables one another. For example, take $x_1,\cdots,x_m$ based on an orthogonal array since the orthogonality is equivalent to independence for discrete random variables and continuous random variables can be in a discrete random variable approximation [8].

In this case, it is well known that the inputs

$$u_1 = F_1(x_1),\cdots,u_m = F_m(x_m),u_\omega = F_\omega(\omega)$$

are independent random variables with the same continuous distribution $U(0,1)$ Assume

$$v_1 = a_1 + u_1\left(b_1 - a_1\right),\cdots,v_m = a_m + u_m\left(b_m - a_m\right),$$
$$v_\omega = a_\omega + u_\omega\left(b_\omega - a_\omega\right),$$

where $b_j > a_j, j = 1,\cdots,m, b_\omega > a_\omega$. Then the inputs $v_1,\cdots,v_m,v_\omega$ are independent random variables with continuous distributions $U(a_1,b_1),\cdots,U(a_m,b_m),U(a_\omega,b_\omega)$. In this case, the function

$$h(v_1,\cdots,v_m,v_\omega) = f\left(F_1^{-1}\left(\frac{v_1 - a_1}{b_1 - a_1} \right), \cdots, \right.$$

$$\left. F_m^{-1}\left(\frac{v_m - a_m}{b_m - a_m} \right), F_\omega^{-1}\left(\frac{v_\omega - a_\omega}{b_\omega - a_\omega} \right) \right)$$

can replace f as a new system function since each of both h and f are all not known. Therefore, without loss of generality, we always consider $x_1,\cdots,x_m,\omega$ as independent random variables with continuous distributions

$$U(a_1,b_1),\cdots,U(a_m,b_m),U(a_\omega,b_\omega)_.$$

On the other hand, the function f is considered as continuous, in order to ensure that condition expectations and partial derivative of existence and make the conventional mathematics method has significance.

To the complex system f, we need to decide an energy goal t, make y more close to target, the greater the function of the system. In general, the Target t is the maximum energy of y.

Image mathematics in TCMath considers the complex system stability problem, because the core problem of any complex system is stability. The stability can only through the fixed program to observe to do a test or experiment since the function f is not known. In general, the human wants to find a testing or experimental center $x^0 = (x_1^0,\cdots,x_m^0)$ and testing or experimental tolerance $\Delta x = (\Delta x_1,\cdots,\Delta x_m)$, $\Delta_{max} = max(\Delta x_1,\cdots,\Delta x_m)$, for the observed function

$$g(x_1,\cdots,x_m) = E\left(f(x_1,\cdots,x_m,\omega) \big| x_1,\cdots,x_m \right)$$

under some similar conditions, such that 1) $(Eg(x_1,\cdots,x_m)-t)^2$

$$= \left(Eg(x_1^0,\cdots,x_m^0)-t \right)^2 + O(\Delta_{max}) \to min;$$

2) $E(y-t)^2 \to min;$

where the controllable inputs $x_1,\cdots,x_m$ are independent random variables and

$$x_j \sim U(a_j,b_j) = U(x_j^0 - \Delta x_j, x_j^0 + \Delta x_j), j = 1,\cdots,m.$$

In order to solve the stability problem, we get easily the following theorem [29]:

Theorem 4.1. Suppose that f is continuous and

$$g = g(x_1, \cdots, x_m) = E\left(f(x_1, \cdots, x_m, \omega) \big| x_1, \cdots, x_m\right).$$

Then

$$E(y-t)^2 = (Eg-t)^2 + Var(g) + Var(\varepsilon_g),$$

where $Var(\varepsilon_g) = E\sigma^2_{x_1, \dots, x_m}$. #

1)   In image mathematics, the index

$U_g^2 = Var(g(x_1, \cdots, x_m))$ is called Distortion Degree. It belongs to the wood (X) subsystem of the complex system f since it cognizes the structure function g of the complex system f which is the beginning or birth stage of all things, just like in the Spring of a year. In mathematics,

$$U_g^2 = Var\left(g(x_1, \cdots, x_m)\right)$$

$$= \sum_{j=1}^{m} \left(\frac{\partial g(x^0)}{\partial x_j}\right)^2 Var(x_j) + o(\Delta_{max}) = U_g^2(\Delta_{max})$$

where $\Delta_{max} = \max(\Delta x_1, \cdots, \Delta x_m)$. It is thought as the wood (X) image of generally complex system f since it is an objective constant independent of human observations, expressing birth, although the maximum and controllable experimental tolerance Δ_{max} can be choose by human.

The index $\varphi(X) = Var\left(g(x_1, \cdots, x_m)\right)^{-1} = U_g^{-2}$ can be taken as the energy function of the subsystem wood (X) since the smaller the distortion degree U_g^2, the better the stability of the complex system f.

Although the distortion degree U_g^2 is unknown for freedom model speaking, it can be easily estimated by using orthogonal arrays $L_n(p_1, \cdots, p_m) = (a_{ij})_{n \times m}$,

$0 \le a_{ij} \le p_j - 1, j = 1, \cdots, m$ (see [29]). For example, take

$$x_{ij} = x_j(a_{ij}) = x_j^0 + C_j(a_{ij})\Delta x_j$$

where

$C_j(a_{ij}) = -1 + 2/(p_j - 1)$ then $Var(x_j) = W(p_j)\Delta x_j^2$, where

$$W(p_j) = \frac{1}{p_j} \sum_{k=0}^{p_j-1} C_j(k)^2$$

And for the experiment data $y_i = f(x_{i1}, x_{im}, \omega_{i\omega})$, $i = 1, \cdots, n$, we have

$$\frac{\widehat{\partial g(x^0)}}{\partial x_j} = \frac{\frac{1}{p_j} \sum_{k=0}^{p_j-1} C_j(k)\hat{\mu}_{jk}}{W(p_j)\Delta x_j},$$

$$\hat{\mu}_{jk} \triangleq \frac{1}{r_{jk}} \sum_{s \in H_{jk}} y_s, \quad r_{jk} \triangleq |H_{jk}|,$$

where $H_{jk} = \{s : a_{sj} = k, s = 1, \cdots, n\}$. Thus, we can obtained the estimation of the distortion degree $U_g^2 = Var(g(x_1, \cdots, x_m))$ as

$$\widehat{U_g^2} = \widehat{Var(g)} = \sum_{j=1}^m \frac{\left(\frac{1}{p_j} \sum_{k=0}^{p_j-1} C_j(k)\hat{\mu}_{jk} \right)^2}{W(p_j)}.$$

The $\Delta_{max} = max(\Delta x_1, \cdots, \Delta x_m)$ smaller, more exact estimate.

2) The index $\gamma_f^2 = Var(f) = Var(g) + Var(\varepsilon_g)$ is called Disturb Degree. It belongs to the fire (X_S) subsystem of the complex system f since it controls the fluctuations of the complex system f which is the development and growth stage of all things, just like in the Summer of a year. In mathematics,

$$\gamma_f^2 = Var(f(x_1, \cdots, x_m))$$

$$= \sum_{j=1}^m \left(\frac{\partial g(x^0)}{\partial x_j} \right)^2 Var(x_j) + E\sigma_{x_1, \cdots, x_m}^2 + o(\Delta_{max})$$

$$= \gamma_g^2(\Delta_{max})$$

where $\Delta_{max} = max(\Delta x_1, \cdots, \Delta x_m)$. It is thought as the fire (X_S) image of generally complex system f since it is also an objective constant independent of human observations, expressing growth, although the maximum and controllable experimental tolerance Δ_{max} can be choose by human. The index $\varphi(X_S) = \gamma_f^2$ can be taken as the energy function of the subsystem fire (X_S) since the smaller the disturb degree γ_f^2, the better the stability of the complex system f.

Although the disturb degree γ_f^2 is unknown for freedom model speaking, it can be easily estimated by using experiment data $y_1, \cdots, y_n$ of orthogonal arrays (see [29]). For example, a good estimation of $\gamma_f^2 = Var(f)$ is the data standard variance

$$\hat{\gamma}_f^2 = \widehat{Var(f)} = S_n^2 = \frac{1}{n-1}\sum_{s=1}^{n}(y_s - \bar{y})^2 \quad \text{where} \quad \bar{y} = \frac{1}{n}\sum_{s=1}^{n} y_s .$$

3) The index $\eta_f = (Ef)^2 / Var(f)$ is called Information Decomposition Ratio (or Signal to Noise Ratio). It belongs to the soil (X_K) subsystem of the complex system f since it makes the coordination of the center and fluctuation in the complex system f which is the continuous development and combined stage of all things, just like in the Long-Summer of a year. In mathematics,

$$\eta_f = (Ef)^2 / Var\big(f(x_1, \cdots, x_m)\big)$$

$$= \frac{(Eg)^2}{\sum_{j=1}^{m}\left(\frac{\partial g(x^0)}{\partial x_j}\right)^2 Var(x_j) + E\sigma_{x_1,\ldots,x_m}^2 + o(\Delta_{max})}$$

$$= \eta_g(\Delta_{max})$$

where $\Delta_{max} = \max(\Delta x_1, \cdots, \Delta x_m)$. It is thought as the soil (X_K) image of generally complex system f since it is an objective constant independent of human observations, expressing combined, although the maximum and controllable experimental tolerance Δ_{max} can be choose by human. The index $\varphi(X_K) = \eta_f$ can be taken as the energy function of the subsystem soil (X_K) since the bigger the information decomposition ratio η_f, the better the stability of the complex system f.

Although the information decomposition ratio η_f is unknown for freedom model speaking, it can be easily estimated by using the experimental data $y_1, \cdots, y_n$ of orthogonal arrays (see [29]). For example, a good estimation of η_f is

$$\hat{\eta}_f = \frac{n\bar{y}^2}{S_n^2}, \ S_n^2 = \frac{1}{n-1}\sum_{s=1}^{n}(y_s - \bar{y})^2, \ \bar{y} = \frac{1}{n}\sum_{s=1}^{n} y_s .$$

Note: In data analysis situation, often taking $\eta_f = nE\bar{y}^2 / \gamma_g^2$ instead of $\eta_f = (Ef)^2 / Var(f)$. Limited to data observation point of view, they are in the statistical meaning equivalent, but the former is more advantageous to the statistical analysis.

4) The index $\rho_f^2 = (Ef - t)^2$ is called Deviation Degree. It belongs to the metal (K_X) subsystem of the complex system f since it makes function characteristics (or data center and the expected goal deviation) in the complex system f which is the getting-results and accepted stage of all things, just like in the Autumn

of a year. In mathematics,

$$\rho_f^2 = \left(Ef - t\right)^2 = \left(Eg - t\right)^2$$

$$= \left(g\left(x^0\right) - t\right)^2 + o\left(\Delta_{max}\right) = \rho_g^2\left(\Delta_{max}\right)$$

where $Ey = Ef = Eg = g\left(x^0\right) + o\left(\Delta_{max}\right)$. It is thought as the metal (K_x) image of a generally complex system f since it is an objective constant independent of human observations expressing accepted, although the controllable inputs $x_1, \cdots, x_m$ of the observed function

$$g\left(x_1, \cdots, x_m\right) = E\left(f\left(x_1, \cdots, x_m, \omega\right) \middle| x_1, \cdots, x_m\right)$$

can be choose by human. The index $\varphi(K_x) = \rho_f^{-2}$ can be taken as the energy function of the subsystem metal (K_x) since the smaller the deviation degree ρ_f^2, the better the stability of the complex system f.

Although the deviation degree ρ_f^2 is unknown for freedom model speaking, it can be easily estimated by using the experimental data $y_1, \cdots, y_n$ of orthogonal arrays (see [29]). For example, a good estimation of ρ_f^2 is $\widehat{\rho_f^2} = \left(\bar{y} - t\right)^2, \bar{y} = \frac{1}{n}\sum_{y=1}^{n} y_y$.

5) The index $R_f^2 = E(y-t)^2$ is called Risk Function (or Risk, or Loss Function). It belongs to the water (S_x) subsystem of the complex system f since it makes the expected value of loss $L(y,t) = (y-t)^2$ between each data y and the goal t in the complex system f which is the risk and hiding stage of all things, just like in the winter of a year. The best condition is $R_f^2 = E(y-t)^2 = 0$, but generally this is very hard to achieve, because right now all the data are equal to the target t and the system f is a simple system $f = t$. In mathematics,

$$R_f^2 = E\left(y-t\right)^2$$

$$= \rho_g^2\left(\Delta_{max}\right) + U_g^2\left(\Delta_{max}\right) + E\sigma_{x_1, \cdots, x_m}^2$$

$$= R_g^2\left(\Delta_{max}\right).$$

where $E\sigma_{x_1, \cdots, x_m}^2 = E\left(\varepsilon_g^2 \middle| x_1, \cdots, x_m\right)$. It is thought as the water (S_x) image of generally complex system f since it is an objective constant independent of human observations, expressing hiding, although the observed function $g\left(x_1, \cdots, x_m\right) = E\left(f\left(x_1, \cdots, x_m, \omega\right) \middle| x_1, \cdots, x_m\right)$ can be choose by human. The index $\varphi(S_x) = R_f^{-2}$ can be taken as the energy function of the subsystem water (S_x) since the smaller the risk R_f^2, the better the stability of the complex system f.

Although the risk R_f^2 is unknown for freedom model speaking, it can be easily estimated by using the experimental data $y_1, \cdots, y_n$ of orthogonal arrays (see [29]). For example, a good estimation of R_f^2 is

$$\widehat{R_f^2} = \frac{1}{n} \sum_{s=1}^{n} (y_s - t)^2.$$

In image mathematics, each of the rows of orthogonal arrays is called one gua which is independent of the unknown system function f (model-free). The state space V of a multilateral system $(V, \Re)$ based the unknown system function f is the set of mathematical indexes of the unknown system function f. By Theorem 2.4, the set V can be divided into five categories. Corresponding to every kind of five categories, each of mathematical indexes of the unknown system function f is called an image. Each of images must shows the complex system certain characteristics, such as, wood, fire, soil, metal, water.

For given each of the gua or each the rows of orthogonal arrays and for any unknown continuous system function f, it can be proved easily that each of true image mathematical indexes above can be obtained if the experiments or observations can be repeated (law of averages of great numbers). The way or calculation method of the image indexes is also independent of the unknown system function f (model-free).

In image mathematics, each of mathematical indexes represents an "Axiom system", called a class. For each of classes, all theories and methods are in order to increase the energy of class or to make the corresponding mathematical index becomes to better. There are the loving and hating (or killing) relations among all images or mathematical indexes of classes. Generally speaking, close is love, alternate is hate.

In every category of internal, think that they are equivalent relationship, between each two of their elements there is a force of similar material accumulation of each other. It is because their pursuit of the goal is the same, i.e., follows the same "Axiom system". It can increase the energy of the class if they accumulate together. All of nature material activity follows the principle of maximizing so energy. In general, the force of similar material accumulation of each other is smaller than the loving force or the killing force in a stable complex system. The stability of any complex system first needs to maintain the equilibrium of the killing force and the loving force. For a stable complex system, if the killing force is large, i.e., ρ_1 becomes larger, then the loving force is large and the force of similar material accumulation of each other is also large. They can make the complex system more stable. If the killing force is small, i.e., ρ_1 becomes smaller, then the loving force is small and the force of similar material accumulation of each other is also small. They can make the complex system becoming unstable.

It has been shown in Theorems 2.1 - 2.4 that the classification of five subsets is quite possible based on the mathematical logic. As for the characteristics of the five subsets is rational or not, it is need more research work. It has been also shown in Theorems 3.1 - 3.4 that the logical basis of image mathematics is a steady multilateral system. The vigor energy (or, Chi, Qi) of image mathematics means the energy function in a steady multilateral system.

There are two kinds of mathematical diseases in image mathematics: Mathematical real disease and mathematical virtual disease. They generally means the subsystem is abnormal, its energy is too high for mathematical real disease or too low for mathematical virtual disease.

The intervening method of image mathematics is to "xie Chi" which means to rush down the energy if a mathematical real disease is treated, or to "bu Chi" which means to fill the energy if a mathematical virtual disease is treated. Like intervening the subsystem, decrease when the energy is too high, increase when the energy is too low.

Both the capability of intervention reaction and the capability of self-protection of the multilateral system are equivalent to the Immunization of image mathematics. This capability is really existence for a mathematical complex system. Its target is to protect other mathematical subsystem while treating one mathematical subsystem. It is because if the capability is not existence, then $\rho_1 = \rho_2 = \rho_3 = 0$. In this time, the energy of the system will be the sum of energy of each part. Thus the mathematical system will be a simple mathematical system which is not what we consider range.

Intervening Principle if Only One Subsystem of the Mathematical Complex System Falls Ill

If we always intervene the abnormal subsystem of the mathematical complex system directly, the intervention method always destroy the balance of the mathematical complex system because it is having strong side effects to the mother or the son of the subsystem which may be non-disease of mathematical subsystem or non-intervened subsystem by using Theorem 3.2. The intervening method also decrease the capability of intervention reaction because the method which don't use the capability of intervention reaction makes the ρ_1 and ρ_2 near to 0. The state $\rho_1 = \rho_2 = 0$ is the worst state of the mathematical complex system, namely mathematical intervenetion failure. On the way, the mathematical intervening resistance problem will be occurred since any mathematical intervening method is possible too little for some small P_1 and P_2.

But, by Corollary 3.2, it will even be better if we intervene subsystem X itself directly $P_2 = P_1 P_3$, $P_3 = P_1$ and $P_1 < P_0$. In this case: $P_1 + P_2 < 0.9375648971$ It can be explained that if a multilateral system which has a poor capability of intervention reaction, then it is better to intervene the subsystem itself directly than indirectly. But similar to above, the intervention method is always to destroy the balance of multilateral systems such that there is at least one of side effects occurred. And the intervention method also have harmful to the capability of intervention reaction making the mathematical interveneing resistance problem also occurred by Theorem 3.2. Therefore the intervention method directly can be used in case $P_1 < P_0$ but should be used as little as possible.

If we always intervene the abnormal subsystem of the mathematic complex system indirectly, the intervention method can be to maintain the balance of the mathematical complex system because it has not any side effects to all other subsystems which are not both the mathematical disease subsystem and the mathematical intervened subsystem by using Theorem 3.2. The intervening method also increase the capability of intervention reaction because the method of using the intervention reaction makes the P_1 and P_2 near to 1. The state $P_1 = P_2 = 1$ is the best state of the mathematical complex system. On the way, it is almost none mathematical intervening resistance problem since any mathematical intervening method is possible good for some large P_1 and P_2.

For example, in China, many mathematical engineers generally only care about risk R_f^2 (or, income, water (S_x)) and interfere γ_f^2 (or, disturb, fire (X_s)) as two mathematical indexes which have the killing relations for a complicated engineering problem based on the freedom model (3). They frequently used method is: If only the risk R_f^2 (i.e., water (S_x) is not normal (or, the bigger, virtual disease), then they reduce deviation degree ρ_f^2 (i.e., increase the energy $\varphi(K_x) = \rho_f^2$ of metal (K_x)) with linear optimization method since metal (K_x) is the mother of water (S_x); If only the interfere degree γ_f^2 (i.e., fire (X_s)) is not normal (or, the bigger, virtual disease), then they reduce distortion degree U_f^2 (i.e., increase the energy $\varphi(X) = U_f^2$ of wood (X)) of system function f with the method of adjusting stable center x^0 of the system function since wood (X) is the mother of fire (X_s) (see [29]). The idea is precisely "Virtual disease is to fill his mother" if one subsystem of mathematical complex system falls virtual ill.

All in all, the mathematical complex system satisfies the intervention rule and the self-protection rule. It is said a healthy mathematical complex system when the intervention reaction coefficient ρ_1 satisfies $\rho_1 \geq \rho_0$. In logic and practice, it's reasonable $\rho_1 + \rho_2$ near to 1 if the input and output in a complex system is balanced, since a mathematical output subsystem is absolutely necessary social other subsystems of all consumption. In case: $\rho_1 + \rho_2 = 1$, all the energy for intervening mathematical complex subsystem can transmit to other mathematical complex subsystems which have neighboring relations or alternate relations with the intervening mathematical complex subsystem. The condition $\rho_1 \geq \rho_0$ can be satisfied when $\rho_2 = \rho_1 \rho_3$ and $\rho_3 = \rho_1$ for a mathematical complex system since $\rho_1 + \rho_2 = 1$ implies $\rho_1 = \left(\sqrt{5} - 1\right)/2 \approx 0.618 > \rho_0$ and $\rho_2 = 1 - \left(\sqrt{5} - 1\right)/2 \approx 0.382$. If this assumptions is set upthen the intervening principle: "Real disease is to rush down his son and virtual disease is to fill his mother" based on "Yin Yang Wu Xing" Theory in image mathematics, is quite reasonable. But, in general, the ability of self-protection often is insufficient for an usual mathematical complex system, i.e., ρ_3 is small. A common standard is $\rho_3 = \frac{1 - \rho_1}{2\rho_2} \approx \frac{1}{2}$, i.e., there is a principle which all losses are bear in mathematical complex system. Thus the general condition often is $\rho_1 \approx 0.618 \geq \rho_3 \approx 0.5 \geq \rho_2 \approx 0.382$. Interestingly, they near to the golden numbers.

On the other hand, in image mathematics, mathematiccal real disease and mathematical virtual disease have their reasons. Mathematical real disease is caused by the born subsystem and mathematical virtual disease is caused by the bear subsystem. Although the reason cannot be proved easily in mathematics or experiments, the intervening method under the assumption is quite equal to the intervening method in the intervention indirectly. It has also proved that the mathematical intervening princeple is true from the other side.

Intervening Principle If Only Two Subsystems with the Loving Relation of the Mathematical Complex System Encounter Sick

Suppose that the two subsystems X and X_S of the mathematic complex system are abnormal (mathematical virtual disease or mathematical real disease). In the mathematical complex system of relations between two noncompatible with the constraints, by Theorem 3.2, only two situations may occur:

1) X encounters mathematical virtual disease, and at the same time, X_S befalls mathematical virtual disease, i.e., the energy of X is too low and the energy of X_S is also too low. It is because X bears X_S. The mathematical virtual disease causal is X.

2)   X encounters mathematical real disease, and at the same time, X_S befalls mathematical real disease, i.e., the energy of X is too high and the energy of X_S is also too high. It is because X_S is born by X. The mathematical virtual disease causal is X_S.

It can be shown by Theorem 3.3 that when intervenetion reaction and self-protection coefficients satisfy $\rho_2 = \rho_1 \rho_3$, $\rho_3 = \rho_1$ and $\rho_1 \geq \rho_0$, if one wants to treat the abnormal subsystems X and X_S, then 1) For mathematical virtual disease, the one should intervene subsystem X directly by increasing its energy. It means "mathematical virtual disease is to fill his mother" because the mathematical virtual disease causal is X;

2) For mathematical real disease, the one should intervene subsystem X_S directly by decreasing its energy. It means "mathematical real disease is to rush down his son" because the mathematical real disease causal is X_S. For example, in China, many factories and enterprises generally only care about cost R_f^2 (or, risk, water (S_x)) and quality ρ_f^2 (or, deviation degree, metal (K_x)) as two mathematical indexes which have the loving relation for a complicated economical problem based on the freedom model (3). They frequently used method is: If both cost and quality are all not normal (or, virtual diseases, i.e., high cost and poor quality), then they reduce deviation degree ρ_f^2 (i.e., improve quality, increase the energy $\varphi(K_x) = \rho_f^2$ of metal (K_x)) with linear optimization method since metal (K_x) is the mother of water (S_x) (see [29]). The idea is precisely "Virtual disease is to fill his mother" if one subsystem of mathematical complex system falls virtual ill.

The intervention method can be to maintain the balance of the mathematical complex system because only two mathematical virtual disease subsystems are treated, by using Theorem 3.2, such that there is not any side effect for all other subsystems. And the intervention method can also be to enhance the capability of intervention reaction because the method of using intervention reaction makes the ρ_1 and ρ_2 greater and near to 1. The state $\rho_1 = \rho_2 = 1$ is the best state of the mathematical complex system. On the way, it almost have none mathematical intervening resistance problem since any mathematical intervening method is possible good for some large ρ_1 and ρ_2.

Intervening Principle If Only Two Subsystems with the Killing Relation of the Mathematical Complex System Encounter Sick

Suppose that the subsystems X and X_K of a mathematical complex system are abnormal (real disease or virtual disease). In the mathematical complex system of relations between two non-compatible with the constraints, only a situation may occur: X encounters mathematical virtual disease, and at the same time, X_K befalls mathematical real disease, i.e., the energy of X is too low and the energy of X_K is too high. The disease is serious because the X_K has harmed the X by using the method of incest, i.e. damaged the king relation of X and X_K.

It can be shown by Theorems 3.3 and 3.4 that when intervention reaction and self-protection coefficients satisfy $\rho_2 = \rho_1\rho_3$, $\rho_3 = \rho_1$ and $\rho_1 < \rho_0$, if one wants to treat the abnormal subsystems X and X_K, the one should intervene subsystem X directly by increasing its energy, and at the same, intervene subsystem X_K directly by decreasing its energy. It means that "Strong inhibition of the same time, support the weak".

For example, in China, many sociologists generally only care about social and economic benefits $\varphi(S_X) = R_f^{-2}$ (or social benefits, water (S_X)) and social equity extent $\varphi(X_K) = \eta_f$ (or, information decomposition ratio, soil (X_K)) with two mathematical killing indexes for a complicated social system based on the freedom model (3). If both social benefit $\varphi(S_X) = R_f^{-2}$ and social equity extent $\varphi(X_K) = \eta_f$ are not normal, often appear serious problem is that social benefit $\varphi(S_X) = R_f^{-2}$ is too good (or, R_f too low, real disease) but social equity extent $\varphi(X_K) = \eta_f$ is too bad (or, η_f too low, i.e., virtual disease). The disease is serious because the X_K has been harmed by the S_X with the method of incest such that the soil (X_K) cannot kill water (S_X). For the serious disease, many mathematical sociologists of China now frequently used method is: to increase social equity extent η_f (or, to increase the information decomposition ratio η_f which is the energy of the soil (X_K)) but to decrease social and economic benefits $\varphi(S_X) = R_f^{-2}$ (i.e., to decrease the energy of water (S_X)) at the same time since soil (X_K) is the bane of water (S_X). The idea is "Strong inhibition of the same time, support the weak" if X_K falls virtual disease and at the same time, S_X befalls real disease.

For another example, thirty years ago, China's social coordination function $\varphi(X_K) = \eta_f$ is very good, the complex system is rife with average socialist, but both social benefit $\varphi(S_X) = R_f^{-2}$ and social structure $\varphi(X) = U_f^{-2}$ are all very poor. In addition, the mathematical complex system is not rich. In other words, both S_X and X fall virtual diseases and at the same time, X_K befalls real disease based on the freedom model (3). The disease is very serious because not only the wood (X) has been harmed by the soil (X_K) with the method of incest such that wood (X) cannot kill soil (X_K), but also there are 3 subsystems falling ill. Generally speaking, there are three or more than three of disease of the subsystem of a complex system, it is very difficult to cure. In order to cure the very serious disease, Deng Xiao-Ping's taking method is to break the "iron bowl" and to develop the economy (to fill up the energies $\varphi(X) = U_f^{-2}$ and $\varphi(S_X) = R_f^{-2}$ of both wood (X) and water (S_X) at the same time, i.e., strengthen social structure and social benefit), and to allow a few people to get rich (to rush down the energy $\varphi(X_K) = n_f$ of soil (X_K), abate the coordinated ability). The idea is, at the same time, to use both "Strong inhibition of the same time, support the weak" and "Virtual disease is to fill his mother", if both S_X and X fall virtual diseases and at the same time, X_K befalls real disease.

The intervention method can be to maintain the balance of mathematical complex system because only two mathematical virtual disease subsystems are treated, by using Theorems 3.2 and 3.4, such that there is none side effects for all other subsystems. And the intervention method can also be to enhance the capability of intervention reaction and self-protection because the method of using intervention reaction and self-protection makes the ρ_3 and ρ_1 greater and near to 1. The state $\rho_3 = \rho_1 = 1$ is the best state of the steady multilateral system. On the way, it almost have none mathematical intervening resistance problem since any mathematical intervening method is possible good for some large ρ_3 and ρ_1.

CONCLUSIONS

This work shows how to treat the mathematical diseases (real or virtual) of a mathematical complex system in image mathematics and three methods are presented.

If only one subsystem falls ill, mainly the intervening method should be to intervene it indirectly for case: $\rho_2 = \rho_1\rho_3$, $\rho_3 = \rho_1$ and $\rho_1 \geq \rho_0$, according to the intervening principle of "Real disease is to rush down his son but virtual disease is to fill his mother". The intervention method directly can be used in case $\rho_2 = \rho_1\rho_3$, $\rho_3 = \rho_1$ and $\rho_1 < \rho_0$ but should be used as little as possible.

If two subsystems with the loving relation encounter sick, the intervening method should be intervene them directly according to the intervening principle of "Real disease is to rush down his son but virtual disease is to fill his mother".

If two subsystems with the killing relation encounter sick, the intervening method should be intervene them directly also according to the intervening principle of "Strong inhibition of the same time, support the weak".

Other properties, such as balanced, orderly nature of Wu-Xing, mathematical forecast, and so on, will be discussed in the next articles.

ACKNOWLEDGEMENTS

This article has been repeatedly invited as reports, such as Shanxi University, Xuchang College, and so on. The work was supported by Specialized Research Fund for the Doctoral Program of Higher Education of Ministry of Education of China (Grant No. 200802691021).

REFERENCES

1. Y. S. Zhang, "Theory of Multilateral Matrices," Chinese Stat. Press, Beijing, 1993.

2. Y. S. Zhang, "Theory of Multilateral Systems," 2007. http://www.mlmatrix.com

3. Y. S. Zhang, "Mathematical Reasoning of Treatment Principle Based on 'Yin Yang Wu Xing' Theory in Traditional Chinese Medicine," Chinese Medicine, Vol. 2, No. 1, 2011, pp. 6-15.doi:10.4236/cm.2011.21002

4. Y. S. Zhang, "Mathematical Reasoning of Treatment Principle Based on 'Yin Yang Wu Xing' Theory in Traditional Chinese Medicine(II)," Chinese Medicine, Vol. 2, No. 4, 2011, pp. 158-170. doi:10.4236/cm.2011.24026

5. Y. S. Zhang, "Mathematical Reasoning of Treatment Principle Based on the Stable Logic Analysis Model of Complex Systems," Intelligent Control and Automation, Vol. 3, No. 1, 2012, pp. 6-15. doi:10.4236/ica.2012.31001

6. Y. S. Zhang, "Mathematical Reasoning of Economic Intevening Principle Based on 'Yin Yang Wu Xing' Theory in Traditional Chinese Economic(I)," Modern Economics, Vol. 3, No. 2, 2012, pp.

7. Y. S. Zhang, S. S. Mao, C. Z. Zhan and Z. G. Zheng, "Stable Structure of the Logic Model with Two Causal Effects," Chinese Journal of Applied Probability and Statistics, Vol. 21, No. 4, 2005, pp. 366-374.

8. C. Luo, X. D. Wang and Y. S. Zhang, "Orthogonality and Independence—New Thinking of Dealing with Complex Systems Series Three," Journal of Shanghai Institute of Technology (Natural Science), Vol. 10, No. 4, 2010, pp. 271-277.

9. C. Luo, X. P. Chen and Y. S. Zhang, "The Turning Point Analysis of Finance Time Series," Chinese Journal of Applied Probability and Statistics, Vol. 26, No. 4, 2010, pp. 437-442.

10. Y. S. Zhang, X. Q. Zhang and S. Y. Li, "SAS Language Guide and Application," Shanxi People's Press, Taiyuan, 2011.

11. Y. S. Zhang and S. S. Mao, "The Origin and Development Philosophy

Theory of Statistics," Statistical Research, Vol. 12, 2004, pp. 52-59.

12. N. Q. Feng, Y. H. Qiu, F. Wang, Y. S. Zhang and S. Q. Yin, "A Logic Analysis Model about Complex System's Stability: Enlightenment from Nature," Lecture Notes in Computer Science, Vol. 3644, 2005, pp. 828-838. doi:10.1007/11538059_86

13. N. Q. Feng, Y. H. Qiu, Y. S. Zhang, F. Wang and Y. He, "A Intelligent Inference Model about Complex System's Stability: Inspiration from Nature," International Journal of Intelligent Technology, Vol. 1, 2005, pp. 1-6.

14. N. Q. Feng, Y. H. Qiu, Y. S. Zhang, C. Z. Zhan and Z. G. Zheng, "A Logic Analysis Model of Stability of Complex System Based on Ecology," Computer Science, Vol. 33, No. 7, 2006, pp. 213-216.

15. C. Y. Pan, X. P. Chen, Y. S. Zhang and S. S. Mao, "Logical Model of Five-Element Theory in Chinese Traditional Medicine," Journal of Chinese Modern Traditional Chinese Medicine, Vol. 4, No. 3, 2008, pp. 193- 196.

16. X. P. Chen, W. J. Zhu, C. Y. Pan and Y. S. Zhang, "Multilateral System," Journal of Systems Science, Vol. 17, No. 1, 2009, pp. 55-57.

17. C. Luo and Y. S. Zhang, "Framework Definition and Partition Theorems Dealing with Complex Systems: One of the Series of New Thinking," Journal of Shanghai Institute of Technology (Natural Science), Vol. 10, No. 2, 2010, pp. 109-114.

18. C. Luo and Y. S. Zhang, "Framework and Orthogonal Arrays: The New Thinking of Dealing with Complex Systems Series Two," Journal of Shanghai Institute of Technology (Natural Science), Vol. 10, No. 3, 2010, pp. 159-163.

19. J. Y. Liao, J. J. Zhang and Y. S. Zhang, "Robust Parameter Design on Launching an Object to Goal," Mathematics in Practice and Theory, Vol. 40, No. 24, 2010, pp. 126-132.

20. Y. S. Zhang, S. Q. Pang, Z. M. Jiao and W. Z. Zhao, "Group Partition and Systems of Orthogonal Idempotents," Linear Algebra and Its Applications, Vol. 278, No. 1-3, 1998, pp. 249-262. doi:10.1016/S0024-3795(97)10095-7

21. J. L. Zhao and Y. S. Zhang, "The Characteristic Description of Idempotent Orthogonal Class System," Advances in Matrix Theory and Its Applications, Proceedings of the 8th International Conference on Matrix and its applications, Taiyuan, Vol. 1, No. 1, 16-18 July 2008, pp. 445-448.

22. X. P. Chen, C. Y. Pan and Y. S. Zhang, "Partitioning the Multivariate

Function Space into Symmetrical Classes," Mathematics in Practice and Theory, Vol. 39, No. 2, 2009, pp. 167-173.

23. C. Y. Pan, X. P. Chen and Y. S. Zhang, "Construct Systems of Orthogonal Idempotents," Journal of East China University (Natural Science), Vol. 141, No. 5, 2008, pp. 51-58.

24. C. Y. Pan, H. N. Ma, X. P. Chen and Y. S. Zhang, "Proof Procedure of Some Theories in Statistical Analysis of Global Symmetry," Journal of East China Normal University (Natural Science), Vol. 142, No. 5, 2009, pp. 127-137.

25. X. Q. Zhang, Y. S. Zhang and S. S. Mao, "Statistical Analysis of 2-Level Orthogonal Satursted Designs: The Procedure of Searching Zero Effects," Journal of East China Normal University (Natural Science), Vol. 24, No. 1, 2007, pp. 51-59.

26. Y. S. Zhang, Y. Q. Lu and S. Q. Pang, "Orthogonal Arrays Obtained by Orthogonal Decomposition of Projection Matrices," Statistica Sinica, Vol. 9, No. 2, 1999, pp. 595-604.

27. Y. S. Zhang, S. Q. Pang and Y. P. Wang, "Orthogonal Arrays Obtained by Generalized Hadamard Product," Discrete Mathematics, Vol. 238, No. 1-3, 2001, pp. 151- 170.doi:10.1016/S0012-365X(00)00421-0

28. Y. S. Zhang, L. Duan, Y. Q. Lu and Z. G. Zheng, "Construction of Generalized Hadamard Matrices $D\left(r^m(r+1), r^m(r+1); p\right)$," Journal of Statistical Planning, Vol. 104, 2002, pp. 239-258. doi:10.1016/S0378-3758(01)00249-X

29. Y. S. Zhang, "Data Analysis and Construction of Orthogonal Arrays," East China Normal University, Shanghai, 2006.

30. Y. S. Zhang, "Orthogonal Arrays Obtained by Repeating- Column Difference Matrices," Discrete Mathematics, Vol. 307, No. 2, 2007, pp. 246-261.doi:10.1016/j.disc.2006.06.029

31. X. D. Wang, Y. C. Tang, X. P. Chen and Y. S. Zhang, "Design of Experiment in Global Sensitivity Analysis Based on ANOVA High-Dimensional Model Representation," Communications in Statistics—Simulation and Computation, Vol. 39, No. 6, 2010, pp. 1183-1195. doi:10.1080/03610918.2010.484122

32. X. D. Wang, Y. C. Tang and Y. S. Zhang, "Orthogonal Arrays for the Estimation of Global Sensitivity Indices Based on ANOVA High-Dimensional Model Representation," Communications in Statistics—Simulation and Computation, Vol. 40, No. 9, 2011, pp. 1324-1341. doi:10.1080/03610918.2011.575500

33. J. T. Tian, Y. S. Zhang, Z. Q. Zhang, C. Y. Pan and Y. Y. Gan, "The Comparison and Application of Balanced Block Orthogonal Arrays and Orthogonal Arrays," Journal of Mathematics in Practice and Theory, Vol. 39, No. 22, 2009, pp. 59-67.

34. C. Luo and C. Y. Pan, "Method of Exhaustion to Search Orthogonal Balanced Block Designs," Chinese Journal of Applied Probability and Statistics, Vol. 27, No. 1, 2011, pp. 1-13.

35. Y. S. Zhang, W. G. Li, S. S. Mao and Z. G. Zheng, "Orthogonal Arrays Obtained by Generalized Difference Matrices with g Levels," Science China Mathematics, Vol. 54, No. 1, 2011, pp. 133-143. doi:10.1007/s11425-010-4144-y

36. Lao-tzu, "Tao Te Ching," In: S. Mitchel, Transl., 2010. http://acc6.its.brooklyn.cuny.edu

37. M.-J. Cheng, "Lao-Tzu, My Words Are Very Easy to Understand: Lectures on the Tao Teh Ching," North Atlantic Books, Richmond, 1981.

38. Research Center for Chinese and Foreign Celebrities and Developing Center of Chinese Culture Resources, "Chinese Philosophy Encyclopedia," Shanghai People Press, Shanghai, 1994.

Chapter 10

WAVELET DENSITY ESTIMATION OF CENSORING DATA AND EVALUATE OF MEAN INTEGRAL SQUARE ERROR WITH CONVERGENCE RATIO AND EMPIRICAL DISTRIBUTION OF GIVEN ESTIMATOR

Mahmoud Afshari

Department of Statistics, College of Science, Persian Gulf University, Bushehr, Iran

ABSTRACT

Wavelet has rapid development in the current mathematics new areas. It also has a double meaning of theory and application. In signal and image compression, signal analysis, engineering technology has a wide range of applications. In this paper, we use wavelet method, for estimating the density function for censoring data. We evaluate the mean integrated squared error, convergence ratio of given estimator. Also, we obtain empirical distribution of given estimator and verify the conclusion by two simulation examples.

INTRODUCTION

One of data types, which researchers are extremely interested in, is caring to the time interval till the occurrence of certain events such as death etc. Any process waiting for a specific event produces survival data. Survival function, which is shown by $S(t)$, indicates the ratio of people who survived since the base time which is the point they enter the experiment. Failure in survival analysis means the occurrence of the event we were waiting for. The time, where survival is measured after that point, is called the start time. The failure time is the time that failure occurs for each individual which is denoted by T_i for $i = 1, 2, 3, \cdots$. The failure time is occurred from the base time up to when the failure occurs and it's known as T_i. It's not always possible to observe the failure time for each individual. In such cases, censorship occurs. The rate of occurrences of an event (failure) in a specific short period of time providing that no failure occurred before that time is the concept which is discussed by the name hazard function in survival analysis. Hazard function for the failure time line is as follows:

$$h_t(t) = \lim_{\Delta t \to 0} \frac{P(t \le T \le t + \Delta t | T \ge t)}{\Delta t} = \frac{F'(t)}{1 - F(t)} = -\frac{S'(t)}{S(t)}.$$

Wavelets can be used for transient phenomena analysis or functions analysis which sometimes changes rapidly, and they are symmetrical and have limited period unlike rugged Sine waves, thus the signals with radical changes are analyzed better. The close relationship between wavelet coefficients and some spaces, wavelet bases being orthogonal and also useful properties of them in wavelet issues simplify the computational algorithms.

Wavelets theory was proposed by Alfred Harr [1] for the first time in 1910. He showed that a continuous function can be approximated as follows:

$$f_n(x) = \langle f, \phi_0 \rangle \phi_0(x) + \langle f, \phi_1 \rangle \phi_1(x) + \cdots + \langle f, \phi_n \rangle \phi_n(x). \tag{1}$$

Such that $\langle f, \phi_i \rangle = \int f(x)\phi_i(x)\,dx,$

$$\phi_n(x) = I_{\frac{k}{2^j} \le x < \frac{k + \frac{1}{2}}{2^j}} - I_{\frac{k + \frac{1}{2}}{2^j} \le x < \frac{k+1}{2^j}} \quad , \quad j \ge 0 \quad k = 0, 1, \cdots, 2^j - 1 \quad , \quad n = 2^j + k$$

Also for mother wavelet and father wavelets the following:

$$\phi_{j,k}(x) = \phi(2^j x - k), \quad \psi_{j,k}(x) = \psi(2^j x - k).$$

Definition 1-1: Assume that $V_j = \overline{\text{span}}\{\varphi_{j,k} : k \in z\}$; $\{\varphi_{j,k}, k \in z\}$ is an orthogonal unit base for V_j and V_j contains all sectionally constant functions and their exact length is twice the interval length of V_{j+1}.

Spaces $\{V_j, j \in Z\}$ are called multiresolatio analysis or scale function φ, if it satisfies the following conditions:

1- $V_j \subset V_{j+1} \forall j \in Z$, 2- $\overline{\bigcup_{j \in z} V_j} = L^2(R)^2$, 3- $\bigcap_{j \in z} V_j = \{0\}$ 4- $j \in z.$ $f(x) \in V_j$, $f(x) \in V_j$, 5-' $-k \in z. \Leftrightarrow f(x) \in V_0$, $f(x - h) \in V_0$.

6- $\exists \varphi(x) \in V_0$ in condition that $\{\varphi(x - k) : k \in z\}$ is an orthogonal base for V_0.

If we consider the scale function in the interval $[0,1]$, then the image of f on the space V_j is defined as $P_{V_j}^f = \sum_{k \in z} \alpha_{j,k} \varphi_{j,k}(x)$ which is a function with the resolution, 2^j and because of the fact that $\overline{\bigcup_{j \in z} V_j} = L^2(R)$

thus $P_{V_j}^f$ is a good approximation of function f for large amounts of j.

Let the nested sequence of closed subspaces; ... $V_{j-1} \subset V_j \subset V_{j+1} \subset \cdots, j \in Z,$ be a multiresolutuon approximation to $L^2(R)$. Define W_j, $j \in Z$ to be orthogonal complement of V_j in V_{j+1}.

The term wavelets are used to refer to a set of basis functions with very special structure. The special of wavelets basis for function $f \in L^2(R)$ as scaling function φ and mother wavelet ψ such that $\{\varphi(x-k)\}_{k \in z}$ forms an orthogonal basis for V_0 and $\{\psi(x-k)\}_{k \in z}$ forms an orthonormal basis for W_0. Other wavelets in the basis are then generated by translation of the scaling function and dilations of the mother wavelet by using the relationships:

$$\varphi_{m_0,k}(x) = 2^{m_0/2} \varphi(2^{m_0}x - k), \quad \psi_{j,k}(x) = 2^{j/2} \psi(2^j x - k) \tag{2}$$

Given above Wavelet basis, a function $f \in L^2(R)$ can be written a formal expansion:

$$f = \sum_{k \in Z} \alpha_{m_0,k} \varphi_{m_0,k} + \sum_{j=m_0}^{\infty} \sum_{k \in Z} \beta_{j,k} \psi_{j,k} \tag{3}$$

where $\alpha_{j_0,k} = \int f(x) \varphi_{j_0,k}(x) dx, \quad \delta_{j,k} = \int f(x) \psi_{j,k} dx$

As for general orthogonal series estimator, Daubechies [2] , density estimator can be written as:

$$\hat{f} = \sum_{k \in Z} \hat{\alpha}_{m_0,k} \phi_{m_0,k}(x) + \sum_{j \geq m_0} \sum_{k \in Z} \hat{\beta}_{j,k} \psi_{j,k}(x) = \mathbf{P}_{m_0} f + \sum_{j \geq m_0} \sum_{k \in Z} \hat{\beta}_{j,k} \psi_{j,k}. \tag{4}$$

where the obvious coefficient estimator can be written:

$$\hat{\alpha}_{m_0,k} = E\left[\varphi_{m_0,k}(X)\right] = \frac{1}{n} \sum_{i=1}^{n} \varphi_{m_0,k}(X_i), \quad \hat{\beta}_{j,k} = E\left[\psi_{j,k}(X)\right] = \frac{1}{n} \sum_{i=1}^{n} \psi_{j,k}(X_i) \tag{5}$$

We divide time axis into two parts, the intervals and the number of events in each interval. We determine number of events and hazard function according to the observations. Then we flatten them separately via linear wavelet density estimation on the whole time and then we calculate the function estimator and evaluate the asymptotic distribution. In this paper we obtain estimator density for censoring data by using wavelet method and evaluate mean integral square error with convergence ratio and empirical distribution of given estimator.

ESTIMATOR OF DENSITY BY USING WAVELET METHOD

Wavelets can be used for transient phenomena analysis or functions analysis which sometimes changes rapidly, and they are symmetrical and have limited period unlike rugged Sine waves, thus the signals with radical changes are analyzed better. The close relationship between wavelet coefficients and some spaces, wavelet bases being orthogonal and also useful properties of them in wavelet issues simplify the computational algorithms. As a result, numerous articles have been published about density function estimation. The mathematical theorem of wavelets and their application in statistics have been studied as a technique for nonparametric curve estimators by Antoniadys [3] .

Afshari [4] -[6] have done some researches about density function estimator, the density functional derivative and the nonparametric regression function for the mixing random variables. Donohu [7] , kyacharyan, Picard [8] , Malat [9] , Meyer [10] , and some articles have been published in this field. Hall and Patil [11] have found a formula for the Mean Integrated Squared Error of Nonlinear Wavelet based on density estimators. Antoniadys et al. [12] achieved the density function estimator and the hazard function for right-censored data with the wavelets. In this section we obtain estimator of density function for censoring data by using wavelet method.

Suppose $X_1, X_2, X_3, \cdots, X_n$ are failure time of n tests that are studied. They are non-negative, independent, identically distributed, with the density function f and distribution function F and $C_1, C_2, C_3, \cdots, C_n$ are corresponding to censored times, non-negative, independent, identically distributed, with the density function g and distribution function G.

Assuming independency of failure times and censored time of the observed random variable, Z_i and the function δ_i and Hazard function are shown as below:

$$Z_i = \min\left(X_i, C_i\right), \quad \delta_i = I_{\left(X_i \le C_i\right)}, \quad h(t) = \frac{f(t)}{1 - F(t)}, \quad F(t) < 1.$$

Such that $I_{(A)}$ is indicator function of A . For data censoring, if $G(t) < 1$ then we have as the following:

$$h(t) = \frac{f(t)(1-G(t))}{(1-F(t))(1-G(t))}, \quad F(t) < 1.$$

Also we definite as follows:

$$L(t) = P(Z_i \le t) = 1 - P(Z_i > t) = 1 - P(X_i > t, C_i > t) = 1 - (1-F(t))(1-G(t)).$$

$$f^*(t) = f(t)(1-G(t)). \quad h(t) = \frac{f^*(t)}{1-L(t)}, \quad L(t) < 1.$$

To estimate $f^*(t)$, we divide the time axis into two parts of small intervals and the amounts of events (0 or 1) in each interval, and then we divide these values to the length of intervals.

Estimation procedures of $f^*(t)$ can be summarized as the following:

Select $\Delta > 0$ and collect the observed failures in $k+1$ intervals with the length Δ and using wavelet estimation on the collected data. We find an estimate of sub density. This means that we calculate the collected wavelet coefficients data on the scale of $j(n)$ by choosing the decomposition level $j(n)$ and then we estimate $f^*(t)$. It is necessary to state the following symbols to show the details:

$$T_F = \sup\{t : F(t) < 1\}. \quad T_G = \sup\{t : G(t) < 1\}. \quad T_L = \sup\{t : L(t) < 1\} = \min\{T_F, T_G\}.$$

We figure estimators on the finite interval $[0, \tau]$ in which $\tau < T_L$. Note that if $Z_{(i)}$ is the ordinal order statistic i of the sequence Z_i then $T_{l_n} = Z_{(n)} \xrightarrow{n \to \infty} T_L$. In fact we suppose $\tau = Z_{(n)}$.

Suppose that N is an integer that could be dependent to n and the estimated points are as follows:

$$t_k = \frac{k\tau}{2^N}, \quad k = 0, \cdots, K = 2^N - 1.$$

Suppose that $\Delta = \tau 2^{-N}$ and we divide the interval $[0, \tau]$ of time axis to $k+1$ intervals with Δ long

$$\tau_0 = -\frac{\Delta}{2}, \quad \tau_k = t_k - \frac{\Delta}{2}, k = 1, \cdots, K, \quad \tau_{K+1} = \tau.$$

The k-th interval is marked by J_k so: $J_k = [\tau_K, \tau_{k+1})$ for $k = 0, \cdots, K-1, \quad J_K = [\tau_K, \tau]$.

Now we define the following indicator function that indicates the number of uncensored failures in the time interval

$J_k : Y_{ik} = I_{J_k}(Z_i)\delta_i, \quad i = 1, \cdots, n \quad k = 0, \cdots, K.$ We assume that U_k the observed failures ratio in the interval J_k n other words: $U_k = \frac{1}{n}\sum_{i=1}^{n} Y_{ik}, \quad k = 0, \cdots, K.$

Theorem 2-1: Suppose that the sub density f^* is a continuous function on $[0,\tau]$ and it's m times differentiable, then if $v^\Delta \to 0$ or $n \to \infty$, we have:

$$Var\left(\frac{U_k}{\Delta}\right) = \frac{f^*(t_k)}{n\Delta} + O\left(\frac{1}{n}\right), E\left(\frac{U_k}{\Delta}\right) = f^*(t_k) + O(\Delta)$$

$$Cov\left(\frac{U_k}{\Delta}, \frac{U_\ell}{\Delta}\right) = -\frac{1}{n}f^*(t_k)f^*(t_\ell) + O\left(\frac{\Delta}{n}\right), \quad k \neq \ell.$$

Proof: see [13].

We smooth the data $\dfrac{U_k}{\Delta}$ by an appropriate wavelet smoother to find the estimation of f^*.

We can write,

$$f^*(t) = \sum_{k=0}^{2^{j_0}-1} \langle f^*, \phi_{j_0,k}\rangle \phi_{j_0,k}(t) + \sum_{j \geq j_0}\sum_{\ell=0}^{2^j-1} \langle f^*, \psi_{j,\ell}\rangle \psi_{j,\ell}(t).$$

(6)

where, $\langle f, g\rangle \equiv \int_0^\tau f(t)g(t)\,dt$

The complex structural polymorphism analysis causes an efficient tree construction algorithm for analysis of functions in V_N with theoretic scale wavelet coefficients $\langle f^*, \varphi_{N,k}\rangle$. However, the integral scale $\langle f^*, \varphi_{N,k}\rangle$ is not well available and we need an initial value for a fast wavelet transform. Antonyadys [4] suggested the following initial amount:

$$\langle f^*, \phi_{N,k}\rangle = 2^{-\frac{N}{2}}f^*t_k + O\left(2^{-\frac{N}{2}}2^{-Nm}\right), \quad 0 \leq k \leq 2^N - 1$$

As a result a reasonable estimate for image of f^* with clarity N is:

$$\tilde{f}_N^*(t) = 2^{-\frac{N}{2}}\sum_{k=0}^{K}\frac{U_k}{\Delta}\varphi_{N,k}(t).$$

(7)

If we assume that the collected values U_k which are equal to the estimators of $\tilde{f}_N^*(t)$, are in Sobolev space $W^m([0,\tau])$ and φ is regular of degree m. We estimate the unknown function f as follows to level the data with a better rate for the sample size n and the sequence $j(n) < N$:

$$\hat{f}_n = P_{V_{j(n)}}^{j_N^*}$$

(8)

That it is the orthogonal image of $\tilde{f}_N^*(t)$ on the leveler approximation space $V_{j(n)}$.

Theorem 2-2: Suppose that the sub density f is a continuous function on $[0,\tau]$ and it's m times differentiable, then if $\Delta \to 0$ for $n \to \infty$ we have:

$$E\left[\tilde{f}_N^*(t)\right] = P_{V_N}^{f^*(t)} + O(\Delta), Var\left[\tilde{f}_N^*(t)\right] = O\left\{(n\Delta)^{-1}\right\} + O\left(n^{-1}\right).$$

Proof: by using theorem (2-1) we can write:

$$\left[\tilde{f}_N^*(t)\right] = 2^{-\frac{N}{2}} \sum_{k=0}^{K} E\left[\frac{U_k}{\Delta}\right] \phi_{N,k}(t) = 2^{-\frac{N}{2}} \sum_{k=0}^{K} \left[f^*(t_k) + O(\Delta)\right] \phi_{N,k}(t)$$

$$= 2^{-\frac{N}{2}} \sum_{k=0}^{K} f^*(t_k)\phi_{N,k}(t) + O(\Delta)2^{-\frac{N}{2}} \sum_{k=0}^{K} \phi_{N,k}(t)$$

(9)

Since, $\sup_t \sum_k |\phi(t-k)| = M$, then $\varphi_{N,k}(t) = 2^{\frac{N}{2}} \varphi(2^N t - k)$ and we can write as the following:

$$\sup_t 2^{-\frac{N}{2}} \sum_k |\phi_{N,k}(t)| = M.$$

So Equations (9) can be written as follows:

$$E\left[\tilde{f}_N^*(t)\right] \leq 2^{-\frac{N}{2}} \sum_{k=0}^{K} f^*(t_k)\phi_{N,k}(t) + O(\Delta)M$$

(10)

By using Equation (1) we have:

$$2^{-\frac{N}{2}} \sum_{k=0}^{K} f^*(t_k)\phi_{N,k}(t) + O(\Delta)M = \sum_{k=0}^{K} \left\{\langle f^*, \phi_{N,k}\rangle + O\left(\Delta^{m+\frac{1}{2}}\right)\right\} \phi_{N,k}(t) + O(\Delta)$$

$$= \sum_{k=0}^{K} \langle f^*, \phi_{N,k}\rangle \phi_{N,k}(t) + O\left(\Delta^{m+\frac{1}{2}}\right) + O(\Delta).$$

(11)

By using Equations (10) and (11) we have:

$$E\left[\tilde{f}_N^*(t)\right] = P_{V_N}^{f^*(t)} + O(\Delta).$$

$$Var\left[\tilde{f}_N^*(t)\right] = 2^{-\frac{N}{2}}\sum_{k=0}^{K} Var\left(\frac{U_k}{\Delta}\right)\phi_{N,k}^2(t) + 2^{-N}\sum_{k=0}^{K}\sum_{l=0,l\neq k}^{K} Cov\left(\frac{U_k}{\Delta},\frac{U_l}{\Delta}\right)\phi_{N,k}(t)\phi_{N,k}(t)$$

By using theorem (2-1) we can writhe as follows:

$$Var\left[\tilde{f}_N^*(t)\right] = 2^{-N}\sum_{k=0}^{K}\left[\frac{f^*(t_k)}{n\Delta}+O(n^{-1})\right]\phi_{N,k}^2(t)$$

$$+2^{-N}\sum_{k=0}^{K}\sum_{l=0,l\neq k}^{K}\left[-\frac{1}{n}f^*(t_k)f^*(t_l)+O(\Delta n^{-1})\right]\phi_{N,k}(t)\phi_{N,l}(t)$$

Using this fact that f^* is uniformly bounded on $[0,\tau]$ and $\Delta=O(2^{-N})$, we have:

$$Var\left[\tilde{f}_N^*(t)\right] \leq \frac{1}{n}C_1\sum_{k=0}^{K}\phi_{N,k}^2(t)+O\left(\frac{\Delta}{n}\right)\sum_{k=0}^{K}\phi_{N,k}^2(t)+C_2\frac{\Delta}{n}\sum_{k}\sum_{k\neq l}\left|\phi_{N,k}(t)\phi_{N,l}(t)\right|$$

$$+O\left(\frac{\Delta^2}{n}\right)\sum_{k}\sum_{k\neq l}\left|\phi_{N,k}(t)\phi_{N,l}(t)\right|.$$

$$(12)$$

Since φ is regular in order m we can write:

$$\sum_{k=0}^{K}\phi_{N,k}^2(t)=O(2^N),\quad \sum_{k}\sum_{k\neq l}\left|\phi_{N,k}(t)\right|\left|\phi_{N,l}(t)\right|=O(2^{2N})$$

$$(13)$$

According Equation (13), we can write: $Var\left[\tilde{f}_N^*(t)\right]=O\{(n\Delta)^{-1}\}+O(n^{-1})$, complete the proof.

EVALUATE OF MEAN INTEGRAL SQUARE ERROR WITH CONVERGENCE RATIO

In this section we evaluate mean integral square error and convergence ratio is investigated.

Definition 3-1: The mean integrated square error (MISE) of kernel estimator of a density function f is given $MISE \approx C_1(nh)^{-1}+C_2 h^{2r}$. In this formula $\approx$ denotes the right and left convergence, when $n\rightarrow\infty$, n denotes the sample size, h denotes the estimator bandwidth core, r denotes core level and C_1, C_2 denote kernel dependent quantities with unknown density.

Theorem 3-1: Suppose that the sub density f^* is a continuous function on $[0,\tau]$ and it's m times differentiable, then if $\Delta\rightarrow 0$ for $n\rightarrow\infty$ and $j(n)\rightarrow\infty$, then $n2^{-j(n)}\rightarrow\infty$,

$$\mathrm{MISE}\left(\hat{f}_n\right) = E\left[\int_0^\tau \left\{\hat{f}_n(t) - f^*(t)\right\}^2 dt\right] \le O\left(2^{-2j(n)m}\right) + O\left(\Delta^2\right) + O\left(\frac{2^{j(n)}}{n}\right)$$
(14)

Proof:

$$E(t) = \hat{f}_n(t) - f^*(t) = P_{V_{j(n)}}^{\tilde{f}_N^*(t)} - f^*(t) = P_{V_{j(n)}}^{\tilde{f}_N^*(t)} - P_{V_{j(n)}}^{f^*(t)} + P_{V_{j(n)}}^{f^*(t)} - f^*(t) = S(t) + A(t).$$
(15)

By using Equation (15) and theorem (2-2) for $m \ge 1$, we can write as the following:

$$E\left[E(t)\right] = E\left[S(t)\right] + A(t) = P_{V_{j(n)}}^{f^*(t)} + O(\Delta) - P_{V_{j(n)}}^{f^*(t)} + A(t)$$

Because $V_{j(n)} \subset V_N$ we can write as the following:

$$E\left[E(t)\right] = O(\Delta) + A(t)$$
(16)

$$\left\|P_{V_{j(n)}}^{f^*(t)} - f^*(t)\right\|_2 = O\left(2^{-mj(n)}\right)$$
(17)

So by using Equations (16) and (17), we can write:

$$\mathrm{MISE}\left(\hat{f}_n\right) = O\left(\Delta^2\right) + O\left(2^{-mj(n)}\right).$$
(18)

For evaluate $Var\left[\mathrm{MISE}(\hat{f}_n)\right]$, we can write:

$$P_{V_{j(n)}}^{\tilde{f}_N^*(t)} = \sum_{k=0}^{2^{j(n)}-1} \left\langle \tilde{f}_N^*, \phi_{j(n),k}\right\rangle \varphi_{j(n),k}(t) = \sum_{k=0}^{2^{j(n)}-1} \left\{\sum_{l=0}^{2^N-1} 2^{-\frac{N}{2}} \left\langle \frac{U_l}{\Delta} \varphi_{N,l}, \varphi_{j(n),k}\right\rangle\right\} \varphi_{j(n),k}(t)$$

Also we can write:

$$P_{V_{j(n)}}^{\tilde{f}_N^*(t)} - P_{V_{j(n)}}^{E\left(\tilde{f}_N^*(t)\right)} = \sum_k 2^{-\frac{N}{2}} \sum_l \frac{U_l - E(U_l)}{\Delta} \left\langle \phi_{N,l}, \phi_{j(n),k}\right\rangle \phi_{j(n),k}(t)$$

then,

$$\int \left\{P_{V_{j(n)}}^{\tilde{f}_N^*(t)} - P_{V_{j(n)}}^{E\left(\tilde{f}_N^*(t)\right)}\right\}^2 dt = 2^{-N} \sum_l \left\{\sum_k \frac{U_l - E(U_l)}{\Delta} \left\langle \phi_{N,k}, \phi_{j(n),l}\right\rangle\right\}$$
(19)

By using theorem (2-1) and expectation of Equation (19), we can write as the following:

$$Var = \sum_{l} 2^{-N} \sum_{k} E\left\{\frac{\left(U_k - E(U_k)\right)^2}{\Delta^2}\right\}\left\langle \varphi_{N,k}, \varphi_{j(n),l}\right\rangle$$

$$+ \sum_{l} 2^{-N} \sum_{k} \sum_{h \neq k} Cov\left(\frac{U_k}{\Delta}, \frac{U_h}{\Delta}\right)\left\langle \varphi_{N,k}, \varphi_{j(n),l}\right\rangle \left\langle \varphi_{N,h}, \varphi_{j(n),l}\right\rangle^2$$

(20)

By using theorem (2-1) we have:

$$\sum_{l} 2^{-N} \sum_{k} E\left\{\frac{\left(U_k - E(U_k)\right)^2}{\Delta^2}\right\} = \Delta \sum_{k,l}\left\{\frac{f^*(t_k)}{n\Delta} + O\left(n^{-1}\right)\right\}\left\langle \phi_{N,k}, \phi_{j(n),l}\right\rangle^2$$

$$= \Delta \sum_{k,l}\left\{\frac{f^*(t_k)}{n\Delta} + O\left(n^{-1}\right)\right\}\left\langle \varphi_{N,k}, \varphi_{j(n),l}\right\rangle^2$$

(21)

$$\sum_{k}\left\langle \varphi_{N,k}, \varphi_{j(n),l}\right\rangle^2 = \left\|P_{V_N}^{\varphi_{j(n),l}(t)}\right\|_2^2 = \left\|\varphi_{j(n),l}\right\|_2^2 = \int \varphi_{j(n),l}(t)dt = \int 2^{j(n)}\phi^2\left(2^{j(n)}t - l\right)dt = 1$$

(22)

By using Equation (22) and this fact that f^* is uniformly bounded, we can write as the following:

$$\sum_{l} 2^{-N} \sum_{k} E\left\{\frac{\left(U_k - E(U_k)\right)^2}{\Delta^2}\right\}\left\langle \phi_{N,k}, \phi_{j(n),l}\right\rangle^2 = O\left(\frac{2^{j(n)}}{n}\right) + O\left(\frac{\Delta 2^{j(n)}}{n}\right) = O\left(\frac{2^{j(n)}}{n}\right).$$

The second part of Equation (20) can be written as the following:

$$\sum_{l} 2^{-N} \sum_{k} \sum_{h \neq k} Cov\left(\frac{U_k}{\Delta}, \frac{U_h}{\Delta}\right)\left\langle \varphi_{N,k}, \varphi_{j(n),l}\right\rangle \left\langle \varphi_{N,h}, \varphi_{j(n),l}\right\rangle$$

$$= \sum_{l} 2^{-N} \sum_{k} \sum_{h \neq k}\left\{-\frac{1}{n}f^*(t_k)f^*(t_h) + O\left(\Delta n^{-1}\right)\right\}\left\langle \varphi_{N,k}, \varphi_{j(n),l}\right\rangle \left\langle \varphi_{N,h}, \varphi_{j(n),l}\right\rangle$$

$$= \sum_{k} \sum_{h \neq k}\left\langle \varphi_{N,k}, \varphi_{j(n),l}\right\rangle \left\langle \varphi_{N,h}, \varphi_{j(n),l}\right\rangle$$

$$\leq \sum_{k} \sum_{h}\left|\left\langle \varphi_{N,k}, \varphi_{j(n),l}\right\rangle\right|\left|\left\langle \varphi_{N,h}, \varphi_{j(n),l}\right\rangle\right| = \left(\sum_{k}\left|\left\langle \varphi_{N,k}, \varphi_{j(n),l}\right\rangle\right|\right)^2$$

$$\leq 2^N \sum_{k}\left\langle \phi_{N,k}, \phi_{j(n),l}\right\rangle^2 = 2^N\left\|\phi_{j(n),l}\right\|_2^2 = 2^N$$

By using $O\left(\frac{2^{j(n)}}{n}\right) + O\left(\frac{\Delta 2^{j(n)}}{n}\right) = O\left(\frac{2^{j(n)}}{n}\right)$, the proof is complete.

EMPIRICAL DISTRIBUTION OF PURPOSE ESTIMATOR

In this section we investigate empirical distribution of estimator under some condition.

Theorem 4-1 Suppose that the sub density f^* is a continuous function on $[0,\tau]$ and it's m times differentiable, for $n \to \infty$, $n\Delta \to \infty$, $n\Delta^3 \to 0$, $n\Delta 2^{-j(n)(2m-1)} \to 0$, then for interval $[0,\tau]$, we have:

$$\sqrt{n\Delta}\left(\hat{f}_n(t) - f^*(t)\right) \to N\left(0, f^*(t) \sum_{k=-\infty}^{\infty} \varphi^2(k)\right).$$

Proof:

$$\sqrt{n\Delta}\left(\hat{f}_n(t) - f^*(t)\right) = \sqrt{n\Delta}\left(\hat{f}_n(t) - E\left[\hat{f}_n(t)\right]\right) + \sqrt{n\Delta}\left(E\left[\hat{f}_n(t)\right] - f^*(t)\right)$$

By using theorems (2-1) and (2-2), we can write as the following:

$$E\left[\hat{f}_n(t)\right] - f^*(t) = O(\Delta) + \left(P_{V_{j(n)}}^{f_N^*(t)}\right) - f^*(t),$$

$$\sup_{f \in W^m}\left\|P_{V_{j(n)}}^{f(t)} - f(t)\right\|_\infty = O\left(2^{-j(n)\left(m-\frac{1}{2}\right)}\right), \left(P_{V_{j(n)}}^{f^*(t)} - f^*(t)\right) = O\left(2^{-j(n)\left(m-\frac{1}{2}\right)}\right).$$

$$\tag{23}$$

$$E\left[\hat{f}_n(t)\right] - f^*(t) = O(\Delta) + O\left(2^{-j(n)\left(m-\frac{1}{2}\right)}\right).$$

$$\tag{24}$$

So by using equation of (23) and (24) we can write as the following:

$$\sqrt{n\Delta}\left(\hat{f}_n(t) - E\left(\hat{f}_n(t)\right)\right) = \sqrt{n\Delta}\left(\hat{f}_n(t) - \tilde{f}_N^*(t)\right) + \sqrt{n\Delta}\left(\tilde{f}_N^*(t) - E\left(\tilde{f}_N^*(t)\right)\right)$$
$$+ \sqrt{n\Delta}\left(E\left(\tilde{f}_N^*(t)\right) - E\left(\hat{f}_n(t)\right)\right) = I + II + III.$$

We prove that II has asymptotically normal distribution and also I, III tend to zero when $n \to \infty$

First, we show that I, III tend to zero when $n \to \infty$. According to Equation (24) we have:

$$\hat{f}_n(t) - \tilde{f}_N^*(t) = P_{V_{j(n)}}^{\hat{f}^*(t)} \tilde{f}_N^*(t) = O\left(2^{-j(n)\left(m-\frac{1}{2}\right)}\right).$$

$$E\left(\tilde{f}_N^*(t)\right) - E\left(\hat{f}_n(t)\right) \le \left| E\left(\tilde{f}_N^*(t)\right) - E\left(\hat{f}_n(t)\right) \right| \le E\left| \tilde{f}_N^*(t) - \hat{f}_n(t) \right| = O\left(2^{-j(n)\left(m-\frac{1}{2}\right)}\right).$$
(25)

By using Equation (23) we have:

$$\left| P_{V_{j(n)}}^{f^*(t)} - f(t) \right| \le K 2^{-j(n)\left(m-\frac{1}{2}\right)}$$
.

So by using Equation (24) and (25), the phrase I, III tend to zero when $n \to \infty$, and finally we have:

$$\sqrt{n\Delta}\,\tilde{f}_N^*(t) = \sqrt{n\Delta}\sum_{k=0}^{K} 2^{-\frac{N}{2}}\frac{U_k}{\Delta}\cdot 2^{\frac{N}{2}}\phi\left(2^N t - k\right) = \frac{\sqrt{n}}{\sqrt{\Delta}}\sum_{k=0}^{K} U_k \phi\left(2^N t - k\right)$$

$$= \frac{1}{\sqrt{n\Delta}}\sum_{k=0}^{K} nU_k \phi\left(2^N t - k\right) = \sum_{i=1}^{n}\frac{1}{\sqrt{n\Delta}}\sum_{k=0}^{K} Y_{ik}\phi\left(2^N t - k\right).$$

So we have:

$$\sqrt{n\Delta}\left(\hat{f}_n(t) - E\left(\hat{f}_n(t)\right)\right) = \sum_{i=1}^{n}\frac{1}{\sqrt{n\Delta}}\sum_{k=0}^{K}(Y_{ik} - p_k)\phi\left(2^N t - k\right) = \sum_{i=1}^{n} Z_{ni}$$
(26)

Such that for each fixed k, while $i = 1, 2, \cdots, n$, Y_{ik} is defined as an independent and identically distributed random sample with the mean as follows:

$$p_k\left| \tilde{f}_N^*(t) - E\left(\tilde{f}_N^*(t)\right) \right| \le \frac{1}{n\Delta} 2^{-\frac{N}{2}}\left(\sum_{k=0}^{K}\left\{ \left|\phi_{N,k}(t)\right|\left|\sum_{i=1}^{n}(Y_{ik} - p_k)\right| \right\} \right).$$

By using cushy Schwartz inequality:

$$\left| \tilde{f}_N^*(t) - E\left(\tilde{f}_N^*(t)\right) \right|^2 \le \frac{1}{n^2\Delta^2} 2^{-N}\left(\sum_{k=0}^{K}\left|\phi_{N,k}(t)\right|^2 \right)\left(\sum_{k=0}^{K}\left|\sum_{i=1}^{n}(Y_{ik} - p_k)\right|^2 \right) \le \frac{M^2}{n^2\Delta^2}\left(\sum_{k=0}^{K}\left|\sum_{i=1}^{n}(Y_{ik} - p_k)\right|^2 \right).$$
(27)

So we can write as the following:

$$E\left[\sup_{0 \le t \le \tau} \left| \tilde{f}_N^*(t) - E\left(\tilde{f}_N^*(t)\right) \right|^2 \right] \le \frac{M^2}{n^2\Delta^2}\left(\sum_{k=0}^{K} Var(Y_{ik}) \right).$$

Using this fact that f^* is uniformly bounded and, $Var(Y_{ik}) = \Delta f^*(t) + O(\Delta^2)$, $k = 1, 2, \cdots, 2^N - 1$, we can write:

$$Var(Y_{ik}) = \Delta f^*(t) + O(\Delta^2)$$
,

$$E\left[\sup_{\circ \le t \le \tau}\left|\tilde{f}_N^*(t)-E\left(\tilde{f}_N^*(t)\right)\right|^2\right]\le \frac{M^2}{n^2\Delta^2}\left(\sum_{k=0}^{K}\left(f^*(t_k)+c\right)\right)\le \frac{c}{n\Delta^2}$$

Thus, the Equation (26) state is convergent in L^2 and thus in the distribution. Also by using Theorem (2-2), we have:

$$Var[Z_{ni}]=\frac{1}{n}\left(\sum_{k=0}^{K}f^*(t)\phi^2\left(2^N t-k\right)+O(\Delta)\right).$$

Thus we have:

$$Var\left[\sqrt{n\Delta}\left(\tilde{f}_N^*(t)-E\left(\tilde{f}_N^*(t)\right)\right)\right]\to f^*(t)\sum_{k=0}^{K}\phi^2(k).$$

We control the Lindberg condition in order to prove that II is asymptotically normal. For this purpose, we set: $U_{ni}=\frac{Z_{ni}}{\sqrt{Var Z_{ni}}}$, and we show that $E\left\{U_n^2 \, I_{|U_n|>\varepsilon\sqrt{n}}\right\}\to 0.$

By using cushy Schwartz inequality:

$$E\left(U_{ni}^4\right)=O\left(n^2\right)E\left(Z_{ni}^4\right),\ E\left(Z_{ni}^4\right)=O\left(\frac{1}{n^2\Delta}\right),\ E\left\{U_{ni}^2 \, I_{|U_{ni}|>\varepsilon\sqrt{n}}\right\}\le\left\{E\left(U_{ni}\right)^4\right\}^{\frac{1}{2}}\left(\varepsilon\sqrt{n}\right)^{-1}$$, So we can

write as the following:

$$E\left\{U_n^2 \, I_{|U_n|>\varepsilon\sqrt{n}}\right\}=O\left(\frac{1}{\sqrt{n\Delta}}\right),$$ and complete the proof.

SIMULATION AND NUMERICAL COMPUTATION FOR TARGET ESTIMATOR

In this section we simulate, $\hat{f}_n(t_k)$ on the data of size n by using Semlayt's wavelet. We consider convergence ratio of given estimator by computing of average mean square error of given estimators. We use R software and wavelet package for simulation.

Example 1: We generate $X_1, X_2, X_3, \cdots, X_n \sim \Gamma(5,1)$ and $C_1, C_2, C_3, \cdots, C_n \sim E(6)$ from the Samples of size $n=400$ and $n=600$ with $K=8$, $K=16$, $K=32$ and $\Delta=0.05$ for optimal surface $j=2$.

The results in Table 1 displays the average mean square errors of subdensity function estimator for sample sizes $n=400$ and $n=600$.

The panel in Figure 1 displays the wavelet estimator of subdensity $\hat{f}_n(t_k)$ of observed failures for a traditional censoring data. The solid line is the density estimator and the dotted line is the true density.

Example 2: Suppose that $X_1, X_2, X_3, \cdots, X_n \sim f=0.6Y+0.4W$, where

$Y \sim LN(0,1)$ and $W \sim N(3,.04)$. We generate $C_1, C_2, C_3, \cdots, C_n \sim E(3)$ from sample size of $n = 400$ and $n = 600$ with $K = 8$, $K = 16$, $K = 32$ and $\Delta = 0.05$.

The results in Table 2 displays the average mean square errors of subdensity function estimator for sample sizes $n = 400$ and $n = 600$.

The panel in Figure 2 displays the wavelet estimator of subdensity of observed failures for a traditional censoring data. The solid line displays the subdensity estimates based actual data and the dotted line is the true density.

Table 1: The average mean square errors of subdensity function estimator by wavelet method.

$$AMSE(f^*) = \frac{1}{K}\sum_{k=0}^{K}\left(\hat{f}_n(t_k) - f^*(t_k)\right)^2$$

K	$n = 400$	$n = 600$
8	26.1	17.9
16	19.2	10.1
32	18.6	7.2

Table 2: The average mean square errors of subdensity function estimator by wavelet method.

$$AMSE(f^*) = \frac{1}{K}\sum_{k=0}^{K}\left(\hat{f}_n(t_k) - f^*(t_k)\right)^2$$

K	$n = 400$	$n = 600$
8	680	610
16	420	275
32	379	278

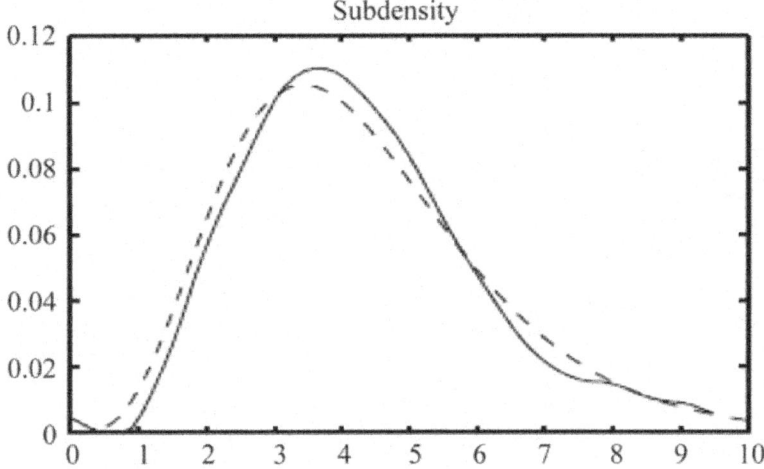

Figure 1: The wavelet subdensity and true density estimator.

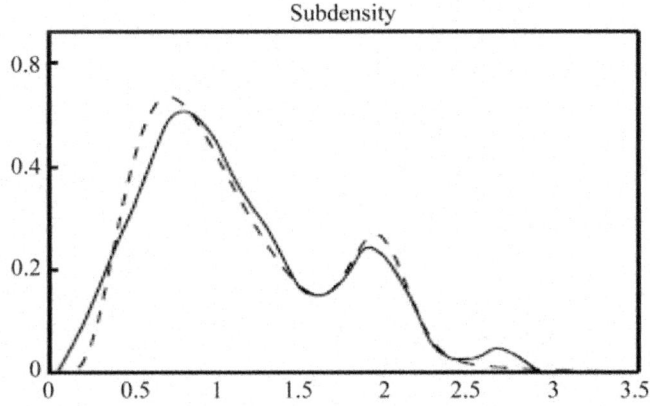

Figure 2: The wavelet subdensity and true density estimator.

CONCLUSION

In this paper we obtain density estimation for censoring data by using wavelet method and evaluate mean integral square error. We show that convergence ratio is acceptable and empirical distribution of given estimator under some condition is normal.

ACKNOWLEDGEMENTS

The support of Research Committee of Persian Gulf University is greatly acknowledged.

REFERENCES

1.   Harr, A. (1910) Zur Theorie der Orthogonalen Funktionen. Mathematische Annalen, 69, 331-371.

2.   Daubechies, I. (1988) Orthogonal Bases of Compactly Supported Wavelets. Communication in Pure and Applied Mathematics, 41, 909-996.

3.   Antoniadis, A. (1996) Smoothing Noisy Data with Tapered Coiflets Series. Scandinavian Journal of Statistics, 23, 313-330.

4.   Afshari, M. (2013) A Fast Wavelet Algorithm for Analyzing of Signal Processing and Empirical Distribution of Wavelet Coefficients with Numerical Example and Simulation. Communication of Statistics-Theory and Methods, 42, 4156-4169.

5.   Afshari, M. (2014) Estimation of Hazard Function for Censoring Random Variable by Using Wavelet Decomposition and Evaluate of MISE, AMSE With Simulation. Journal of Data Analysis and Information Processing, 2, 1-5. http://dx.doi.org/10.4236/jdaip.2014.21001

6.   Afshari, M. (2008) Wavelet-Kernel Estimation of Regression Function for Uniformly Mixing Process. Word Applied Sciences Journal, 4, 605-609.

7.   Donoha, D.L. and Johnstone, I.M. (1994) Ideal Spatial Adaptation by Wavelet Shrinkage. Biometrika Journal, 81, 425-455. http://dx.doi.org/10.1093/biomet/81.3.425

8.   Kerkyacharian, G. and Picard, D. (1993) Density Estimation Bykernel and Probability. McGraw-Hill Science, New York, 327-336.

9.   Mallat, S.G. (1989) A Theory for Multiresolution Signal Decomposition: The Wavelet Representation. Transformations on Pattern Analysis and Machine Intelligence, 11, 674-693.

10.   Meyer, Y. (1990) On de lettes et operateurs. Hermann, Paris.

11.   Hall, P. and Patil, P. (1995) Formula for Mean Integrated Squarederror of Non-Linear Wavelet Based Density Estimators. Annals of Statistics, 23, 905-928. http://dx.doi.org/10.1214/aos/1176324628

12.   Antoniadis, A., Gregoire, G. and Nason, P. (1999) Density and Hazard Rate Estimation for Right Censored Data Using Wavelet Methods.

Journal of Royal Statistical Society Series B, 23, 313-330.

13. Vidakovik, B. (1999) Statistical Modeling by Wavelets. Wiley, New York.http://dx.doi.org/10.1002/9780470317020

CITATION

CHAPTER 1

Shi, N. , Wu, C. and Yang, J. (2015) Mathematical Approach to the Platonic Solid Structure of MS2 Particles. Applied Mathematics, 6, 655-662. doi: 10.4236/am.2015.64059.

CHAPTER 2

Dipankar Chakraborty, Dipak Kumar Jana, and Tapan Kumar Roy, "A New Approach to Solve Intuitionistic Fuzzy Optimization Problem Using Possibility, Necessity, and Credibility Measures," International Journal of Engineering Mathematics, vol. 2014, Article ID 593185, 12 pages, 2014. doi:10.1155/2014/593185

CHAPTER 3

Bin Xu and Man-Qing Xu, "Numerical Analysis of Vibration Isolation Using Pile Rows against the Vibration due to Moving Loads in a Viscoelastic Medium," International Journal of Engineering Mathematics, vol. 2014, Article ID 810525, 12 pages, 2014. doi:10.1155/2014/810525

CHAPTER 4

Alecsandru Simion, Leonard Livadaru and Adrian Munteanu (2012). Mathematical Model of the Three-Phase Induction Machine for the Study of Steady-State and Transient Duty Under Balanced and Unbalanced States, Induction Motors - Modelling and Control, Prof. Rui Esteves Araújo (Ed.), ISBN: 978-953-51-0843-6, InTech, DOI: 10.5772/49983.

CHAPTER 5

Wayne S. Kendal and Bent Jørgensen, A Scale Invariant Distribution of the Prime Numbers, doi:10.3390/computation3040528

CHAPTER 6

Daniel Law, Jennie D'Ambroise , Panayotis G. Kevrekidis and Detlef Ki, Asymmetric Wave Propagation Through Saturable Nonlinear Oligomers, doi:10.3390/photonics1040390doi:10.3390/photonics1040390

CHAPTER 7

Richard A. Guinee (2012). Mathematical Analysis for Response Surface Parameter Identification of Motor Dynamics in Electric Vehicle Propulsion Control, New Generation of Electric Vehicles, Prof. Zoran Stevic (Ed.), ISBN: 978-953-51-0893-1, InTech, DOI: 10.5772/54483.

CHAPTER 8

Banerjee, A. , Ghosh, A. and Das, M. (2015) High Performance Novel Square Root Architecture Using Ancient Indian Mathematics for High Speed Signal Processing. Advances in Pure Mathematics, 5, 428-441. doi: 10.4236/apm.2015.58042.

CHAPTER 9

Y. Zhang and W. Shao, "Image Mathematics—Mathematical Intervening Principle Based on "Yin Yang Wu Xing" Theory in Traditional Chinese Mathematics (I)," Applied Mathematics, Vol. 3 No. 6, 2012, pp. 617-636. doi:10.4236/am.2012.36096.

CHAPTER 10

Afshari, M. (2014) Wavelet Density Estimation of Censoring Data and Evaluate of Mean Integral Square Error with Convergence Ratio and Empirical Distribution of Given Estimator. Applied Mathematics, 5, 2062-2072. doi: 10.4236/am.2014.513200.

INDEX